HALLAM UNIVERSI.
LEARNING & IT SERVICES
ADSETTS CENTRE CITY CAMP
SHEFFIELD S1 1WB

102 024 313 9

SHEFFIELD HALLAM UNIVERSITY
LEARNING CENTRE
WITHDRAWN FROM STOCK

Urban Regions

Ecology and Planning Beyond the City

Natural systems and their human uses are of central importance in urban regions, where diverse greenspaces and built spaces of essentially equal value spatially intertwine. With land planning, socioeconomics, and natural systems as foundations, this book combines urban planning and ecological science in examining urban regions. Writing for graduate students, academic researchers, planners, conservationists, and policy makers, and with the use of informative urban-region color maps, Richard Forman compares 38 urban regions from 32 nations, including London, Chicago, Ottawa, Brasilia, Cairo, Beijing, Bangkok, Canberra, and a major case study of the Greater Barcelona Region. Alternative patterns of urbanization spread (including sprawl) are evaluated from the perspective of nature and people, and land-use principles extracted from landscape ecology, transportation, and hydrology are stated. Good, bad, and interesting spatial patterns for creating sustainable land mosaics are pinpointed, and urban regions are considered in broader contexts, from climate change to biodiversity loss, disasters, and sense of place.

RICHARD T. T. FORMAN is Harvard University's Professor of Advanced Environmental Studies in Landscape Ecology. Previously at the University of Wisconsin and Rutgers University, he is an American Association for the Advancement of Science Fellow and recipient of the Lindback Award for Teaching Excellence as well as of honorary degrees and medals in the USA and internationally.

D1628103

Urban Regions

Ecology and Planning Beyond the City

RICHARD T. T. FORMAN

Harvard University

CAMBRIDGE
UNIVERSITY PRESS

CAMBRIDGE UNIVERSITY PRESS
Cambridge, New York, Melbourne, Madrid, Cape Town,
Singapore, São Paulo, Delhi, Mexico City

Cambridge University Press
The Edinburgh Building, Cambridge CB2 8RU, UK

Published in the United States of America by Cambridge University Press, New York

www.cambridge.org
Information on this title: www.cambridge.org/9780521670760

© Cambridge University Press 2008

This publication is in copyright. Subject to statutory exception
and to the provisions of relevant collective licensing agreements,
no reproduction of any part may take place without the written
permission of Cambridge University Press.

First published 2008
Reprinted with corrections 2010

A catalogue record for this publication is available from the British Library

ISBN 978-0-521-85446-7 Hardback
ISBN 978-0-521-67076-0 Paperback

Cambridge University Press has no responsibility for the persistence or
accuracy of URLs for external or third-party internet websites referred to in
this publication, and does not guarantee that any content on such websites is,
or will remain, accurate or appropriate. Information regarding prices, travel
timetables, and other factual information given in this work is correct at
the time of first printing but Cambridge University Press does not guarantee
the accuracy of such information thereafter.

SHEFFIELD HALLAM UNIVERSITY
WL
307.12 16
FO
ADSETTS LEARNING CENTRE

Dedicated to Sabrina Forman Heim
and her family,
Karl, Brianna, Matthew

Contents

Color plate section can be found between pages 122 and 123.

Foreword

Encountering the title of this book by Richard Forman, my first reaction was one of surprise. It is commonplace, of course, that cities are embedded in natural systems, but the modern city seems such a triumph of modern technology over the constraints of nature that one can easily understand why urban planners have rarely found it necessary to spend much time talking with urban ecologists – or, to state it another way, why ecology and urban planning have remained quite separate domains of inquiry and action in the modern division of labor.

Ecology, a branch of biological sciences, strives to understand relationships of interdependence in the natural world, and (while not at heart activist) at times to devise strategies for their preservation. As such, of course, it informs environmental regulators and thereby places some constraints on development activity. Urban planning, on the other hand, exists to provide analysis in the service of action, and its principal concerns historically have been economic – to pursue and facilitate development while striving as well to preserve and enhance the market value of existing property investments. Planners have other concerns as well, to be sure, such as improved public health, social equity, and an attractive public realm – all of which have vital ecological dimensions. So one would be hard-pressed indeed to find a planner who disagreed with the proposition that good plans must be ecologically sound. This agreement has traditionally had a ritualistic quality, however, in that, with rare exceptions, planners have viewed ecological values mainly as constraints – to be addressed late in their analyses, particularly at the behest of environmental regulators – rather than at the very core of their mission. And they have rarely viewed ecologists as indispensable participants in their deliberations from the outset.

Richard Forman would change all that, and the argument he lays out in this volume is compelling. Though modern technologies are dazzling, he observes, having enabled us to separate urban residents from their sources of nourishment, potable water, and even jobs, by greater distances than would ever have been imaginable in earlier times, the "tsunami" of urban growth now threatens widespread disaster. With three billion people living in urban areas, and two billion more expected within the next quarter-century, with global warming, with energy demand rising more rapidly than energy production (the latter, moreover, often with devastating environmental effects), and with the continuing depletion of fresh water supplies and biodiversity – we seem to be racing beyond the capacities of our technological ingenuity to shield us from the natural limits of our environment. It is past time, in short, for urban planners and policy makers to recognize ecological health as the single most urgent value to be served by urban planning – without which all the others are likely to prove illusory before too many more decades pass.

Forman's analysis is global, greatly enhancing its power. He examines 38 regions in 32 countries, representing most of the variety of large cities and regions throughout the world, and reports as well on a detailed case example of ecologically focused urban planning in Barcelona, a pioneering effort that he personally led in 2001–2002. The latter provides a truly eye-opening example of big-picture planning, carried out at the behest of Barcelona's mayor and chief planner, to preserve the critical natural assets of that region and direct its development for generations to come in ways that minimize environmental degradation. What emerges clearly from this exercise is that there need not be a major conflict between the objectives of ecological health and economic development, but that one had better focus on the ecology early on if there is to be much hope of reconciling them in the end.

In brief, though written by an ecologist, this is very much a book for urban planners, policy-makers, and all others who care seriously about the future of urban life on this planet. Moving to implement Forman's ideas will be a formidable challenge indeed, even in those very rare enlightened jurisdisctions with planning traditions comparable to Barcelona's, and vastly more so everywhere else. But global transformations invariably begin in the realm of ideas. And Forman here lays out a very big one. Planners and urban policy-makers everywhere, take heed!

Alan A. Altshuler
Ruth and Frank Stanton Professor in Urban Policy
*and Planning, Harvard University**

* Also: formerly Secretary, Massachussetts Department of Transportation; formerly Director,Taubman Center for State and Local Government, Kennedy School of Government; formerly Dean of the Faculty of Design, Harvard University.

Foreword

When Daniel Burnham exhorted the planners of the early twentieth century to "make no small plans," ecology was not something that very many urban planners knew much about. Indeed, the ecology of that time probably seemed irrelevant to planners, because it had little to say about humans in ecosystems, or even about the structure and function of broad-scale landscapes. These concerns were late inventions in ecology, but they have finally emerged and come together to generate a sound and growing body of knowledge that relates to urban mosaics as ecological systems. The new knowledge and perspectives of landscape ecology and urban ecological studies bring the science of ecology, the practice of urban planners, and the needs of dwellers in urban regions to an unprecedented threshold of truly ecological planning at regional scales. Richard Forman has written a satisfyingly original and compelling book to carry us over that threshold.

As an ecologist I find several things particularly exciting about this book. It identifies the big issues and concerns about urbanization at the beginning of the first urban century – the century in which humans become numerically an urban species. The growth, intensification, and global spread of urbanization are staggering. Ecology must find a way to engage with this wave, and not retreat in its face. With the rapid changes in urban systems, they take on new forms and establish new interactions with their regions. The old assumptions about the forms of cities, and the ecological implications of those forms will not support effective ecological research or interaction between ecologists and planners in the future. Forman recognizes the significance of the amazing transformations our urbanized world is experiencing, and translates them into a conceptual language that can help bring ecological knowledge to bear in the design and management of the Earth's changing urban face. Ecologists have much to learn from this, and it provides a way to interact with planners who are confronting and dealing with the novel urban patterns around the world.

Perhaps most pleasing to an ecologist is the fact that the ecology here is up-to-date and sound. This is no mere urban ecology by analogy with the patterns of processes outside the human realm. Rather this book articulates, in clear and relevant ways, the major principles that must guide the application of real scientific ecology to cities and their regions. Also important to see here is the fact that generalizations are couched within taxonomies of urban regions that can identify the constraints governing their applicability. These generalizations are couched so that they should not be inappropriately applied, or over extended.

This book will be a key tool in the important and widely recognized work of bridging between contemporary ecology and urban planning. The "worked example" of plans for the Barcelona region in the context of examining the ecological opportunities and constraints for 38 metropolises around the world is a very powerful guide for truly ecologically based regional planning. Ecology has too often been a weak or small tool in planning. Forman shows us how strong and central it can be. At the same time, the reader will find respect for both ecology and planning, based on long experience in education and research in both. This is a brave and necessary book, which does its work with both scientific clarity and the poetry or keen observation and sensitivity to the humanities and the social realm.

I'm finishing my notes for this piece while descending into the Johannesburg, South Africa airport. Below me sprawls a region of immense natural resources, a beacon of social hope, a locus of economic power, a cultural engine. It contains an old center, quiet suburbs from the last century, gleaming new suburbs with their business and entertainment districts, and the crowded townships. There are the mine shafts, spoils, and cooling towers of the mineral industry; there are farms, and the green leafy canopy of jacaranda blooming over some neighborhoods. Forman's book tells us how to truly bring ecology and the built and social mosaics together to envision how a metropolis such as this can evolve sustainably in the future. This book also suggests how to deal with the dynamism of Baltimore, Maryland, a very different region where much of my ecological work now takes place. But it also gives us an important way to deal with the different dynamics of the next new city – perhaps not yet named – to be established in China. The deep regional perspective and sound ecological principles articulated and put into action in this book make us think about Burnham's exhortation in a different light. Whatever one thinks about the value or success of big plans over this last 100 years, the vision of this book suggests that it is time to make big plans again, but to do so on a regional, landscape ecological base – the real ecology – encapsulated in this book.

Steward T. A. Pickett

Distinguished Senior Scientist, Institute of Ecosystem Studies, Millbrook, New York
Director, Baltimore Ecosystem Study, Long Term Ecological Research

Preface

I grew up in a world of rural nations where forests and farmland seemed infinite. Now most are urban nations. Indeed, at this moment the globe zooms over a threshold; half of us (three billion people) now aggregate in and around cities.

Yet the big change lies just ahead. In a single generation two billion more people are expected to join the urban population. Where will they live their lives? In much bigger and more numerous urban regions? Next door? Like an urban tsunami, easily visible today, we sweep swiftly and powerfully across our finite land.

For years I asked people in audiences to visualize the place where they grew up, and indicate whether it is better or worse now. Minds instantly left the room, speeding through images of memorable neighborhoods, glorious experiences, tough times, meaningful spots, and inspiring nature. Upon rapid return, virtually all audiences on different continents agreed: 80 to 90 % of their formative landscapes are worse today. Yet this trend could be turned around. Incremental solutions crowd our plate, while promising big-picture solutions increasingly appear, often ready for serious evaluation or action.

One of the great challenges of history has appeared, the giant urban region. At the center, a huge city population depends fundamentally and daily on resources that are out of sight, out of the city. An engineering and architectural marvel, the city expands at its edges or along transportation corridors or dispersed as sprawl. Too often expansion devours the city's closest and best resources, impoverishing both the land and the people. Proximity is value, as transportation cost, scarce clean water, local food sources, and tourism/recreational access emphasize. Aquifers supply clean water, greenways support walkers and wildlife, and floodplain vegetation reduces flood damage. Natural systems, from groundwater and wetlands to riparian zones and wooded parks, provide these valuable resources to society.

Four motifs with ever-changing harmonies cascade through the book's pages: (1) urban regions, rather than cities or all-built metropolitan areas, are the key big objects today and in our future; (2) natural systems, or simply nature, and human uses of them in an urban region are of major importance; (3) all regional characteristics are changing, driven by growing populations, more cities, and diverse urbanization patterns; (4) using principles and a rich array of existing solutions, society can significantly improve every distinctive urban region.

The book title provides further clues to content. "Urban region" highlights the 150 to 200 km diameter (90 to 125 mile) area where a major city and its surroundings interact to effectively form a functional region. "Ecology" refers to interactions between plants/animals and the physical environment, though often the slightly broader concept of natural systems is used. "Planning," as used here, is about product (rather than process), the tangible arrangement of human pieces and natural systems that forms the big picture. "Beyond the city" highlights patterns in the ring around the city. This book does not focus on the city, or all-built metropolitan area, or urban history, or socioeconomic dimensions, or mainstream urban planning, or town planning, or housing developments, or the methods of developing plans, or the implementation of plans, though, of course, bits of each appear. Finally, the book's perspective is global.

What are the benefits, and costs, of creating a globe with a scatter of huge growing urban powerhouses? Concentrating people helps protect natural and agricultural resources elsewhere. Economic growth often occurs in growing cities. Specialized resources such as the opera house and biotechnology center appear. But seemingly intractable problems multiply for cities, surrounding towns, villages, and farms. Natural systems are degraded, even eliminated. Floods and air pollutants are harder to control. As cities grow outward, alas, we keep traveling further and importing more of our needs, at greater cost.

A prominent sign adorns my office: "Think Globally, Plan Regionally, and *Then* Act Locally." As a philosophic foundation for a 1995 piece on land management and planning, I was thinking mainly of geographic regions. I now realize the vision especially applies to urban regions.

Big ideas – nationalism, hard-work-makes-productive-land, economic growth, environmentalism – evolve, dominate, and are transformed or replaced over time. Will urban-region planning inevitably appear in this overlapping sequence? If so, where will the giant solutions be found? Unfortunately today most planners avoid emphasizing natural systems, and most ecologists avoid studying urban regions. That leaves a near void of directly useful models. Yet both ecologists and urban planners, along with economists, engineers, and architects, are well-equipped to contribute. Who would want to live in a major area planned or designed by only one of the group? Lack of planning might be better. But the full range of expertise would be best. Indeed, fitting together small pieces makes incremental progress, while overall success or failure depends fundamentally on addressing the big picture and the long term.

Why would I, with roots in ecology, write this book? In essence three foundations suddenly came neatly together. One is ecology, especially landscape ecology's spatial scientific focus on land and natural systems at the human scale. The second is more than two decades of Harvard teaching and learning from planners, designers, engineers, social scientists, humanists, and scientists. The third foundation is an intensive 15-month project developing a land-mosaic plan for the Greater Barcelona Region. Highlighting low-profile big problems, offering tangible steps for improvement, and outlining robust frameworks for real solutions are what ecologists in their finest hour do for society.

Imagine, one afternoon the head planner for Barcelona, a major European city, telephoned saying surprisingly that he had been reading my books and articles, and then two weeks later appeared during a family Thanksgiving holiday. We talked and sensed mutual respect. He asked me, in effect, to do an ecological and conceptual plan analysis for his whole urban region based on my recent book, the only model that made good sense to him. "But I'm a scientist, not a planner." Good. "I've hardly ever been to Spain." No problem. "I've *never* been to Barcelona." Fine. "I'll have to think about it." In three weeks you should meet the Mayor and get started. After an awkward seven-month dance I started. But how does one start? I never found a model or a real city plan highlighting natural systems and their human uses as major components, though valuable pieces did accumulate over time. With an impressive team in a magical place, the outlines of a promising land mosaic emerged.

Writing this book for the wide range of people interested in urban areas, plus the equally diverse array interested in ecological science, is tough. Ecologists are overwhelmingly rural, natural systems, plants/animals, water, and management oriented. In contrast, planners are overwhelmingly urban, economics, social, people, and policy oriented. Success also means reaching perceptive educated citizens who will live in, care about, and depend on tomorrow's urban region.

The chapters flow cascade-like through the book, until expanding with broader visions at the end. An unusual array of important foundations (Chapters 1 to 4) launches the reader into a close-up of 38 urban regions of large-to-small cities worldwide (Chapter 5). Numerous characteristics of nature, food, and water, plus built systems, built areas, and whole regions, are highlighted (Chapters 6 to 7). Then alternative urbanization models, also using many assays, identify good and bad patterns of change (Chapter 8). More pieces are added – a set of basic principles, the detailed Barcelona case study – and gathered together as key components for a land mosaic (Chapters 9 to 11). Finally, urban-region ecology and planning beyond the city is analyzed in the context of broader big-picture perspectives (Chapter 12).

Lots of promising patterns and trends, plus pitfalls to avoid, emerge for the thousands of distinctive urban regions worldwide. But no single solution is proposed, other than urban-region planning for natural systems and us. Rather, a richness of spatial patterns and principles are portrayed together, ready for readers to arrange, add their own ingredients, and create a land mosaic framework or vision for an urban region. The patterns represent handles for wise planning.

With fire in the belly and a dash of optimism, I expect to see more of the world on a trajectory meshing nature and people so they both thrive . . . and I can't wait.

In addressing the great urban-region challenge of history, this book simply helps get the window open a crack to grasp broader horizons. Insights, solutions, big problems, and surprises lie in wait for the reader.

Land as capital, heritage, nature,
    as investment, inspiration, home.
All finite,
    all requiring care.
So, add planning and ecology as wisdom,
    for nature's future, our future.

Richard T. T. Forman

Acknowledgments

I deeply appreciate an impressive group of colleagues and friends who provided key insights that significantly shaped and broadened my thinking on the ecology and planning of urban regions: *scientists*, Sito Alarcon, Michael W. Binford, Mark Brenner, Barney Foran, David R. Foster, Robert L. France, David Lindenmayer, Xavier Mayor Farguel, Robert I. McDonald, Mark J. McDonnell, Marc Montlleo, Steward T. A. Pickett, Ferran Roda, and Ross W. Wein; *planners and designers*, Josep Acebillo, Alan Berger, Paul Cote, Patrick Curran, Susan Feinstein, Anna M. Hersperger, Peter Pollock, Eva Serra, Carl Steinitz, Patrick Troy, François Vigier, Christina von Haaren, Yang Rui; *social scientists and humanists*, Alan Altshuler, Joan Clos, Jose Gomez-Ibañez, Jerald Kayden, John H. Mitchell, John C. V. Pezzey, Kate Rigby, Peter P. Rogers, Sam Bass Warner, Jr.; and *engineers*, Daniel Sperling, Hein van Bohemen.

I am also grateful to many people who have graciously aided my attempts to understand specific cities and urban regions: Arthur Adeya, Christian Albert, Javier A. Arce-Nazario, John Beardsley, Alessandra Cazzola, Caroline Chen, Susan Curtin, Thomas Curtin, Gareth Doherty, Guido Ferrara, Erica Field, Susan L. Forman, Janet Franklin, Randy Gragg, Wolfgang Haber, Bryant E. Harrell, Elizabeth F. Harrell, Jock Herron, John P. Holdren, Michael C. Houck, Huang Guoping, Dorothee Imbert, Hussein A. Isack, Mikiko Ishikawa, Marianne Jorgensen, Hong Joo Kim, Young Min Kim, Wayne Klockner, Jennifer Leaning, Robert L. Liberty, David Lindenmayer, Roy A. Lubke, Taco Iwashima Matthews, Mark J. McDonnell, Yoshiki Mishima, Summer Montacute, Emanuela Morelli, Orhun Muratoglu, Elizabeth Nguyen, Laurie Olin, Grant Pearsell, George F. Peterken, Anchalee Phaosawasdi, Narcis Prat, Jennifer Claire Provost, Rodolpho Ramina, Maria Rieradevall, Ferran Roda, Carme Rosell Pages, Brooke Rosenthal, Peter G. Rowe, Sumeeta Srinivasan, Denis A. Saunders, Mario Schjetnan, Mack Scoggin, Simon Shaw, Robbert Snep, Carl Steinitz, Harry Stelfox, Frederick J. Swanson, François Vigier, Kotchakorn Voraakhom, Ross W. Wein, Joseph Wheeler,

Verona Wheeler, Emily Wilson, Wu Yue, Yang Lu (Gavin), Yang Rui, Yu Kongjian, and Zhu Yu-Fan and Yao.

It is a pleasure to acknowledge students and colleagues at the Harvard University Graduate School of Design, The Harvard Forest, and the Centre for Resources and Environmental Studies at Australian National University for the intellectual catalysis from which this book emerged. Also, I thank the leaders of two impressive organizations, Barcelona Regional and Editorial Gustavo Gili, for permission to reproduce material from my 2004 book, *Mosaico territorial para la region metropolitana de Barcelona*.

Five key people have made especially important contributions to the pages of this book. Patrick Curran produced scores of large, high-quality satellite images for urban regions around the world, including those that underlie much of this book. Eva Serra de la Figuera offered planning experience and reworked the rich geographic-information-system maps of Barcelona until they were just right. Taco Iwashima Matthews once again contributed her designer magic to create the wonderful information-rich multicolor maps of the 38 urban regions explored in this book. Finally, Lawrence Buell and Barbara L. Forman have contributed in countless ways, from fount of ideas to problem-solver, friendly reviewer, and facilitator, par excellence. I warmly thank each of you.

1

Regions and land mosaics

Imagine a group of rhinos rampaging through a restaurant, while we concentrate on adjusting the napkins, filling a glass, and brushing up some crumbs. So it seems on land, we focus on our house lots, our housing developments, sometimes our towns, while giant forces are degrading, even transforming, our valuable land. These are new giants, unseen in history. We notice their fingers, an ear, a heel, but rarely see them. Who are they? What's happening to the land? Should we keep fixing the little pieces and hand our land to the giants? Or could we raise our vision . . . and do something?

This leadoff chapter provides a set of unusual regional and land lenses through which to view urban regions, a key analytic foundation for later chapters. Chapters 2 to 4 add the other major foundations: land planning, socio-economics, and natural systems. The resulting synthesis uses three motifs: (1) urban regions; (2) natural systems; and (3) human uses of nature, to open windows and to pinpoint ecological and planning insights ready for use.

A framework

As a student and insatiable traveler, my idealism colored problems and offered ready solutions. But also as a budding scientist I learned to look more deeply, analyze the internal elements of a problem, and try to expunge opinions from my science. Generally, problems were narrow, at my scale of vision. Those were exciting times.

Big pictures were all around, but as solvable problems I missed them. Big wars were leaving scarred lands and people. Waterways were heavily polluted. Traffic and accidents grew. Road building accelerated through the terrain. Distinctive spread-out suburbs were just appearing. Many national populations were growing at 3 % per year, doubling in a bare 23 years. Now, a generation or two later, most land problems seem much bigger and also widely recognized (McNeill 2000). Yet hardly anyone seems to have a real solution.

In spots, problems have been solved. Many waterways are cleaner but others dirtier. Some war-torn areas have partly healed, while new ones have appeared. Some population growth rates have dropped, yet the total population and its proportion of pre-reproductive people remain high. Road building has decreased here and increased there. New big problems have emerged. Megacities have mushroomed. Rapidly growing poor areas mark most cities around the globe. Sprawl has blanketed some valuable land areas. Freshwater has become scarce and expensive over large areas. Topsoil for food production has thinned with wind and water erosion.

So what can be done? The so-called "paradox of management" is useful. Focus on a solution that is big enough to have some chance of continued success, and small enough that your efforts are visible (Forman 1995, Seddon 1997). For instance, it is hard to have an effect on the globe which is likely to muddle along in similar form, no matter what you do. But also, whereas it is easy to affect your garden, over decades the plants there are likely to fluctuate widely, never reaching any semi-stable sustainable state. So, to solve big problems, address the middle spatial scales such as landscapes and regions, which are most promising for combining the visible effects of your effort and a reasonable chance of success.

Or, to solve big problems, break them into parts, and address enough to tip the balance toward solution (Gladwell 2000). Or establish a promising trend, and wait (Ozawa 2004). Or do not wait; keep adaptively adjusting the trajectory. In all the cases, of course, a key first step is to recognize big problems as tractable, rather than hopeless or too complex.

Urban regions have half the world's population, three billion people. Consider some big problems at the urban-region scale such as megacities, rapidly growing poor areas, and outward urbanization (*State of the World's Cities* 2006). Then add overwhelmed sewage wastewater systems, threatened water supplies, public health, traffic jams, and growing urban air pollution. Worldwide all of these patterns are worsening. Yet a city's urban region is a useful scale for addressing such problems and offering solutions that last.

What does the future promise? No one knows, but, according to the United Nations Population Division, population trends point to nearly 200 000 people added daily to the urban population (70 million people per year). In the onrushing year 2030 (hopefully both author and reader will be here then), some five billion people, about 60 % of the world's population, are expected to live in urban areas. So, today's urban problems are big. How about tomorrow?

Figure 1.1 "Natural" disaster. In this area earthquakes are natural and frequent, while the "disaster" resulted from a bridge in this location which was unable to withstand the earthquake. Gavin Canyon, Los Angeles County, California; earthquake 6.8 on Richter scale. The absence of housing on these slopes near Los Angeles prevented worse effects. Photo courtesy of US Federal Highway Administration.

City populations grow over time as a consequence of births and immigration exceeding deaths and emigration. Economic fluctuations may especially affect immigration and emigration rates, producing short-term population rises and drops. Urban population drops, usually short term, may also result from human conflict (Leningrad, 1930s; Hiroshima, 1940s; Bujumbura, Burundi, 1970s–80s) or so-called "natural" disasters (Kobe, Japan earthquake, 1990s; Aceh, Indonesia tsunami, 2000s; New Orleans, USA hurricane, 2000s) (Figure 1.1). Still, cities usually grow in population, today commonly at a 3–5 % annual growth rate, with some sections or municipalities growing at 5–10 % annually. With cities as the major central portion of urban regions, many of these trends also apply widely to urban regions.

Consider the land surface of the urban region. Land is home and heritage, and therefore a source of sustenance and inspiration to be cared for. Land is also capital and investment to be bought, used, and sold. Furthermore, nature depends on land, and we depend on nature. Yet curiously, "We're wasting land!" (Josep Acebillo, Chief Architect for the Mayor of Barcelona, 2000), particularly in the urban region.

So, focusing the lens on patterns and processes within an urban region reveals a dynamic mosaic of people and nature (Forman 2004a). Nature varies from some relatively large natural pieces to many highly degraded pieces. Society is arranged in a single huge central aggregation plus numerous dispersed places. The region works as a system, with flows and movements across the mosaic. Also, the great mosaic changes over time, especially as human pieces expand and natural pieces shrink. This leaves nature further degraded, and the fundamental human dependence on nature's resources riskier, less sustainable. Plato even described what his ancestors did to Greece, leaving him only a late stage of this process, a skeleton of the once-rich land and water.

Nature's flows and movements across the land are particularly important in the urban region, partly because they are so buffeted by human activities. Surfacewater flowing in streams and rivers supports many human needs, from clean drinking water to recreation, wastewater treatment, and aesthetics. Groundwater flows create "underground reservoirs" that support wells, agriculture, and diverse natural plant communities. Wildlife disperses and migrates across the land, a key value for recreation and even human culture. In effect, important natural flows inexorably permeate the region.

Meanwhile urbanization spreads across the same region. Traffic jams increase. Energy efficiency drops, leaving less-sustainable built areas. Clean unpolluted water becomes scarce and expensive. Highways form barriers that subdivide the remnants of nature. Appealing recreational and tourism sites degrade. Hard surfaces spread and flood pulses get worse. Productive agriculture and family farms shrink. Forestry withers. Biodiversity is threatened and erodes. All so familiar.

People of the region, long dependent on the local resources and benefits of natural systems, must increasingly depend on more distant, more expensive resources. Concurrently the value of natural systems drops, as nature-dependent aesthetics, inspiration, ethics, and resources for future generations erode. This disconnect between nature's fundamental patterns and processes and current development trends could lead to crises, forcing prompt costly actions. Irrespective, it calls for new thinking or vision, with the core objective to mesh nature and people so they both thrive (Forman 2004a).

Usually it costs money to do something. Yet also it is costly, and penalizes both citizens and nature, to do nothing. Solutions to quickly address crises are normally expensive. To gradually address a legacy of cumulative impacts or accomplish a major new initiative, solutions are costly, but spread out over time. Finally some solutions cost little to provide significant benefits. Planning that heads off crises or creates positive legacies for a region is good economics.

Economic gains also can be expected from many solutions involving natural systems. Consider: (a) maintaining diverse productive agricultural landscapes on the best soils; (b) concentrating rather than dispersing growth to reduce infrastructure and servicing costs; (c) investing in key areas for nature protection and nature-based tourism; (d) rethinking floodplain design to reduce flood-damage costs; and (e) targeting a handful of pollution sources, plus creating stormwater wetlands, to increase a scarce supply of costly clean water. Such investments in natural systems pay dividends.

Social patterns and municipalities are equally central to planning and natural systems. Towns whose edges have light and medium industry tend to have both nearby jobs and fewer traffic problems and costs. Towns whose edges have parks with nature and recreation may have nearby stable appealing neighborhoods. Housing that is relatively concentrated rather than dispersed, has a much lower impact on natural systems. Strategically focusing population growth and urbanization in areas of low ecological value enhances the regional natural-systems' value. Creating a convenient efficient large-industry center or a truck-transportation center in such a location does too.

These many benefits to both society and natural systems are explored in the pages ahead. Such benefits emphasize that, rather than overwhelmingly concentrating on the traditional socioeconomic aspects of public transit, highways, housing, employment, urbanization, and economic development, which often can be provided in many places across the region, we should begin with best uses for the fundamental distinctive and somewhat fixed land resources for the future of a region. The many specific socio-economic aspects, of course, are also critical and likely to be addressed in most regularly updated planning. Plans for specific issues as well as specific areas can be readily meshed spatially with the land-use frameworks presented in the chapters ahead.

Also by focusing on land use, rather than regulatory and legal approaches that can change "overnight," the approach helps provide a solid long-term future for a region. Political leaders with foresight, along with planners, engineers, economists, ecologists, and others who can think big, collaborate, and effectively mesh regional land uses, can accomplish a vision. They hold and will mold the future of a region in their hands.

Urban planning often highlights the quality of people's life and promotes intelligent growth (Fainstein and Campbell 1996, Hall 2002), whereas conservation planning highlights the natural systems and nature on which people

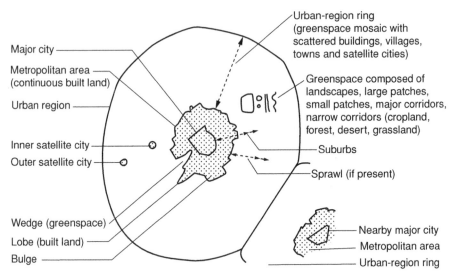

Major city

Metropolitan area
(continuous built land)

Urban region

Inner satellite city

Outer satellite city

Wedge (greenspace)

Lobe (built land)

Bulge

Urban-region ring
(greenspace mosaic with
scattered buildings, villages,
towns and satellite cities)

Greenspace composed of
landscapes, large patches,
small patches, major corridors,
narrow corridors (cropland,
forest, desert, grassland)

Suburbs

Sprawl (if present)

Nearby major city

Metropolitan area

Urban-region ring

Figure 1.2 Concepts and terms for urban regions.

live and depend (Noss and Cooperider 1994, Dale and Haeubner 2001, Marsh 2005). No models were found that provide for sustained viable natural resources and nature around cities. Clearly a new strategic approach is needed to mesh both halves, people and nature, and create a whole.

Terms and concepts to reveal urban regions

Concepts are usefully grouped into three clusters: (1) urban region and its built areas; (2) greenspaces and natural systems; and (3) urbanization.

Urban region and its built areas

A *city* is a relatively large or important municipality. Cities analyzed in this work range from just over 250 000 to over 10 million population, though the basic city concept includes smaller important population centers, even down to 10 000 in the Amazon Basin (Browder and Godfrey 1997). The *urban region* is the area of active interactions between a city and its surroundings (Figure 1.2). Thus the outer boundary of an urban region is determined by a drop in rate of flows and movements as one proceeds outward from the city.

From the eye of a satellite, the boundary delineating a city is normally invisible. Instead the *metropolitan area*, or *metro area*, the nearly continuously built, or all-built, area of the city and adjoining suburbs, is prominent as a visible object. Here, the metro area is not defined as a "commuter-shed," as in the USA (Office of Management and Budget 2000), since extensive commuting beyond the built metro area is limited in most cities worldwide. A *built area* is land with continuous closely spaced buildings, as on small properties or (p)lots.

Suburbs are mainly residential municipalities, such as towns, close to a city. A suburb may be entirely within, partially within, or altogether outside a metropolitan area. *Suburbia* (or *the suburbs* or the *suburban landscape*) refers to adjoining, or all, suburbs around a city. The area on both sides of a metro-area border, where built and unbuilt areas intermix, is the *peri-urban area* (though also some scholars use the term, peri-urban, in the more general sense of the area around a city).

The *urban-region ring* is the "ring-around-the-city" outside the metro area and inside the urban-region boundary (Figure 1.2). This variable-width ring is a mosaic of greenspace (or unbuilt) types of land interwoven with built systems and relatively small built areas. Towns and villages are distributed over the urban-region ring. Also *satellite cities* (here, <250 000 population) are normally present, *inner satellites* and *outer satellites* in the inner and outer portions, respectively, of the urban-region ring. Major highways, railroads, and powerline corridors are the prominent built systems criss-crossing the urban-region ring.

Urban regions have a city-center nucleus and are generally rounded. The all-built metropolitan area surrounded by an urban-region (or urb-region) ring is reminiscent of a gargantuan donut, and indeed a *donut model* is later used for analysis.

The term *megacity*, used occasionally, refers to a city with a population of >10 million. The concepts of city size and urban region size are discussed more fully in Chapter 5. The term *megalopolis* refers to a group of adjoining urban regions of major cities (each with >250 000 population), such as Boston to Washington or Amsterdam–Utrecht–The Hague–Rotterdam (Carbonell and Yaro 2005).

Greenspaces and natural systems

Greenspaces are unbuilt areas in an urban region, i.e., areas without continuous closely spaced buildings. Greenspaces (sometimes called open spaces) often have no buildings, but may contain a small number of relatively scattered structures. Numerous important greenspace types are present, including playing fields for sports, wetlands that reduce floods, nature reserves that protect biodiversity, tree corridors providing cool shade in summer, and market-gardening areas that produce fresh vegetables and fruits close to a city. Greenspaces range from tiny city parks to extensive woodland landscapes, and from rounded spots to linear greenways and river corridors. Greenspaces, evident on aerial photos and satellite images, may or may not be protected or have public access. Thus the many types of greenspaces parallel the many types of built areas, such as industrial, commercial, high-rise-apartment, and single-family-home. Most types of both greenspaces and built areas are useful to society.

Nature, natural systems, and natural areas are terms widely used in this book, yet, as noted in numerous treatises, defy easy definition (Peterken 1996, Kowarik and Langer 2005). Here *nature* refers to what humans have not made or strongly altered (Williams 1983, Buell 2005). Normally a frog – or a mud bank, a gust of wind, a woods, an aquifer, or even the Universe – is an example of nature, and therefore is natural. A caveat is important for a world long populated by people, namely that some things like a hedgerow or desertified area, if human-created long ago, may be considered naturalized, or simply nature.

Natural system refers to nature, but focuses on its structure, functioning, and change. Nature has a form or anatomy. Nature works, as energy, material, and species flow and move. And nature changes both its form and functioning over time. The *ecosystem* concept is used where organisms play major roles in the structure, function, and change of the system. In urban regions, the somewhat broader natural system concept is helpful in order to include important aquifer systems, subsoil groundwater flows, earthen- and rubble-fill systems, and wind transport systems, as well as ecosystems. *Natural resources*, both in-place (e,g., for recreation and aesthetics) and extractable (e.g., mineral and wood removal), are characteristics of nature with value to people.

The concepts of *natural area*, *natural habitat*, *natural vegetation*, *natural community* (or assemblage), and *natural land*, on the other hand, denote a type of space, an area unplanted and without intensive human management or use. Thus a *woodlawn* area, as a mowed grassy space often with scattered trees and shrubs, such as a typical golf course, cemetery, or city park, is not a natural area. But as something in between, a *semi-natural area* is commonly dominated by natural vegetation patterns with intensive-human-use unbuilt spaces intermixed. Semi-natural areas are especially characteristic of metropolitan areas, though widely present in the surrounding urban-region ring. To enhance readability in the text, semi-natural areas are commonly lumped under the term natural area or natural land. The idea of "native vegetation," not used here, contrasts with vegetation dominated by non-native species, whereas natural vegetation contrasts with that degraded by human activities.

Degradation is the human-activity process of decreasing natural vertical structure, horizontal pattern, biodiversity, and/or flows in a natural area. Habitat perforation, dissection, fragmentation, and isolation, as well as familiar processes such as polluting and overgrazing, cause habitat degradation (Odum 1982, Forman 1995, 2006, Lindenmayer and Fischer 2006).

A *nature reserve* is an area established mainly to protect nature or biodiversity. *Protected areas* are spaces that have legal constraints or are guarded to maintain valuable resources, such as nature, historic structures, scenic roads, forestry tracts, game, diverse recreational opportunities, flagship features of the land,

and much more, for the long term. Normally each protected area accomplishes multiple functions and objectives for society.

Habitat refers to a relatively distinct area and its physical and biological conditions where an organism, population, or group of species mainly lives. For example, a panda or metasequoia habitat refers to an area with suitable conditions for those populations or species, and an aquatic or grassland habitat applies to the area with suitable conditions for the community of species present. A *multi-habitat species* regularly uses two or more habitats. When used alone, the term habitat means natural habitat.

Biodiversity or *nature's richness* refers to biological abundance. The focus is primarily on native species and secondarily on natural communities, in both cases highlighting their number and the presence of rare ones. Thus a *biodiversity area*, whether protected or not, harbors a large number of native species or natural communities, or supports one or more rare species or community.

Conservation, as long-term protection of natural resources, apparently first focused on water, especially water quantity and to a lesser extent on water quality and fish (Pinchot 1967, Nash 1982, Schrepfer 1983, Robin 1998). Park conservation for stunning natural features and scenic value quickly followed. Also forest conservation focusing on trees, soil, and flooding came to the fore. Soil conservation, emphasizing erosion, sedimentation, and vegetation cover, was next. Finally biological or biodiversity conservation highlighting species and natural communities reached center stage. The long-term protection of the combined interacting components of a natural system, whether of direct value to society or not, is *nature conservation*. The concept of nature conservation (Saunders and Hobbs 1991, Peterken 1996), long familiar and understood by scientists and the public, is therefore used in this book.

Given the rates of urbanization and other land-use changes and the limited resources available for conservation, a site-by-site or species-by-species approach to nature conservation is of limited or local value. Instead, the focus here is on landscape pattern and (multi-species) communities. Not every water body, scenic feature, erosion-free site, or species will be conserved with this approach. However, the bulk of nature and its most important known components should be sustained for the future.

Urbanization

Urbanization is the combination of densification and outward spread of people and built areas. In *densification*, the density of people and building units increases, for example, by infilling greenspaces or by changing from low- to high-rise apartment buildings. In addition or alternatively, the city grows by expanding outward. The *outward urban expansion* or spread may occur in many spatial ways, such as expanding concentric zones or extending out transportation corridors or dispersing small developments outward. Cities may urbanize by rolling over suburbs, and suburbs urbanize by rolling over farmland or natural land. The outward spread of a town or village is sometimes included in the urbanization concept.

Outward urbanization may or may not involve sprawl. Webster's dictionary, consistent with the roots of the word, defines the verb, sprawl, as to spread out or stretch out awkwardly. For urban expansion, awkward is perhaps best translated as unsatisfactory or unsuitable or uncoordinated. This concept is relative to numerous characteristics of importance to society, from transportation, public health, and sense of community to loss of valuable farmland and disruption of nature (Bullard *et al.* 2000, Benfield *et al.* 2001, Lopez 2003, Frumkin *et al.* 2004, Burchell *et al.* 2005). Therefore *sprawl* is the process of distributing built structures in an unsatisfactory spread-out (rather than compact) manner or pattern. The concept can refer to constructing single-family rather than multiple-unit buildings, houses on large rather than small lots, and many rather than few separate developments. (Note that some authors use the term sprawl as essentially a low-density concept without the dictionary dimension of awkward or unsatisfactory [Antrop 2000]). The term *sprawl* also refers to an area with relatively new residential structures in an unsatisfactory spread-out or low-density pattern. In this sense, the process of sprawl produces sprawl as a recognizable form on the land.

An alternative pattern, especially in much of Europe, is effectively *nucleus expansion* or *growth*, where a village or town expands outward with adjacent compact urbanization. This approach capitalizes on an existing central cultural and commercial center and on the people's sense of place. Later, however, near a major city such expanding nuclei may threaten to coalesce, or indeed coalesce, and produce a huge disjointed urbanized landscape, yet which is not a city (Forman 2004a).

Some related terms are usually avoided: (a) "open space," because sometimes it implies a low-value space waiting to be filled or built upon (most types of greenspaces are highly valuable), and sometimes it implies non-forest, which is inappropriate in mainly forested portions of a region; (b) "urban edge," "urban fringe," and "urban-rural fringe," because typically these seem to be lines or narrow zones, nearly equivalent to the metropolitan-area border and considerably narrower than the peri-urban zone described above; (c) "exurban zone," which is similar to the urban-region ring, but with an uncertain inner boundary, an unspecified outer boundary, and the suggestion of an outside void rather than valuable area; and (d) "rural area," since the outer boundary is unspecified, and also because the term refers to the country, usually farmland, whereas in urban regions (where the term seems awkward) predominant land covers vary from cropland to forest or desert.

Regions

We will explore this key topic from three perspectives: (1) regions and urban regions; (2) bioregions and ecoregions; and (3) internal structure and external effects.

Regions and urban regions

Hercules was right when he placed two huge rocks at the mouth of the Mediterranean. Ever since, all ships have had to pass cautiously between these Pillars of Hercules to enter the Mediterranean Region. Once inside, the land and the diverse people are strongly linked. The sea itself serves as a giant tub for the surrounding nations' economies and cultures. The climate of dry summers and moist winters, distinctively called the Mediterranean climate, bathes everyone. Similar vegetation, formally named Mediterranean-type vegetation, covers the region (Grove and Rackham 2001). The region contrasts mightily with the Sahara area to the south, temperate Europe to the north, and a cool, dry region to the east. The Mediterranean Region is distinctive in both physical and human terms.

Two broad characteristics are commonly central to the concept of a region; macroclimate and cultural-social pattern (Forman 1995). In global air circulation, atmospheric "cells" form due to solar energy and the configuration of continents, oceans, and mountain ranges. Each cell has a common *macroclimate*, i.e., the history of weather pattern that covers a relatively large area and differs from that in surrounding areas. Hence a geographic region typically corresponds spatially with, or is a subset of, the atmospheric cell. The Mediterranean Region, Southwestern USA, Eastern Queensland, and Scandinavia are regions with distinctive macroclimates corresponding to atmospheric cells. Southern England, Southern Ontario, and Northeastern China are recognized regions, but subsets of larger macroclimate areas.

Culturally determined human activities on the land, as in the idea of *regionalism*, determine the subset-macroclimate regions. Often a single large city is the major hub, though a number of linked cities may characterize a region. A transportation network connected to the city typically ties the region together socially and economically. Mountain ranges and coastlines often form boundaries of the region. A closer look usually reveals clear evidence of a common culture or cultures across the region, as in architecture, town/village form, language, and arts. So, a *region* has both a common macroclimate and a common sphere of human activity and interest.

The urban region is a distinctive and increasingly important type of region. In this case a single major city is of central importance and surrounding land is closely linked with the city. The predominant role of the city is the main difference between an urban region and a *geographic region*. An added cultural dimension is also present in an urban region. Typically an "urban culture" exists in the city, promulgated by a particular population formed by and committed to city life. People of an urban culture can move rather seamlessly from city to city, where they may thrive and contribute. However, normally they have little commitment to the urban region surrounding a city. A sense of place, either for a particular city or its urban region, may be limited.

Certain cities are well known for a long history of regional studies and planning. Certainly London, Chicago, and Berlin are among the leaders in this group. London's greenbelt, nearby Letchworth's earlier greenbelt, London's intense dependence on resources of the surrounding countryside, and many more characteristics have been grist for leading urban scholars and planners (Munton 1983, Turner 1992, Parsons and Schuyler 2000, Hall 2002). Berlin, encompassed in the Brandenburg Region, has undergone dramatic physical changes, and with the meticulous German planning tradition has also merited much scholarly attention (Sukopp *et al.* 1995, Breuste *et al.* 1998, Kuhbler 2000, von Krosigk 2001, Bahlburg 2003).

Chicago is of special note for several reasons: (a) a City Beautiful Movement launched in Chicago in the 1890s which influenced American city centers and architecture; (b) an influential broad-scale 1909 Plan of Chicago; (c) a group of social scientists, including Robert Park, who drew in part from ecological principles to understand urban dynamics; (d) a subsequent sequence of plans with greenways including Cook and DePage counties; and (e) a recent push for a "green" Chicago with numerous parks, green roofs, and natural areas (Schmid 1975, Cronon 1991, Nowak 1994, Cityspace 1998, Greenberg 2002, Daley 2002, Platt 2004). Of course many other cities including Tokyo, New York, Paris, Melbourne, Beijing, Moscow, and Mexico City have had major regional emphases, in some cases including important environmental dimensions (Sit 1995, Pezzoli 1998, Ishikawa 2001, Hall 2002).

The literature of urban history, urban studies, and urban planning focuses on the city, yet is continuously marked by a series of leaders who highlight the importance of the urban region. Illustrative are the works of Geddes (1915), Cronon (1991), Fainstein and Campbell (1996), Warren (1998), Ravetz (2000), Soja (2000), Bullard *et al.* (2000), Hall (2002), and Dreier *et al.* (2004). Much less common are cases that emphasize natural systems, in addition to the traditional economics, transportation, and housing, as key factors in urban regions

(Warren 1998, Atkinson *et al.* 1999, Ravetz 2000, Steinitz and McDowell 2001, White 2002, Register 2006, Moore 2007).

Many challenges remain, though, to create a body of literature and work useful to society that combines the urban-region scale with major environmental dimensions and their human uses. Ecologists are challenged by the idea that ecological conditions within an urban region really matter, much less that they are of major importance (Sukopp *et al.* 1995, McDonnell *et al.* 1997, Pickett 2006, Grimm *et al.* 2003, Musacchio and Wu 2004, Kowarik and Korner 2005). Architects may highlight the importance of greenspaces for amenity and aesthetic benefits, but are challenged to deal with the many powerful environmental forces at broad scales (Norberg-Schulz 1980, Calthorpe 1993, Duany *et al.* 2000, Register 2006). Engineers are challenged to recognize the significant construction, maintenance, and repair benefits of designing and building with, rather than against, nature (van Bohemen 2004). And on and on. As cities explode with people and roll outward over valuable land, a powerful regional and ecological perspective is needed. Indeed environmental and recreational resources often require the broadest spatial perspective for the urban planner (Robert Yaro, personal communication).

Bioregions and ecoregions

The idea of linking the bio-physical and cultural dimensions in regionalism is well-illustrated and strengthened in literature and art (Buell 2005). Wordsworth was especially a poet of England's Lake District and Henry David Thoreau an interpreter of New England (USA). The American Southwest came alive in the art of Georgia O'Keefe and the landscape-detective eyes of J. B. Jackson. Grant Wood's regionalism art portrays a US region of cornfields, tree groves, farms, and hedgerows (Corn 1983). Tom Roberts' and Arthur Streeter's late nineteenth-century paintings, unencumbered by English landscape forms, revealed real Australian landscapes (Radford 1996).

In this way *bioregionalism* integrates the geographical terrain and the terrain of consciousness (Berg and Dasmann 1977). A major drainage basin is the big picture and one portion of it is, e.g., colored green or yellow by the inhabitants who sink in roots there over time. As a place-based sensitivity (Buell 2005), the meshing of ecology and culture at this spatial scale provides a bioregion dimension, usefully grounded between local culture and thinking globally.

The bioregion concept applies well to the city and its region (Snyder 1990). A culturally and economically diverse populace congregates in a spot. Is its sense of place the city, or the city with its surroundings? In a place-based bioregion, people care about and care for the region. The combined threads of culture and ecology run deeply in both space and time.

The ecoregion concept, in contrast, specifically highlights biological distributions over a large area. The *ecoregion* is a large unit of land and water typically characterized and delineated by climate, geology, topography, and associations of plants and animals. Hence it divides the land surface up biophysically rather than by political boundaries. This is the basic framework currently used by The Nature Conservancy to protect biodiversity (Groves *et al.* 2002, Anderson 2003, Magnusson 2004). It also has been used for planning by the US Environmental Protection Agency (Omernik 1987), USDA Forest Service (Bailey 1995, 1998), and World Wildlife Fund (Olson *et al.* 2001). In general, ecoregions are unfamiliar and difficult for policymakers and the public. Normally urban regions are much smaller than an ecoregion, though the location of an urban region may have considerable impact on processes across an ecoregion.

Internal structure and external effects

All the regions discussed share the same basic type of internal structure. Many landscapes, e.g., from suburban to forested and industrial to cropland, are present, and their spatial arrangement is a key to understanding and planning a region (Forman 1995). The Greater Yellowstone Region in the Rocky Mountains is a good example (Keiter and Boyce 1991, Hansen and Rotella 2001, Hansen 2002). Cattle ranchlands, river floodplains, pine forests, spruce-fir forests, alpine tundra zones, and built areas are well intermixed. In addition, many processes tie these landscapes together: fast-moving wildfires, streams of tourists, moving livestock, horseback riders, streamflows and floodwaters, tree harvesting and logging trucks, grizzlies, elk migration, bison herds, hikers, local economic activities, rafters and fishermen, vehicles on road networks, and more. The linked landscapes work as a region, and are occasionally planned as a region.

A region is larger and inherently more stable than a landscape within it (Forman 1995). Therefore planning a region as a sustainable environment or place provides a higher probability of achieving success.

The suburban landscape sandwiched between city and, for instance, cropland/woodland surroundings plays a huge role in how the urban region works. Suburbia is source, sink, and filter. As a *source*, which gives off objects, suburbia provides commuters, manufactured goods, and suburban species to the city. Suburbs also provide recreationists, commercial products, and suburban species to the outer cropland/woodland areas. As a *sink*, which soaks up objects, the suburban landscape absorbs air pollutants, water pollution, and non-native species from the city. Also suburbia absorbs food products, dispersing woodland and cropland species, and farmland dust and chemicals from the outer zone.

These flows emphasize that the city is also both a source and a sink. The city's economic activity may depend on the flows of commuters, and its semi-natural parks and greenspaces depend on continued native species dispersing in from outside. Put another way, a city is swamped by commuters, bathed by outside air pollutants, protected from flooding by suburban wetlands, nourished by market-gardening food products, and enriched by outside recreational opportunities. And of course the outer cropland/woodland zone is also a source and sink. These numerous flows and movements among city, suburbs, and surroundings represent a regional system with many feedbacks. An urban region is eternally working.

An outward expanding city pushes these flows outward. However, the areas and the spatial arrangement of city, suburbia, and cropland/woodland also change. A larger city means bigger inward and outward flows. If the suburban landscape noticeably widens, its inward and outward flows also increase, but the linkage between city and outer cropland/woodland ring becomes more tenuous. People in the city are further divorced from natural and agricultural landscapes. When an outer cropland/woodland area shrinks significantly, the flows do too. In effect, outward urbanization and its spatial arrangement become critical in determining how the urban region of the future will work.

The change in width of suburbia and the spatial pattern of urbanization point to another important little-analyzed role of the suburban landscape. It serves as a *filter*, selectively reducing flows between city and cropland/woodland outside. Wind may blow agricultural dust inward and city air pollutants outward, sometimes unaffected by suburbia. But streams and rivers that flow from outer areas into the city pass through suburbia. Suburbs may have a high or low impermeable-surface cover (Arnold and Gibbons 1996, Forman *et al.* 2003). With considerable hard surface, much rainwater is added to the streams, increasing flood hazard in the city (Jared 2004). Alternatively, ample wetlands and other natural vegetation in suburbs absorb rainwater, helping to protect the city from flooding.

Moving in the opposite direction, city residents crowd highways on weekends to recreate in outer woodland areas. Narrow commercial suburban highways squeeze the traffic flows. However, providing a richness of small recreational locations across suburbia that attract and are used by city residents would reduce the congested inward and outward weekend flows. As for any filter, the degree of filtering depends on the amount of input and the prevention of clogging. The suburban landscape varies in width and the greatest filtering may occur in the widest portions. In short, adjustments in the suburbs and in channeling urbanization spread can make an urban region work much better.

Like any large area, an urban region is tightly linked to surrounding regions and to distant regions. These linkages often strongly affect spatial patterns and processes within the region of interest (Forman 2004a, Forman *et al.* 2004). One set of patterns might be called *boundary issues* because their origin is near the urban-region boundary, either just inside or just outside. Boundary issues often warrant careful watching, because they can rapidly affect the urban region, or the adjoining region, and often change over time.

Inputs from an adjacent region that affect a major portion of an urban region are typically of greatest concern. Examples include a major water supply from an adjacent region's aquifer, people entering for recreation or the city's cultural resources, and industrial air pollutants blown in. *Outputs from a region* to its adjoining regions may be equally significant though lower profile. People and goods enter and leave by car, truck, rail, sea, and air, so each of those routes warrants evaluation. For example, holiday traffic is often channeled between the metropolitan area and coastal or mountain areas.

Distant changes also affect regional inputs and outputs. A high-speed rail line, new ski recreation areas, changes in immigration policy, and government policy changes elsewhere may significantly affect a region. Across a continent effects may involve migratory birds, livestock disease spread, changing crops, Nature-based tourism, new markets, and international policy changes. In short, land use in a region is tightly linked in both directions to other regions.

Land-mosaic perspective and landscape ecology

Urban planning, city planning, regional planning, natural resource planning, and conservation planning are all reasonably well known fields with textbooks, journals, academic programs, professional societies, and leading scholars and practitioners. All contribute extremely important knowledge and insight to planning the future of a region. The *land-mosaic perspective* that has emerged from landscape ecology and related fields in the past two decades builds from these and other foundations. It provides a body of theory and principles focusing on the spatial arrangement of land uses for meshing and sustaining both natural systems and people (Forman 1995, 2004a).

In essence, *landscape ecology* focuses on analyzing and understanding land mosaics, large heterogeneous areas with important natural systems viewed at the human scale, such as landscapes, regions, or the area seen from an airplane window or in an aerial photograph (Hobbs 1995, Dramstad *et al.* 1996, Burel and Baudry 1999, Farina 2005, Decamps and Decamps 2001, Turner *et al.* 2001, Wu and Hobbs 2007, Ingegnoli 2002, Anderson 2003, Decamps and Decamps 2004, Wiens and Moss 2005, Turner 2005). Spatial arrangement is a core analytic

Figure 1.3 Landscape structure, function, and change altered by outward urbanization from a metropolitan area. Relative to regional urbanization patterns, the central patch of multi-unit housing is compact development, the older group of house lots on right was sprawl when built, and, at a broader scale, the residential developments in the landscape represent a sprawl, rather than compact or contiguous, arrangement. Northwest of Baltimore, Maryland (USA). Photo courtesy of US Department of Agriculture.

approach. Landscape ecology is at exactly the right spatial scale for effective planning. It explicitly integrates nature and people. Its principles work in any landscape, from urban to forest and cropland to desert. Its spatial language is simple, facilitating easy communication among land-use decision-makers, professionals, and scholars of many disciplines. Centered on spatial pattern at the human scale, landscape ecology is directly usable.

Like a cell or human body, the landscape exhibits three broad characteristics, structure, function, and change (Forman 1995). Landscape *structure* or *pattern* is simply the spatial arrangement of the elements present, the natural areas and human land uses (Figure 1.3). Landscape *functioning* is the movement or flows of water, materials, species, and people through the pattern (Harris *et al.* 1996, Forman 1999, 2002b). And *change* is the dynamics or transformation of pattern over time, somewhat analogous to sequential images seen by turning a kaleidoscope.

Conveniently, the land mosaic or structural pattern may be modeled or understood using only three types of elements: patches, corridors, and a background matrix (Forman 1979a, Forman and Godron 1981). These universal elements are the handle for comparing highly dissimilar landscapes and developing basic principles. This *patch–corridor–matrix model* is also the handle for land-use planning, since spatial pattern strongly controls movements, flows, and changes of both natural systems and people.

The simple spatial language is further highlighted when considering how patches, corridors, and the matrix combine to form the variety of land mosaics on Earth, either existing or planned (Forman and Godron 1981, Forman 1995, Lindenmayer and Franklin 2002, Ingegnoli 2002, Hilty *et al.* 2006). What are the key attributes of *patches*? They are large or small, smooth or convoluted, round or elongated, few or numerous, dispersed or clustered, and so on (Figure 1.3). What are the properties of *corridors*? Narrow or wide, straight or curvy, continuous or disconnected, etc. The *matrix* is single or subdivided, variegated or relatively homogeneous, perforated or dissected, and so forth. These spatial descriptors are close to dictionary definitions and familiar to all.

Adding a housing development, a nature reserve, or a highway, for example, changes the mosaic pattern. Consequently the diverse flows and movements – of water, materials, species, and people – are altered in generally predictable ways. Basic form-and-function principles help (e.g., why rabbit ears are short in the arctic and long in the tropics). Round patches protect internal resources, whereas convoluted patches enhance flows across the boundary. Negative environmental impacts often emerge from unplanned human alterations, or from changes designed overwhelmingly for people. On the other hand, beneficial results, especially for the long term, often follow planned changes that highlight both natural systems and people.

In short, landscape ecology brings to the table simplicity and clarity, a focus on spatial arrangement, a broad-scale perspective, easy communication among users, a meshing of natural systems and people, and application to any landscape. It becomes increasingly central as society begins to seriously address the question of creating sustainable environments. Enlightened, sustainable, visionary, economically and ecologically viable, or glorious land mosaics are a worthy target for planning and society. As a vision or product of planning, a land mosaic is effectively a spatial arrangement so nature and people both thrive long term. In urban regions built spaces are meshed with green spaces.

Spatial scales and their attributes

Ecological studies and planning projects overwhelmingly focus on spaces smaller than a region. Although these fine-scale areas and sites are not the focus of this book, they are important here from three perspectives. First, many types of the spaces are repeated by the hundreds or thousands in an urban region. Consequently, if good models or generic solutions were determined for the small spaces, their cumulative effect could be measurable or even quite significant at the regional scale.

A second reason to focus inward on small spaces relates to *hierarchy theory* (Forman 1964, O'Neill *et al.* 1986). To understand or manage something of interest, three levels of scale are especially important (Freemark *et al.* 2002). The scale just above or broader than the area or object of interest exerts effects on the area. Second, other areas at the same scale as the area of interest exert competitive or collaborative effects. And finally, the scale just below or finer than the area of interest affects the area. This finer scale is where most people look for answers. How does the internal structure and functioning affect the larger object of interest? All three scales are important for urban regions.

The third reason to look at fine-scale patterns emanates from human perception and policy. Unless one goes up in a balloon or analytically looks down from airplane windows or pores over satellite images and maps, one does not really see an urban region. Rather, the public mainly sees and relates to small spaces. Thus translating public preferences into public policy and planning generally means dealing with small spaces (Nassauer 1997, Johnson and Hill 2002).

The urban region is a hot spot of highly diverse small spaces packed together. So, rather than considering numerous internal urban-region patterns here, a few key ones are illustrated in a sequence of scales from broad to fine. Areas or patches are first presented, followed by linear features or corridors. Then a close-up of four types of spaces is presented, pinpointing their spatial or unusual attributes, along with some interesting types of planning options.

Patches and corridors at a sequence of scales

Areas or patches, as well as strips or corridors, are conspicuous and important at each spatial scale in a region (Freemark *et al.* 2002). Repeatedly using a giant zoom lens, we first view patches and then the corridors. At the broadest scale, the urban region is composed of a metropolitan area and an urban-region ring (Figure 1.2).

Focusing the lens in a bit, typically a city is composed of districts: a central business district, other commercial districts, industrial areas, various multi-unit residential areas, major city parks, and so forth (Lynch and Hack 1996, Warren 1998, LeGates and Stout 2003, Wong 2004). The suburban landscape in turn is composed of towns and municipalities, commercial/industrial land, residential land, agricultural areas, natural areas, and more. The metropolitan area is the city and the inner, continuously built, portion of suburbia. The urban-region ring consists of the outer, incompletely built, portion of suburbia, natural land, agricultural land, towns, villages, and satellite cities.

Focusing in still further puts the preceding somewhat out of focus, but sharpens up new patterns. A town or municipality is covered with various residential densities or (p)lot sizes, commercial types, light and heavy industry, mixed-use areas, plus similar numbers of important greenspace land-covers, such as school land, municipal land, water-protection land, farmland, ballfields/playgrounds, nature-based recreation land, and so on. Continuing to turn the lens, a residential neighborhood (composed of housing developments, parks, fields, and small shopping areas) appears (Figure 1.3), next a housing development (composed of house lots), and then a house lot (composed of front yard/garden, house, backyard, etc.).

Analogous patterns for prominent corridors or strips appear at these different scales (Forman 1995, Warren 1998, Bennett 2003, Ahern 2002, Vos *et al.* 2002, Jongman and Pungetti 2004, Hilty *et al.* 2006). Mountain ridges and river corridors, along with major valleys, stream corridors, highways, and railroads predominate in urban regions. The unusual greenbelt may be conspicuous here. The metro-area and city scales manifest river corridors bulging with infrastructure, occasional stream corridors (most streams are in underground pipes), highways, rail lines, and greenways. Continuing to focus the lens inward reveals in sequence, a town or municipality (with water-protection and walking/wildlife movement corridors, roads/railroads, and pipelines/powerlines), housing development (with streets, sidewalks, street-tree lines, and continuous back-lot lines), and finally a house lot (with driveway, shrub/tree rows especially along side-lot and back-lot lines, and open view-lines in front and back). In short, patches and corridors in a sequence of scales usefully describe an urban region.

Spatial attributes and planning options illustrated

The prevalence of patches and corridors at all scales highlights the importance of using landscape ecology in analysis and planning of the urban region. However, let us first look more closely at some attributes, and associated planning options, for four of the patches and corridors at different scales: city parks, road networks, stream corridors, and house lots. These four spatial features are abundant in urban regions.

City parks

As patches or areas, city parks are scattered over a matrix of densely built area, with the density of residents varying widely from place to place. The number of nearby residents per park and the average distance of residents from a park are useful attributes for planning (Turner 1992, Cityspace 1998, Beatley

2000, Ishikawa 2001). One could add the attractiveness and safety of local routes to and from a park as important to park planning. The focus here, however, is on parks themselves.

Is it better for a city park to be relatively homogeneous and different from other parks, or for it to be quite heterogeneous and similar to other parks (Forman 2004a)? In the first case, each park can be relatively large and important for a single land use, say ballfields or semi-natural vegetation. Thus the park system, as a whole, contains a collection of large specialized flagship parks. In the second case, each park has a similar wide diversity of small land-use spaces packed together. Yet the system as a whole is monotonous and missing the large flagship land-use spaces.

Advantages and disadvantages of these extremes are evident. The homogeneous large-land-use park permits specialization, such as unusually high-quality ballfields or a semi-natural area with somewhat rare species. It is apt to draw "specialized" residents from, and have some positive effect on, a larger radius within the city. In contrast, the heterogeneous park of small land-use areas packed together draws "all" residents from, and has a stronger positive effect on, a smaller radius. The homogeneous park is likely to be a source of some uncommon species which disperse through the surrounding neighborhoods (Houck and Cody 2000, Wein 2006). More species may disperse from the heterogeneous parks, but nearly all of the species are common (Boada and Capdevila 2000).

Consider the interactions between parks, and between park and neighborhood. If all parks have similar and diverse small land-use spaces, they attract residents from small circles around them. Therefore a high density of parks is needed to serve everyone. Local residents tend to have pride in and help care for such parks. Common species can be expected to move readily among these similar parks. On the other hand, specialized dissimilar parks draw residents from broadly overlapping circles, but may not engender as much pride and care by local residents. Here the somewhat different species in each park, including some uncommon species, are less likely to move between parks because the land cover in each park is so different.

For the diverse-land-use park, higher maintenance budgets may be needed to deal with the ever-prevalent conflicts among land uses. Confounding the situation, government and especially park-maintenance budgets often fluctuate markedly over time. Consequently, land care by local residents is important to get through low-budget phases without major degradation of a park.

Providing attractive and safe walkways and other transportation modes that radiate from parks should enhance park usage, pride, and care by residents. Also green strips radiating outward from a park provide routes for species to move through and enrich a neighborhood. These corridors may be greenways, tree rows, shrub strips, lines of balcony plants, or even a sequence of green roofs (Hien *et al.* 2007).

Road networks

The form and usage of road networks vary widely, from a regular grid with similar roads and traffic levels to a highly irregular net with a strong hierarchy of roads and traffic flows (Forman *et al.* 2003, Forman 2004b). The latter case has major highways with large traffic flows and tiny lightly used roads. It also has high-road-density portions and low-road-density portions, and intermixed straight and winding roads. Furthermore, community planners often discuss the pros and cons of including cul-de-sac or dead-end roads in the network. All these options overlie a mosaic of land uses in the urban region, and provide to-and-from access for residents and business.

Meeting traffic demand is traditionally the watchword for transportation planners. Sometimes, in response to economic investment or other interests, government builds a road to open up a little-used area for development. Perhaps more frequently, when development and traffic build up, and then demand increases, government builds or widens a road. Thus road building may stimulate development, and development may stimulate road construction. In the former case, the question is whether the value to society of the little-used area, e.g., for protecting an aquifer or recreational opportunities or biodiversity, is greater or less than the value resulting from road building. In the case of development stimulating roadbuilding, the question is whether a viable alternative for moving people or goods exists rather than road construction. Providing alternate modes (types) of transportation, as well as *traffic calming*, the slowing and channeling of traffic by creative road designs and modifications, are widely known planning approaches.

To promote a sense of community in neighborhoods, designing the road networks "for 7-year-olds and 70-year-olds" is sometimes advocated. In other words, provide for attractive, safe walking, playing, bicycling, and meeting places. In addition, this can be combined with road-network design, at least at a limited scale, to provide attractive, safe accessibility for visitors.

Road networks are also of major ecological importance (Forman *et al.* 2003, Forman 2006). The road infrastructure and the traffic on it in an urban region have effects reverberating widely through the natural systems present. A busy highway may degrade the avian communities in natural vegetation for hundreds of meters on both sides, presumably due to traffic noise (Reijnen *et al.* 1995, 1996, Forman *et al.* 2002).

Probably most roads alter the groundwater levels and surface-flowing waters locally, but because the network is so dense around metro areas, hydrology is widely disrupted. This results in flooding, bridge problems, water-supply degradation, loss of fish populations, aquatic habitat loss, wetland loss, and so forth. Stormwater washes a range of pollutants, such as heavy metals and hydrocarbons, from roads and vehicles into water bodies. Thus stormwater-mitigation techniques and structures may be present, but are usually needed in much greater abundance near roads. Basically, instead of accelerating stormwater flows and pollutants in ditches and pipes directly to water bodies where several negative effects result, the stormwater can often be dispersed into the ground. For example, common solutions include vegetated swales and detention basins, where water flows can be slowed and reduced while chemical pollutants are filtered and broken down (France 2002, Brandt et al. 2003, Hough 2004).

Most likely, any road is a barrier or filter to crossing by some animals. Hence the dense road networks in urban regions are particularly disruptive, and divide the land up into little sections, often containing small populations rather than the former large populations. However, busy multilane highways are major barriers to crossing by wildlife, as well as to walkers and local residents. In addition to the traffic-noise degradation or avoidance zone referred to above, the highway structure with moving traffic is intimidating and dangerous.

Thus *wildlife underpasses* and *overpasses* are increasingly built to overcome the highway barrier effect. These reconnect habitats on opposite sides of the road and facilitate wildlife movement across the road (Rosell Pages and Velasco Rivas 1999, Forman et al. 2003, Iuell et al. 2003, Trocme et al. 2003). Large and small crossing structures, with or without water flows, are built in many creative designs to aid different types of animals. The distance between wildlife-crossing structures to reestablish landscape connectivity is a key dimension subject to active research (Clevenger and Waltho 2005). Major underpasses and overpasses in places may also be designed for attractive, safe crossing by people, both local residents and longer-distance hikers.

Managed roadsides cover a huge total land area, and could be used to address a wide range of societal objectives, in addition to stormwater and wildlife-crossing issues (Forman et al. 2003, Forman 2005). Aesthetics, wood products, habitats for biodiversity, carbon sequestration, enhanced traffic safety, enhanced crossing by wildlife, and much more are addressable in roadsides. The overall result would be *variegated roadsides*, visually diverse vegetated strips alongside roads which serve many objectives of society.

Stream corridors

As the third example of a small structure widely repeated across the land, the stream corridor offers special challenges and opportunities in an urban region. In the natural landscapes present, stream corridors with wide continuous vegetation protecting the stream are interconnected into a complete dendritic network (Wetzel 2001, Kalff 2002, Binford and Karty 2006). In the more widespread agricultural landscapes, tiny intermittent channels and small first-order streams may have their strips of covering vegetation entirely removed to create large crop fields requiring large tractors. Considerable wildlife habitat is lost. Perhaps more importantly, the loss of tiny vegetation strips results in extensive stormwater runoff producing downstream floods, plus extensive soil erosion and sedimentation problems. Both the altered hydrology and erosion mean that downstream stretches have degraded water quality, aquatic habitats, and fish populations. Solutions to these problems are tough and must address the intermittent and tiny streams.

Larger streams are often channelized through farmland in urban regions, but with narrow strips of riparian vegetation discontinuously distributed along them. Widening and increasing the connectivity of these vegetated stream corridors is particularly valuable for wildlife habitat and movement.

Streams in built areas suffer different fates (Paul and Meyer 2001). Some continue for stretches, but tend to be channelized or straightened with rock or concrete "rip-rap" sides. In consequence, water velocity increases, channels dry out in dry periods, and aquatic habitat and species diversity are drastically reduced. Other streams disappear for stretches, or entirely, into concrete channels, or into underground pipes, where water rushes directly to a downslope water body. The stream is gone. Sometimes the vegetated stream corridor along a concrete channel or over the pipe also disappears, in this case to development. In some cases the former stream corridor continues as a recreational greenway or a seemingly abandoned green strip between communities. "Daylighting," the conversion of an underground piped water flow into a channelized or somewhat curvy stream, is one solution occasionally achieved. Many designs and plans exist to address this range of stream-corridor-in-built-area issues, but progress is typically slow or negative.

In both the agricultural and built landscapes the surrounding land use normally has a much greater impact on streamwater quality than does the riparian zone or vegetated stream corridor. Thus a range of fine-scale solutions over the land is available and used in spots. Hedgerows, scattered trees, wooded patches, grassy swales along intermittent channels, limited herbicide use, and other practices can noticeably decrease water and pollutant runoff from farmland. They also increase habitat cover, diversity, and connectivity for biodiversity.

In built areas a stream is typically lined with adjacent house or other building lots, which are designed and managed in extremely different ways. Thus dumping, dogs, cats, yard fertilizers, insecticides, other chemicals, vegetation cutting, erosion, septic seepage, trampling, and much more, varies markedly from lot to lot at an extremely fine scale (Matlack 1993). Design and planning solutions here can be at the neighborhood, housing-development, and house-lot scales.

House lots

A house or building (p)lot exists in context. It may be an opening surrounded by wooded lots, a tiny woods or oasis surrounded by open lots, a spot along a distinct gradient say from uphill to downhill, a location in a major corridor such as for snowmelt water flow or migrating elk (*Cervus*), or simply representative of the surrounding lots (Forman 1995). Planning and design starts with such context. Location of the lot relative to road network, shopping, schools, and so on further highlights the importance of context.

Within the lot, a building is often located partly based on cultural tradition, regulatory setbacks from the street or lot-lines, space needed for a septic system, the proximity to wetlands or a flood zone, and so forth. The building's design and outer surface provides few or many microhabitats for a rich assortment of species, from lichens and mosses to lizards and birds (Kellert 2005). A driveway and various other structures may be present, and their location is extremely important, ecologically, for the house lot.

As at broader scales, the patch and corridor approach to understanding and designing the lot is valuable. Often wooded corridors are used, e.g., for privacy along side-lot-lines, as foundation plantings along a wall, to separate sections of a yard, and along a back-lot-line (Owen 1991). The *back-lot-line* may be especially important for biodiversity since, if far enough behind the building, it is likely to be least manicured and most natural. Also the back-lot-line may be aligned with those of surrounding lots, which provides a corridor for wildlife movement through the housing area. Open corridors for unobstructed views are common in front to see up and down a street, and in back for views to the back-lot-line and even to the sides. These corridors provide fine-scale routes for certain wildlife movement.

Numerous small different patches, as in the city park example above, provide habitat heterogeneity for many common species. Alternatively a somewhat-large open patch might attract some open-country species that otherwise would be absent. If it is a well-used lawn, the species will be ephemeral, but if it is a seldom-cut meadow, some species may become resident. Analogously, a somewhat-large wooded patch may attract and provide habitat for some uncommon woodland species (Goldstein *et al.* 1981, Forman 1995). Thus a clump of trees, e.g., in a back-lot-line corner, may attract some species. A large shrubby patch provides both food and cover for ground animals and will normally attract a relatively different set of species. Trees with shrub cover beneath is a still-better combination, also because it provides darker shade and moister conditions in the center at ground level. Having a similar woody clump adjacent on the neighbor's lot, or even four adjacent areas of woody vegetation where four lots intersect, can produce quite a significant habitat for woodland species in a housing development.

Two other design options that have big effects on biodiversity should be mentioned. First, edges or boundaries between land uses or habitats may be hard or soft (Forman 1995). *Hard boundaries* are relatively straight and abrupt and attract a limited number of species. *Soft boundaries*, which normally attract many more species, may be gradual (i.e., wider edges), curvy with lobes and coves, or simply irregular and patchy.

The second broad design option is to artificially enhance or inflate biodiversity by adding human-created resources. The options stretch the imagination: bird feeders, brush piles, limestone walls, east–west soil berms, bat boxes, gradual fish-pool borders, still or splashing water, deer salt-licks, red foliage to attract autumn bird-migrants, and on and on. Concentrating such approaches can artificially raise biodiversity enormously.

In short, city parks, road networks, stream corridors, and house lots are small objects that are typically numerous and widespread across an urban region. Each has a range of spatial attributes suggested above which represent useful handles for planning and design options. More importantly, wise solutions, when multiplied by the hundreds or thousands, are likely to have a major cumulative beneficial effect on the urban region as a whole.

2

Planning land

How much of the land should be planned and how much designed? I would say that remote areas dominated by natural processes should only be subjected to the broadest-scale planning, and certainly should escape any fine-scale design. The rest of the world needs broad-scale planning to identify, protect, and develop best land uses. In contrast, fine-scale design that protects and creates inspiring places for people is extremely valuable in scattered spots, especially in urban regions where people concentrate. How would you answer the question?

Land planning is now considered from three broad perspectives: (1) planning and land management; (2) conservation planning; and (3) urban planning focused on urban regions.

Planning and land management

Leading off the chapter with land management highlights the relatively short-term issues of planning, where adaptively managing existing land and its resources, particularly protected land, is the goal. We begin with some perspectives on the role of planning, and then focus on land management.

The role of planning

Physical planning is the prime concern here, rather than the relatively non-spatial political, economic, social, and policy planning. Both urban planning and conservation planning focus on space. Place is more important in urban planning, and habitat in conservation planning. Both place and habitat have deep ties and meaning to space. *Place* includes the natural and the built, but expresses the human affinity for a space (Norberg-Schulz 1980, Seddon 1997). *Habitat* includes the natural and sometimes the built, but highlights the dependence of natural organisms on a space's set of environmental conditions. So, physical planning, both urban and conservation, provides spatial arrangements. Places and habitats are arranged. Furthermore the spaces arranged vary widely in the relative degree of human and natural influence.

Still, physical planning is directed to a potpourri of targets. For example, planning may focus on biodiversity, hazards/disasters, economic development, public health, water supply, energy, air pollution, climate change, and on and on. A plan often includes many or all of the targets, and if so, a hierarchy of priorities and emphasis is present. Usually one, two, or three targets are primary, with the rest superficially considered. Land planning occurs at national, state/province, and local county/town levels (Babbitt 2005).

Narrow plans focusing on a particular sector or goal may be useful, but then meshing them effectively with diverse land uses becomes problematic. Rather, multiple-goal solutions reflected in a mosaic of best land uses is the subject at hand. Plans may be weak or strong, with a short-term or long-term horizon. The alternative, or control, is lack of overall planning which characterizes most areas of land. Perspectives of the public, of policymakers, and of the planner determine the focus areas of a plan. Usually extreme views, such as considering the city to be an "urban desert," or alternatively, the surrounding land to be simply "hinterland," are filtered out early. Thus in urban regions the traditional primary targets have been economics, transportation, housing, industry, and, to facilitate implementation, public policy (Campbell and Fainstein 2003, Hall 2002, LeGates and Stout 2003, Berke et al. 2006). Commonly water, biodiversity, air pollution, and other ecological dimensions (Atkinson et al. 1999, Ravetz 2000, Tress et al. 2004, Marsh 2005, Register 2006) are not.

The final planning subject introduced here is especially significant. Most people want to leave the world a little better. Yet worldwide, both cities and nature seem to be degrading in the face of huge, almost unstoppable forces. So, in practice, the *incremental approach* implies improving spots one by one within the broad trend. Or, more ambitiously, one may slightly slow the rate of degradation. An alternative approach for planners is to envision a better future. Then make the *vision* spatial, sketching out or outlining, without details, its general form and structure. The policymaker and the public can evaluate and appropriately modify a tangible vision, even lay out possible routes to get there. Which approach is more promising in a downward spiral, incremental steps or striving for a vision?

Land management

All protected lands and resources have one planning and management objective: *prevention of human overuse*. After that, each type of protected land has its own somewhat distinct priority goals. To prevent overuse, consider two useful spatial attributes. Typically people are concentrated outside a protected area, especially in one direction (Figure 2.1). Second, the most valuable resources are primarily in the central portion of a protected area. Therefore management concentrates on controlling people, their movements, and their effects near the boundary on that one side.

Three potential *spatial filters* exist for protected areas (Schonewald-Cox and Bayless 1986, Forman 1995). First the zone just outside the administrative boundary may be planned and managed to reduce somewhat the number of people reaching the boundary. Approaches might include mainly off-limits land uses, roads parallel rather than perpendicular to the boundary, and the presence of "decoy" lands designed to be so appealing and interesting that many people never get beyond them. Second, the boundary itself can be a filter, such as stopping for an entrance fee, creating a linear inhospitable wetland, and maintaining few access points. Third and most useful is to make the edge portion of the protected area wonderful, so very few people head on into the valuable central resource area. For instance, edge portions may have most of the accessible roads, several good loop walking trails, fishing areas, wildlife-viewing towers, and countless other attractive attributes. In effect, these *magnets in the edge* are resources provided to attract the public's interest and appreciation sufficiently to remain near the boundary, thus conserving the center of a protected area (Forman 1995).

Land management also deals with the key resources protected, and often includes lots of planning, monitoring, and research. Nevertheless, spatial arrangements and movements of people are the main key to successful land management (Dale and Haeuber 2001, Karr 2002, Liu and Taylor 2002). Normally it is much easier and less costly to lock an access-road gate, give a portable communication device to a ranger, or provide educational information at one strategic point, than it is, for example, to continually repair eroded-soil areas, artificially manipulate wildlife populations, or restore overused riparian vegetation.

Let us look more closely at three types of protected lands common in urban regions, recreation sites, nature reserves, and wetlands. These illustrate some of the preceding points plus additional ones.

Recreation sites

The challenge in recreation sites is to encourage large numbers of people to enjoy a limited set of types of recreation in a finite space without "loving the place to death," i.e., degrading the site and its resources (Knight and Gutzwiller 1995, Liddle 1997). Separating intensive recreation from nature-based recreation is a major principle. Intensive recreation spots, such as ballfields, playgrounds, and picnic areas, handle large numbers of people, provide a reasonable diversity of recreation types, and tend to have nature obliterated. These intensively used spots are located near access points and far from the most valuable natural resources present.

A trail system providing access for nature-based recreation, such as birdwatching, photography, and quiet walking, channels walkers through natural areas. Planning and maintaining the trail layout to avoid or protect the most valuable natural resources is a key goal. Recreation is an important component widely supported by society. Managing recreation sites requires strategic land planning because of fluctuations in budgets, numbers of people entering, and even types of recreation requested and provided.

Nature reserves

Nature reserves may have limited recreation, but involve other major planning and management issues due to the sensitivity and often rarity of the nature being protected. In the classic question of the relative importance of content and context (Forman 1987), clearly both are important for a nature reserve. The valuable nature is the content. The context surrounding area normally involves multiple land ownership and multiple management goals, many of which remain well beyond the nature reserve planner and manager to affect. Nevertheless, as for all protected lands, controlling inputs from outside is a major focus.

However, planning and managing the internal resource, nature, is often a different challenge. Should we leave it alone and "let nature take its course"? Ecologists no longer believe in the so-called "balance of nature," but rather see a *non-equilibrium nature*, where species and environmental conditions are continually changing (Pickett and White 1985, Peterken 1996, Lindenmayer and Burgman 2005). Therefore, if nature is left alone in the protected reserve, we expect and predict that it will look different in the future. A dominant species may decrease and be replaced by another species that increases and becomes dominant. A rare species, even a rare natural community, may disappear. Pond sediment may noticeably accumulate, stream habitats may increase or decrease, and different insect populations may explode or disappear. That's nature. This non-equilibrium perspective means that the nature reserve will look different, often very different, to people over time. Leaving nature alone may or may not be a worthwhile planning strategy. It is particularly difficult on a small nature reserve in an urban region.

Alternatively, a nature reserve can be more intensively managed to achieve a different goal. Restoration could attempt to return nature to a mimic of a former state, such as old-growth or pre-settlement conditions or how it was when we were young (Primack 2004, Groom *et al.* 2006, Lindenmayer and Burgman

Figure 2.1 Wooded hillslope adjacent to city that provides many natural and societal values. Land management against human overuse protects recreational opportunity, cultural sites, biodiversity, erosion/sedimentation benefits, flood control values, and cool air that helps ventilate the city of heat and air pollutants on still nights. Kyoto, Japan. R. Forman photo.

2005). However, just as for historic building restoration and preservation, determining what the place looked like, the basic goal, and how attainable the goal is remains problematic. Management might, for example, periodically burn an area to reduce colonizing woody plants and produce a grassland with rare grassland species. But maybe the former grassland being mimicked was itself simply a product of human economic activity, such as burning shrubs to stimulate grass for sheep grazing (Foster and Aber 2004). Or management might attempt to minimize the presence of non-native invasive species. As suggested above, in an urban region with a dense human population, manifold widespread activities, and outward urbanization spread, attempting to control non-native species is reminiscent of Cervantes' Don Quixote tilting at windmills. Still, maybe a large nature reserve is one of the few places in the urban region where the goal may be worthwhile. Keeping our eye on habitat loss and nature's richness is far more important.

Wetlands

Wetland planning and conservation is particularly relevant and important in an urban region, in part because wetlands tend to be scarce. Most were drained and filled over history as human activities spread and intensified. Protecting the marshes, swamps, acid peatlands, and other wetlands that remain in an urban region is a valuable societal priority. Wetlands provide ecosystem services, including flood-hazard reduction and stormwater pollutant removal, plus many other values from recreation to aesthetics and biodiversity protection (Keddy 2000, Mitsch and Gosselink 2000, Parsons *et al.* 2002). However, mosquitoes and public health problems may also be present (Robinson 1996).

Wetlands have water at or above ground level for a prolonged period most years. Three wetland characteristics are primary: hydrologic conditions, low-oxygen soils, and so-called wetland vegetation. Planning and management of a wetland must focus on its surroundings, especially in the direction of incoming water.

Wetland restoration may include mitigating degraded wetlands or creating new ones. Establishing wetland vegetation and associated wildlife is the visible part appreciated by society. However, wetland restoration mainly depends on getting the hydrology right (Salvesen 1994, Mitsch and Gosselink 2000, Keddy 2000, France 2003). Three components are essential: establishing the right water level to support wetland vegetation, maintaining the necessary entering flows (and hence water source), and sustaining both inputs and water levels through drought and flood periods over time. Tiny wetlands, particularly seasonal ones, may be created at the ends of stormwater drainage pipes. Several types and sizes of wetlands may be restored at the base of certain hills and mountains, and particularly in floodplains with a high water table. Large wetland complexes may be restored in low elevation areas. Because of the typical scarcity and degraded state of wetlands in an urban region, their successful restoration can be expected to have a noticeable benefit for both nature and the public. Also, wetland species quickly colonize wet spots, so restoration success tends to be rapid.

Finally, successful land planning and management normally requires a reasonable level of public knowledge and support. Land can be managed directly, but is better managed when preceded by planning. Still better is *adaptive management* which incorporates new knowledge and planning on an ongoing basis. Management of a protected area by local people is valuable, because they know and care about the local area. Management by government or experts from afar is valuable, because they bring a broader perspective and expertise, and are less affected by narrowly focused local interests. Probably combined management by local and broad outside expertise is usually optimal for sustaining the resources protected. This combination may also be more likely to provide the political and public support needed for long-term protection.

Furthermore, public perception plays a key role. If the public perceives that a place is beautiful or appealing, that translates into both public support and ultimately political support (Yaro *et al.* 1990, Nassauer 1997, Eaton 1997, Forman 2002a, Johnson and Hill 2002, Berkowitz *et al.* 2003). Aesthetics is embedded in culture, a deep and persistent force providing stability (Seddon 1997, Nassauer 2005). If people are culturally tied to a place, long-term protection is facilitated.

Conservation planning

Conservation planners must be sustained by rare idealism, as every day and in almost every place they are faced by, paraphrasing Aldo Leopold, a world of wounds. The wounds are festering, the land degrading. However, let us start this section with the values, resources, and types of conservation. Then we add the processes of land protection, planning, and management.

In conservation planning, a storied literature lays out the theories, controversies, trends, and successes (Noss and Cooperider 1994, Robin 1998, Dale and Haeuber 2001). Impressive, perceptive, even courageous US leaders here would certainly include Henry David Thoreau, George Perkins Marsh, Harriet Hemenway, John Muir, Theodore Roosevelt, Gifford Pinchot, Aldo Leopold, Marjorie Stoneman Douglas, Wendel Berry, and Edward O. Wilson, among others.

Values and resources of conservation

Without conserving a resource, loss follows and we are incrementally poorer. Without long-term conservation, nature and people become impoverished. Although conservation is certainly important for historic buildings and heritage sites, here we focus on the much bigger picture of conservation of nature and natural resources. Soil, biodiversity, rivers, forests, game, nature as a whole, and much more, require conservation planning from local project to mega-project. Big money and massive land areas are involved.

Consider nature's richness or biodiversity. For society, this provides many ecosystem services (nature's services), such as soil erosion control and pollination for food production, a range of extractable products including medicines and foods, and intrinsic values such as aesthetics, nature recreation, and inspiration (Wilson 1992, Noss and Cooperider 1994, Lindenmayer and Burgman 2005). The phrase, "extinction is forever," also underlines the importance of biodiversity. Rare and representative natural communities, as well as rare and representative species, are to be conserved. Rare species are subdivided into various categories, such as endangered/threatened, and rare at global, national, state/province, and local town/county levels. Conservation of all of these biodiversity categories is planned for, though efforts are highly uneven around the world. One type of rare species is not considered, the non-native species. Usually this is a new immigrant, and the concern is that it may become an invader that could degrade a natural community.

The urban region differs in two important ways from other areas. First it contains a high human density with intense diverse human activities on the land, which threaten the persistence of any sensitive species or natural community present. Second, outward urbanization from a city is widespread and often rapid, which, especially in the case of sprawl, further threatens rare species in the region. These two issues of human density and human expansion highlight a conservation-planning problem for urban regions and society.

How much effort should be placed into protecting an existing rare native species in an urban region? If the species or natural community is globally or nationally rare, conservation is important, even though the long-term probability of success for the species at that location may be low (Beatley 1994). The rare species might persist, and later spread to a more promising location(s) outside the urban region. If rarity of the species is at a finer scale, such as a town or county, local efforts to protect it are appropriate. But society as a whole would better put its finite conservation efforts elsewhere.

Types of conservation

Now consider the big conservation picture. Planning and management for lots of critical resources fall under this umbrella. Soil conservation involves planting native and sometimes tough non-native plants, along with a range of soil modifications and treatments, to control water and wind erosion which occurs over extensive areas of land. Forest resource conservation focuses on minimizing wildfire, soil erosion, road construction, and overcutting, plus addressing lots of other uses including recreation and water supply. Game or wildlife conservation commonly involves land protection and management, plus enhancing populations for hunters and for the long term. River conservation, a particularly challenging issue in urban regions, typically involves riparian-zone and surrounding-land inputs, natural-flooding regimes, fish migration, and simply concrete in the floodplain and river, which forms dams, roadways, bridges, and diverse encroaching structures. Rangeland conservation empha-sizes livestock effects, such as overgrazing, soil erosion, riverbank and river degradation, and obliteration of natural communities around wet spots. Water resources or aquatic-system conservation involves maintaining habitat diver-sity (especially bottom characteristics), waterside zones, fish, hydrology, physi-cal/chemical/biological water-quality attributes, and outside impacts from log-ging, agriculture, livestock, and built areas. Conservation targeted to specific habitat types, such as acid bogs, salt marshes, rainforest, and streams, or to specific species, such as waterbirds, desert plants, and big fish, is common.

Nature conservation includes all of these and is the optimum and prime tar-get of conservation. Also, in contrast to, e.g., game or biodiversity or rangeland, almost everyone relates to and supports the idea of conserving nature. As noted in Chapter 1, natural systems are effectively nature, with the advantage of focus-ing on nature's structural, functional, and change attributes. These attributes link tightly with conservation planning.

A conservationist may find the city or metropolitan area to be hopelessly complex and full of unpredictable people. Ironically, the urbanist may note the hopeless complexity of conservation, focused on so many critical resources and objectives of society spread over such a vast surface of the Earth. The urban planner is quite comfortable with preservation of historic buildings, heritage sites, and even cultural landscapes as a subfield in its own right (Green and Vos 2001). Yet most of the issues in cultural site preservation are quite similar or analogous to those in many of the conservation subfields. Land is protected, an internal resource managed, outside impacts minimized, portions undergo restoration, and costs are high.

Land protection

Four perspectives are particularly valuable in understanding land protection: (1) organizations and results; (2) conservation by The Nature Conser-vancy; (3) large green patches and corridors; and (4) the metapopulation concept. Two types of land are the highest priorities for conservation protection: remote and large.

Organizations and results

Who protects land? Local units such as towns and counties normally protect small parcels. Even individuals may protect small parcels for a period.

Cumulatively a lot of land is protected in this way, but it is highly fragmented. Some resources can be protected in little parcels and some cannot. State and local land trusts and environmental organizations usually have adequate resources to protect only small lands.

National and state/province governments tend to have the most capital and periodically invest in serious land protection. If conservation planning has been done, especially valuable large natural patches or areas, the "emeralds" on the land, can be protected. In large areas or patches almost all resources can be protected. Non-profit or non-governmental organizations (NGOs), also, are major players in land protection. International organizations often direct conservation planning toward education and policy, though many also protect land. Some of the lands protected are large, such as tropical rainforest in Latin America and wildlife parks in Africa, including some crossing the borders of nations. National non-profit organizations, also policy oriented, sometimes protect large areas and sometimes small ones.

However, planning and partnering (collaboration) by various non-profit organizations and/or government bodies can also protect a large area. More commonly, conservation planning by different groups adds land to a nucleus of protected land, in this way creating a large protected tract over time. Such a tract has multiple ownership, objectives, and management practices. Thus different management in each section tends to favor different local resources. Nevertheless large-area-dependent resources (Forman 1995), such as an aquifer or large-home-range vertebrate, usually do well because the different managers clearly see the importance of their section as part of the whole.

Land-use planning occurs at all spatial scales from international and national to local. Some African parks cross national borders to protect migrating wildlife herds. National-level land-use planning is widespread including, for example in the USA, investments in the development of highways and associated land, transfer of water supplies between drainage basins, responses to disasters, dam and irrigation projects, and the dredging of channels and harbors (Babbitt 2005). Still, the week-by-week decisions on tiny spots by local officials and local citizens represent a gargantuan enterprise, which molds the future and fragmented face of the land.

The protection of land occurs in highly diverse ways. Perhaps all cases have two things in common: long-term protection, and protection of resources against human overuse. Long-term is effectively in contrast to short-term, and typically refers to decades, generations, or more, rather than years. Permanent or in-perpetuity protection is often mentioned, but, at least in an urban region, that probably means until urbanization or human pressure and activity degrades the resource. Protection often means strong or not-so-strong legal constraints, which usually can be altered by government, war, or other action. Effectively guarding and managing the resource is usually essential, and may be in combination with, or an alternative to, legal constraints.

Conservation by The Nature Conservancy

Conservation planning by The Nature Conservancy (TNC), the largest private landholder in the world, is highly developed and warrants brief description, particularly because of its use of regions. Ecoregions (Chapter 1) have been mapped worldwide by various sources (Groves *et al.* 2002, Anderson 2003, Magnusson 2004). They are the somewhat distinctive groupings of natural communities, plants, and animals over extensive land areas, and differ dramatically from society's hierarchy of familiar administrative units with mapped political boundaries. The mission of TNC is the long-term protection of all plant and animal species and the habitats needed to support them. Ecoregional conservation identifies and prioritizes a "portfolio" of conservation areas that should collectively conserve the biodiversity of each region delimited. A portfolio encompasses multiple examples of all native species and natural communities in sufficient number, distribution, and quality to hopefully support their existence long term.

Conservation planning then moves to the land protection phase, both by TNC and with "partner" agencies and organizations, to protect the lands identified. Terrestrial ecosystems are addressed at three spatial scales. "Matrix-forming" areas, such as extensive forest or rangeland, are at the scale of thousands to millions of hectares. Large patches are more delimited areas, some 2000 to 20 000 ha (5000 to 50 000 acres), with relatively distinct environmental conditions. Small patches are small sites with rare species dependent on unusual environmental conditions present. Combinations of these three scale types comprise a portfolio to protect the biodiversity of an ecoregion.

Land protection planning for a particular project, of course, is complex and somewhat project-specific. How well a site fits with the organization's mission and with the determined ecoregion portfolio goal is important. Cost, balanced against available and expected financial resources, is important. Threat to the resource and urgency for protection are important. Finally availability of the land for protection is important. Increasingly, broad-scale issues such as urbanization rate, a dropping water table, climate change, highway traffic impact, roadlessness, and invasive species spread, are being considered in the planning.

This TNC conservation planning and action approach has produced impressive results in North America and elsewhere. Still some major biodiversity issues have been little addressed with this approach. Marine ecosystems have been mainly ignored. Stream and river systems, together with their migratory fish, form connected linear networks that cut across these hierarchical area-focused

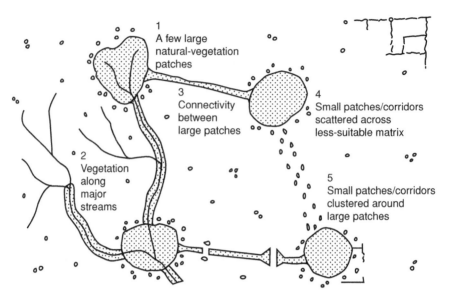

Figure 2.2 Five priorities for nature conservation illustrated with the patch–corridor–matrix model. Patches and corridors are natural or semi-natural vegetation and the background matrix is less-suitable land use. Numbered in the typical order of conservation priority. Adapted from Forman (1995).

lands. Migratory birds depend on and move across many ecoregions. The TNC approach is unique and focused on regions. In contrast, each government agency and NGO and local organization has its own approach to conservation planning.

Large green patches and corridors (emerald network)

For years I took photographs out of airplane windows around the world. Perhaps the most striking and ubiquitous pattern evident is the fragmentation of nature into little parcels. While habitat loss is overwhelmingly the giant cause of nature's problems (Wilson 1992, Forman 1995, Wilcove *et al.* 1998), habitat degradation and habitat fragmentation are the two giants following in the scene.

Most of the principles used by The Nature Conservancy for ecoregions could be used by planners and policymakers for familiar political/administrative units, such as nations, states/provinces, and counties/towns. Thus using the patch–corridor–matrix model (Chapter 1; Forman 1979b, 1995), the largest extensive natural patches can protect an aquifer, connected stream headwaters, large-home-range species (e.g., tigers and wolves), viable populations of interior species, and natural disturbance regimes (Figure 2.2). Analogously, large patches at a town scale can protect some of these resources, and small patches provide generally small, but different benefits.

Connectivity, however, warrants more focus (Forman 1995, Bennett 2003, Fahrig 2003, Lindenmayer and Fischer 2006, Hilty *et al.* 2006). Vegetated corridors, such as water-protection riparian strips, wildlife-movement routes, and walking-trail routes, can connect the large patches. The next best way to provide connectivity is with stepping stones, the small sequential patches used by animals to cross a less-hospitable area. Connectivity and corridors, though much planned and used as greenways and greenbelts in urban areas, have been more slowly incorporated in conservation plans. One historical reason was ecologists' slow recovery after a 1980s controversy about the efficacy of natural corridors, despite an empirical and conceptual literature overwhelmingly supporting their importance. The second more important reason is that large patches are almost universally agreed to be the highest conservation priority for land areas, and connectivity, e.g., by green corridors, a second priority.

In other words, conservationists do not see a network of green corridors, such as a greenway network, as the prime objective. Rather, a *group of large natural patches* or emeralds is the top priority goal. Connecting them with green corridors is a second priority (Saunders and Hobbs 1991, Noss and Cooperider 1994, Forman 1995, Lindenmayer and Burgman 2005). That sequential combination achieves the big conservation objective provided by an *emerald network* (Forman 2004a).

The metapopulation concept

One ecological concept, the metapopulation, has recently emerged as particularly important for land planning and protection. A *metapopulation* is a population subdivided into spatially separate groups, with some movement of individuals among groups (McCullough 1996, Hanski and Gilpin 1997, Lindenmayer and Burgman 2005, Groom *et al.* 2006). For instance, foxes (or deer or kangaroos) on four equal-sized patches with periodic dispersal among the patches represent a metapopulation. In an urban region human activities are constantly removing large or extensive natural habitat, leaving separate small patches as fragments of nature. Thus a large fox population in the former large natural area is converted to four small subpopulations. Over time, small (sub)populations fluctuate more in size, and have more inbreeding and resulting genetic problems, than do large populations. As a result of these two characteristics, demographic and genetic, small populations have a greater probability of disappearing or going locally extinct on a patch. Landscape fragmentation leaves the foxes on each of the four small patches with a dubious future.

An important alternative to four small patches is a metapopulation distributed on one large patch and three small ones. In this case, both demographic fluctuation and genetic inbreeding effects are reduced, because the large patch is a major source of outward-moving foxes. Some animals reach the small patches. That relatively continuous influx helps prevent a small subpopulation from dropping to very small. It also brings new genes to the small patch, reducing the detrimental effects of inbreeding. In effect, the presence of a large natural patch provides a brighter future for foxes in the landscape, and also for foxes in the small fragments of nature surrounded by human land uses. In this way land protection of large patches helps sustain biodiversity as a whole.

Spatial land planning for protection and management can provide additional benefits that improve the persistence of species on small patches. Local species extinction or disappearance from patches followed by recolonization of the patches is referred to as *metapopulation dynamics*. Attributes of the patch primarily affect local extinction rate, whereas attributes between patches mainly affect recolonization. Large patch size and high-quality habitat on the patch reduce the chance of a subpopulation size dropping to zero, i.e., local extinction. Increasing recolonization rate, in turn, benefits from patches being near rather than far apart, from the presence of a connecting natural corridor or row of stepping stones, and from higher-quality habitat conditions between patches. Protecting and managing land to reduce local extinction and increase recolonization provides a brighter future for species, including foxes, on the numerous small natural-habitat fragments in the urban region.

Planned cities

Since planning and urban regions are key themes in this book, we now turn to the ambitious projects of whole cities that have been planned and built, in order to gain insights into urban regions. City and urban planning and its rich intellectual history, theory, diverse approaches, and examples, provide the framework (Sutcliffe 1980, Ravetz 2000, Willis *et al.* 2001, Hall 2002, LeGates and Stout 2003, Campbell and Fainstein 2003, Berke *et al.* 2006). Case studies and giants in the field predominate, though guidelines and theories somewhat emergent from history and persona exist. Still, it would be hard to articulate, for example, US urban planning today without mentioning Frederick Law Olmsted, Daniel Burnham, Jane Jacobs, Robert Moses, Benton MacKaye, Rexford Tugwell, Lewis Mumford, Ian McHarg, and even, perhaps, Britain's Patrick Geddes.

Today's theories to guide planning partly rest on earlier "basic" principles, such as settlement, location, concentric-ring, industry, central-place, neighborhood-unit, and circulation theories (Hall 2002). Olmsted used proto-ecology principles in his work, and Geddes frequently cited biological models, ranging from nature to the human body, for his planning.

Insights from Brasilia and other cities

Probably most cities display the imprint of strong centralized planning and building at times in their history. The prominent square cities of Northern China, medieval rounded hilltop cities of Europe, and early semi-circular cities by major water bodies come to mind. By today's standards these are relatively small areas. Sections of cities produced by central planning are also conspicuous, such as Tokyo's ancient "Edo" design for buildings and mini-greenspaces, Paris' nineteenth-century boulevard area, and Barcelona's eight-sided city-block buildings with inside courtyard and outside mini-neighborhoods.

However, the prime interest here is to gain insights into urban regions from more-recent cities essentially planned from the start. Four are oft discussed in planning circles (Hall 2002): Chandigarth (India), New Delhi, Brasilia, and Canberra. The last three became the national capitals of India, Brazil, and Australia, respectively. The new Indian cities were built near existing population centers, while Brasilia and Canberra were built in relatively remote, low-population areas at the time.

Chandigarth was perhaps more designed than planned. Architectural form and aesthetics played a central role rather than planning for the diverse needs of people. A relatively small area was planned, with glorious pieces within it intensively designed. The surrounding urban region of the time received little emphasis. The city rapidly grew in population, quickly overrunning or expanding beyond the original design.

New Delhi, today a major world city, reflected a more balanced combination of planning and design. People's needs, from transportation and housing to water, wastes, and recreation, were seriously addressed. Stunning structures worthy of a national capital were also incorporated in places. New Delhi's population mushroomed, soon overwhelming some of the original plan's area and expectations. But extracting urban region lessons from New Delhi is difficult because it was built on the outskirts of Delhi, already a large city.

The plan for *Brasilia* also represented a balance between design of structures and their combinations and planning for people's diverse requirements (Hall 2002). Population growth rapidly spread over and beyond the plan. The original urban region received some overall planning, e.g., for two huge parks sandwiching the city, but today's urban region (Color Figure 9) extends well beyond the original vision. The central portion exhibits some of the grandeur appropriate for one of the largest nations of the world. Housing for residents is concentrated in several separate communities, now small cities, and public transport is widespread. Convenient parks provide recreation. Wetlands are relatively protected, an unusual situation in urban regions. A large reservoir adjacent to the city provides aesthetics and recreation, and originally, water supply.

Serious planning shortcomings for Brasilia quickly became evident. Housing for the poor was limited, and informal squatter communities (Perlman 1976, Main and Williams 1994) appeared in numerous locations, most considered inappropriate by planners and existing residents. Some settlements invade and degrade designated park areas. Food and agriculture, a foundation of any community, seem to have been largely overlooked by the planners. Suitable soils for cropland are scarce in the region, and a surprising amount of food is imported at considerable cost. The reservoir initially provided clean water, but soon became polluted, because built areas were designed next to and upstream of it. Water quality has continued to decline. Despite such planning shortcomings, Brasilia, the nation's center of government, has become a new and different star inland on the Cerrados plains and on the map of Brazil.

Many other cities have been planned and built in a rural spot, or have covered a small earlier city, or have almost completely rebuilt an existing city, such as after war or other disaster. Curritiba (Brazil), Adelaide (Australia), Ankara, Washington, DC, and varied German cities are examples. In most cases, however, the vision planned was the city or a portion of the city, rather than its region.

Two cities are mentioned as models of environmental sensitivity. Freiburg (Germany) has an unusual concentration of environmental solutions, including greenways for walking/biking, semi-natural areas with biodiversity, streetside stormwater detention swales and basins, traffic calming, and green roofs.

Curritiba (Brazil) has an environmental reputation partly because of successful implementation of specific projects that work, and partly because of its "marketing of ecology" which has created an image and stimulated people to work together for solutions (Schwartz 2004, Irazabal 2005, Moore 2007; Rodolpho Ramina, personal communication). Projects accomplished include: forest maintained on nearby hillslopes; seasonal linear parks which are mainly water-holding depressions along the five rivers; resistance to stream/river channelization, thus protecting downstream areas; protecting river headwater areas; housing near jobs with a vibrant economy; mass-transit emphasis; extensive tree-planting which includes fruit-bearing species. Urban liveability and recreational access, rather than ecology, were the central goals. No overall city plan guided actions (and apparently no scientist has yet evaluated the perception of environmental success). Now population has grown and expanded outward, so that urban region problems are worsening and new planning, especially regional, seems important.

Canberra, Australia

Canberra, the final example of a planned city, is a contrast. From the air it appears as a few large strange blobs of European trees with sprinkled buildings, all surrounded by paddockland and bush (pastureland and woods), a perception pervasively confirmed on the ground. The city was planned around a central reservoir with buildings largely kept well away from it (Color Figure 12). Aesthetics was important in designing boulevards and views, and axes lead the eye, not to dated human structures, but to the surrounding forested hills.

Four town centers with surrounding residential areas were established, somewhat like close-by satellites surrounding the central city area. The five centers are separated by wide connected semi-natural areas. Each town center provides daily shopping needs and some cultural activity for its adjoining residential areas. The surrounding urban region is little planned except for extensive land protection to the west to maintain a water supply. Unlike the three previous examples, Canberra has grown slowly. Although well over the population envisioned in the plan, growth has mainly occurred as compact development on the outer edges of suburbs, and, in general, the population still "fits" the plan.

The Aborigine community is small though includes many leaders. Rather few Aborigines (Australian residents for some 50 000 years) from the outback have visited Canberra. Probably their reaction would be an analog of "it's from another planet," although it is hard to know how residents from the continent's diverse outback areas would view a city (Layton 1989, Troy 1995, Forman 1995, Rigby 2006). Roads divide up the place into countless squares. Traffic is noisy and dangerous. People put up fences everywhere which block views and movement. Dreaming (lines) may be disrupted or obfuscated. Sacred sites must have been destroyed. Food comes covered with plastic. The surroundings are sterile for walkabouts. Strange deciduous trees from afar corrupt the bush. The place is boring. It has no meaning.

In contrast, most residents and visitors find Canberra to be a pleasant green city with boulevards, parks, gardens, and many other greenspaces in the central city. At the core is a blue lake (reservoir) with clean water, partly protected by the green areas around it. Attractive buildings have appeared over time in the context of the city's plan. Except for one massive communications tower that looms menacingly, development was basically kept off the slopes of the three surrounding nearby mountains. Utilities, including sewers, are connected to a site before houses can be built, which helps prevent sprawl. Some say the city is "very Australian" and residents already manifest a strong sense of place.

The four nearby town centers with residential areas help create neighborhoods and active communities. Walking and bicycle trails seemingly connect everything. The highway network is in the natural area, where it doubtless disrupts nature somewhat, rather than in or by residential areas. Everyone's home is quite close to nature. Wildlife from nearby natural areas enriches the residential areas, as loud, brightly colored birds fly in and wallabies hop in. Residents say, "It's great for raising a family." The close-by bush encourages exploration and imagination beyond the confines of a planned community.

Yet Canberra is no utopia. Visitors constantly run into circular and diagonal streets and get lost. Terrible wildfires occasionally sweep into the city consuming buildings and reducing air quality. Extensive water use (in a dry continent) is required to maintain extensive manicured greenspaces and everyone's tidy gardens. Signs discourage people from swimming in the central lake for two days following a rainstorm, due to runoff of stormwater pollutants. Sprawl close to the city is rare, though it is beginning to occur >20 km out to the southeast in the urban region. The total area of the city is large compared with its limited population. Still, because the centers are separated, very little heat island effect or concentrated air pollution buildup is evident.

Although town centers provide the basics, specialized needs such as a tuba lesson or a particular health clinic generally require considerable driving. Kangaroo–vehicle crashes are frequent, as are "roo-bars" in front of cars. Petroleum use and greenhouse gas production per person is high. Public transport is limited and not exactly rapid. A trolley or light rail system connecting the town centers and city center was envisioned and space provided in the design, but it was never built and remains a dream for some. Rich cultural resources such as museums and historic monuments are conspicuous, and recreational resources as well. Yet residents and critics periodically say the place is "dead" or "doesn't have a soul" or "there's no beach," referring to the perceived limited cultural diversity and nightlife. Some claim the situation results from Canberra being a government city. Others blame it on the original plan which disperses people, thus limiting growth of the central city.

Canberra highlights an intriguing perceptual framework. Hardly anyone can feel emprisoned in the claustrophobic hallways and cells of most major cities. The freedom of greenspace is always but a step away. Yet the city's built areas and greenspaces, the latter largely covered with mowed grass and planted tree lines, are so planned, so tightly fitting, that to some everything seems predictable, dominating, constraining, even boring. Planning and design have permanently snuffed out opportunities for imagination and creativity. Yet take but another step outward, and one finds nature, with rocks and gum trees and butterflies and venomous snakes and koalas. Exploration and imagination and experimentation are available near everyone's doorstep.

For Canberra, the urban region was a marginal part of the planning vision, mainly included to protect the land surrounding its water supply. Yet because of relatively slow population growth, today's Canberra region remains relatively viable. In contrast, Brasilia rapidly outgrew the planner's vision, an ironic measure of success. Today Brasilia's much larger urban region displays rather few marks of effective overall planning.

Urban-region planning

Regional planning, such as a regional rail-transportation system or the TVA dam system (Tennessee, USA), was introduced in Chapter 1. Here planning of urban regions lies center stage (Geddes 1915, MacKaye 1940, Barker and Sutcliffe 1993, Steiner 1994, Simmonds and Hack 2000, Ravetz 2000, Hall 2002). Issues addressed, such as water supply, wastes, and commuter routes to recreation areas, are no longer solvable by cities or even metropolitan areas (Rowe 1991, Simmonds and Hack 2000, Tress *et al.* 2004, Ozawa 2004, Berger 2006). Even where the planning imprint is strong, much of a region's form has resulted from uneven finer-scale plans, and particularly from little-planned or unplanned forces. Consider briefly some urban region examples, from Beijing to Boston.

Examples and approaches

Beijing is unusual because essentially one strong central government controls and plans the entire urban region outward to about 100 km (65 mi) from the center city (Sit 1995, Gu and Kesteloot 1998, Chen *et al.* 2004, Yang 2004). A prominent concentric ring-road form, like progressively larger hula hoops, provides both major benefits and problems. Parts of the seventh ring road are under construction and attempts to stitch in greenways are underway. A huge increase in vehicles and traffic, tree planting, removal of old buildings, soft-coal burning for power, air pollution, and greenhouse gas production characterizes this centrally planned urban region (Color Figure 7). Brisbane (Australia) also has a single centralized government for its region (Troy 1996).

Such strong *centralized planning*, which avoids the multiple-stakeholder process and proverbial least-common-denominator committee-report plan, is faster and better able to produce big change. Yet without checks-and-balances, the result may be good or bad, often depending on the degree of subsequent acceptance by the public.

Moscow, Berlin, and Bucharest illustrate a quite-different conspicuous pattern, whereby essentially only large agricultural fields and large wooded areas cover the urban-region ring, a product of a long strong Soviet-dominated planning process. London's greenbelt and Portland's (USA) urban growth boundary are both products of government policy and planning (Munton 1983, Hall 2002, Avin and Bayer 2003, Ozawa 2004). Stockholm, Copenhagen, Melbourne, and perhaps in the future Nanjing City (China) are notable for prominent greenspace wedges (Geddes 1915, Jim and Chen 2003) projecting into the metropolitan area, which result from government planning and policy (Color Figure 36).

In contrast, the planner, Robert Moses, was the driving force behind a New York regional plan focused on a parkway system of roads and parks (Hall 2002). Still different, and simplifying a bit, Atlanta (USA) is extensively spread out on the land, perhaps largely a result of inexpensive oil and little regional planning (Bullard *et al.* 2000). Chicago's region results, in part, from the combined forces of agribusiness spread, little-planned suburban sprawl, much-planned city sections, and socioeconomic policies (Cronon 1991, Hall 2002). Unplanned immigrant squatter settlements are prominent in the Tegucigalpa (Honduras) and Rio de Janeiro regions (Perlman 1976, Main and Williams 1994). Small farm fields and villages cover the London, Hannover (Germany), and Nantes (France) regions, mainly unplanned by government, but maintained by policies.

Transit-oriented development (TOD) focuses urbanization growth around stations on commuter-rail lines, an important urban-region planning approach (Cervero 1998, Gomez-Ibanez 1999, Ozawa 2004, Dittmar and Ohland 2004, Handy 2005). Compact mixed-use development, including shopping, multi-unit housing, and small-lot single-family homes, is connected by convenient walkways within an 800 m (half-mile) radius of the station (e.g., San Diego, Los Angeles, Sydney). Providing walkable employment opportunities would further enhance a low-vehicle-use community. Transit-oriented development communities contrast with high-vehicle-use sprawl areas and "edge cities" (Garreau 1991). Although transit-oriented development emphasizes development and transportation, it could be noticeably improved by a focus on greenspaces in the community. This design capitalizes on the human need for nature (Wilson 1984, Kaplan *et al.* 1998, Donahue 1999, Hobbs and Miller 2002, Kellert 2005), and avoids the perception of suburban living hemmed in by buildings and concrete.

Frederick Law Olmsted's celebrated late-nineteenth-century Emerald Necklace planning for Boston was largely along the city's edge, rather than over its region (Zaitzevsky 1982, Warner 2001). Shortly thereafter Charles Eliot developed plans for a Boston greenbelt with walking trail well out from the city, a good example of regional planning. The greenbelt was not established, one of countless regional plans that were not implemented. Today a circular walking trail passes through a handful of large greenspace patches. Regional greenway systems or networks are being pieced together around San Francisco, Chicago, Minneapolis/St. Paul (USA), and many other cities by coalitions of interests (Ahern 2002, Jongman and Pungetti 2004, Erickson 2006).

Urban region solutions may emerge from diverse fields (Orr 2002), as the accomplishments of landscape architect Olmsted emphasize. Thinking big, practically, and "outside the box," he successfully integrated recreation, flood control, transportation, vegetation, sewage treatment, and aesthetics. Fortuitously, today some landscape architects, building beyond aesthetics and amenities in small spaces, are elevating serious ecological science, urban planning, and other key fields to the forefront for solutions.

A region could be transformed by planning in a number of ways. First, the *big project*, such as Arthur Morgan's TVA dam-system project in the US South and Robert Moses' New York parkway system (Morgan 1971, Hall 2002), is what many people think about for regional planning. Design a regional plan so brilliantly that, after public review and refinements, government implements it as a whole. A second approach, call it the *planned trajectory approach*, effects a policy change which establishes a trajectory of change. The Portland urban growth boundary is an example, whereby few people could see anything different after five years, small differences were widespread after 10 years, and from 20 years onward Portland looked quite distinct from all other US cities (Avin and Bayer 2003, Ozawa 2004). Another approach, effectively a *land-mosaic* or *puzzle-pieces plan*, provides for implementable changes in small-to-mid-size areas that fit together to form the whole region; no pieces are left out. The Greater Barcelona Region land-mosaic plan illustrates this, whereby planning solutions for each separate portion, as well as for small features repeated across the land, were outlined (Forman 2004a) (Color Figures 41–44).

Planners have long valued, even emphasized, urban greenspaces mainly for people (Rowe 1991, Warren 1998, Ishikawa 2001, Clark 2006). The emphasis on ecology and key environmental dimensions has become prominent more recently (Platt *et al.* 1994, Atkinson *et al.* 1999, Ravetz 2000, Steinitz and McDowell 2001, White 2002, Steiner 2002, Orr 2002, Marsh 2005, Hilty *et al.* 2006). This trend seems partly due to increasing recognition of the central importance of ecological dimensions in the city and its region, and partly to society's interest in sustainability. These approaches use the science of ecology, instead of simply coloring in green bushes and trees on and among city buildings. The ecology used is still introductory or general, but the trend augurs well for the future in planning.

In parallel with this trend, a few urban regions are now being studied as a whole by teams of ecologists and other experts. Prominent among these are Melbourne (McDonnell *et al.* 1997, van der Ree and McCarthy 2005, Hahs and McDonnell 2007), Baltimore (Nilon and Pais 1997, Pickett *et al.* 2001, Pickett 2006), Phoenix (USA) (Jenerette and Wu 2001, Luck and Wu 2002, Grimm *et al.* 2003), and Berlin (Sukopp *et al.* 1995, Breuste *et al.* 1998, Kowarik and Korner 2005). Fortuitously, the predominant scientific paradigms and approaches vary in the different urban regions. The key paradigms include urban-to-rural gradient, landscape ecological pattern, watershed analysis with water and material flows, dynamics of plant and animal communities (biotopes), and the city as an ecosystem with energy and material flows. Though not done for the objective of planning, doubtless these regions will noticeably gain from the pioneering urban-ecology studies.

Major environmental components of plans

A particularly interesting example of ecologically focused regional planning is the *Multiple Species Conservation Plan* (MSCP) for San Diego County, which is roughly the San Diego (California) region (adjacent to the Tijuana, Mexico region) (Color Figure 32). A single environmental goal, protecting biodiversity, was the focus. Three groups, ecological, development/financial, and governmental (federal, state, and local), planned jointly (Beatley 1994, Babbitt 2005, DiGregoria *et al.* 2006). In essence, the results were an agreed-upon map and strategy to protect a system of large greenspaces and connecting corridors, and thereby essentially remove ecological constraints on development elsewhere in the region (Figure 2.3). Ecologists concluded that the greenspace network would sustain the bulk of (but probably not all) the region's biodiversity, and developers were able to invest and build in areas outside the network with fewer uncertainties and time delays. Government played honest broker, also protecting both of the other parties in case new information convincingly showed that the map needed adjustment in spots. Significantly, government put money on the table to help with land acquisition and management.

About a third of the greenspace was protected when the plan was developed. A decade later about two-thirds of the areas were protected. Not surprisingly, most of the difficult protection projects remained for the final third. Even at the two-thirds point though, the MSCP plan is a remarkable success story and model for other urban regions. Also, the network of connected large natural patches (Chapter 1) used for San Diego's biodiversity corresponds closely with the emerald network integrated into a multiple-objective land-mosaic plan for the Greater Barcelona Region (Forman 2004a).

Only one environmental dimension, *air pollution, requires regional planning* in all large US cities. For air quality, a major project in an urban region, such as a new highway or large airport, must be evaluated in a rather lengthy data-collecting and modeling process before approval and construction (Forman *et al.* 2003). The proposer must provide convincing evidence that regional air quality will not significantly decline. Major projects on urban coastal areas also require regional evaluation and planning.

This basic regional-planning concept, using somewhat different models, would apply nicely to other environmental dimensions. Hydrologic flows/flooding and biodiversity are two obvious areas where urban regional evaluation and planning would be of significant benefit to society. Other key environmental dimensions such as wildlife movement, water pollution, and fish migration might be regionally planned, individually or in some combination. Providing

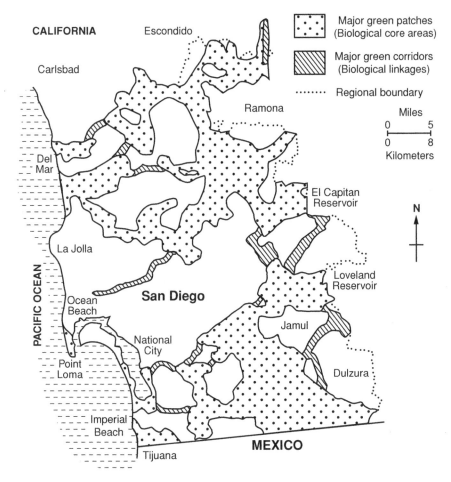

CALIFORNIA   Escondido

Major green patches
(Biological core areas)

Major green corridors
(Biological linkages)

Carlsbad

Ramona

········ Regional boundary

Miles

0   5

0   8

Kilometers

Del
Mar

El Capitan
Reservoir

N

La Jolla

Loveland
Reservoir

PACIFIC OCEAN

Ocean
Beach

San Diego

Jamul

National
City

Point
Loma

Dulzura

Imperial
Beach

MEXICO

Tijuana

Figure 2.3 Major patch-and-corridor system of natural vegetation (emerald network)
in plan for the San Diego Region, California. Adapted from *The New York Times,*
February 16, 1997, page 1.

evidence that regional levels of such environmental dimensions would not be
degraded by a proposed major project would be a tangible measurable manifes-
tation of a sustainable urban region.

Hierarchical, economic, and political problems are familiar constraints on
a regional approach to planning (Forman *et al.* 2004). Hierarchically, an urban
region commonly lies within a broader state or province or nation, while a host
of local political/administrative units such as towns or counties lie within the
region. Land-use issues are routinely handled both at the broader state and narrower local levels (Babbitt 2005), but not at the urban-region level where many
key problems needing solution emerge. Also, both states and local units often
have strong governments with taxing and budgetary authority, as well as political leaders who are likely to have been appointed or elected with public support.

Inserted into this established hierarchy, a regional authority or planning organization has difficulty. It is likely to be seen as a threat, both from above and below. Moreover, it generally has a limited budget, limited political power, and a finite life before disappearing. Despite these handicaps, regional organizations play a valuable role for society. In some places they are the only voice for regional thinking and planning.

3

Economic dimensions and socio-cultural patterns

A friend suggested that a book on urban regions would focus overwhelmingly on economics. But that would unfairly place the ills, delights, and challenges of the world in one corner, whereas a range of human dimensions are central to understanding and solutions. Dividing the human condition into three overlapping categories – economics, social patterns, and culture – is convenient, though admittedly a simplification. Culture is used in its core sense of fundamental aesthetic, intellectual, and moral traditions. This glimpse of the three big subjects, economics in this section, and social patterns and culture in the final section, is obviously incomplete. Still, selected concepts, particularly linked with resource and environmental dimensions, provide useful foundations and insights.

Growth, regulatory, and ecological economics

Key economic systems for considering natural systems and their uses in urban regions are presented as follows: (1) growth economics and regulatory economics, which are familiar and in various combinations currently predominate in urban regions; and (2) ecological economics for resources and the environment, which is growing, because in many ways it complements and addresses the shortcomings of the familiar approaches.

A few background observations are helpful. First, most economic theories are essentially non-spatial. Places for people and habitats for species are basically ignored and unimportant in economic models. Yet since spatial arrangement is so important to understanding and policy in urban regions, linking economics and spatial pattern is included here.

Second, for comparability and analysis, attempts are made to translate "everything" into a *universal currency* with the same units for direct comparability. The most familiar case is ascribing monetary value, such as euros, yen, or dollars, to things, a foundation of most economics. Converting everything to calorie equivalents is another approach (Odum 1973, Odum and Odum 1981). Yet a newer approach spreading in both economics and the public is the ecological footprint. In footprint analyses things are converted to area, such as hectares or acres, as a universal currency (Wackernagel and Rees 1996, Costanza 2000, Rees 2003, Luck *et al.* 2001).

Growth and regulatory economics

Urban regions cut across all the current economic systems worldwide (World Bank 2006). The 38 regions considered in detail in this book cover most systems, varying, for example, from economics driven by strong market forces to strong government controls. The formal constructs of industrialized economies contrast with the informal financial networks and traditions of some developing nations. Thus the current economic system of China applies poorly to France, that of Honduras poorly to Japan, and that of the USA poorly to Chad. Urban regions in all six of these countries are examined in later chapters.

With roots in China, Europe, and elsewhere, economic growth emerged in the early 1940s as one of the "big ideas" of history (McNeill 2000). *Growth economics* is the widespread familiar approach embodied in the phrase, "Let the free market determine it." Three characteristics are central (Dasgupta and Heal 1974, Romer 1990, Aghion and Howitt 1998, Gomez-Ibanez 1999, Jones 2002). Human consumers are the central players. Preferences and tastes are the predominant determining force. And the resource base is essentially limitless, because substitution or technology can overcome a limited resource.

Thus for society, the goal is to sustain economic growth, with the assumption that this can continue forever. Success minimizes the "too high" and "too low" rates over time. For individuals, the goal is often to attain happiness or a high quality of life (Jacobs 1992, Costanza *et al.* 1997a, Kasser 2003, Easterlin 2003, Layard 2005). Broad economic measures of individual success usually focus on high economic status (wealth) or high consumption rate.

The essentially limitless resource base assumes that, if a resource becomes scarce, another resource (a different type or trade from a different area) can be substituted or technologically developed (Dasgupta and Heal 1974, Aghion and Howitt 1998, Pearce and Ulph 1999, Jones 2002). Indeed, human-made capital can be substituted for natural resource inputs to production. Protecting or conserving natural resources reduces the need for substitution or technical progress. But more importantly in growth economics, the limitless resource base means that there is no essential need to protect a particular natural resource.

Before considering the far-reaching implications of this assumption in the next section, we briefly consider another familiar widespread economic system, *regulatory economics*. Government regulations, laws, and other limitations on free markets and freedom of action usually attempt to head off, or result from, crises or problems, or may result from planning for the future (Aghion and Howitt 1998, Jones 2002, Perman *et al.* 2003). The public may demand regulations or the government impose them. Also the regulations are policed by government with varying degrees of effectiveness. Crisis management and the solutions to problems generally involve relatively short-term regulatory actions. On the other hand, planning may involve long-term limitations, such as zoning for appropriate land uses, land protection for a park system, or investment in a transportation system. Government may be particularly suited for multi-sectoral optimization or "what if" analyses, as well as implementation of their results.

Governments often act and invest to meet demand. Such regulations are reactive rather than proactive. Sometimes government acts to open up opportunity. An array of institutional structures, economic instruments, and incentives may be used. Most government actions are relatively short-term and are usually somewhat dependent on changing levels of income.

Regulations intrude on business and the free market, and too often suppress innovation. While they may protect the public against inept or unethical actions, regulations put limits on experimenting with new ideas and may protect mediocrity.

Perhaps all national and urban economies have a combination of free-market growth and government regulation, and mainly differ by their position along the gradient between the two poles. As politics change, the economies slide, normally temporarily, along the gradient to right or left.

Finally, corruption should be mentioned as an economic limitation though not a regulatory one. Corruption, which varies from high to low both spatially and temporally, may put limits on both the free market and government regulation.

The relationship between population growth and economic growth is especially important around growing cities. In general, if suitable pre-conditions, such as various financial institutions, are in place, population growth commonly leads to economic growth and development (Cheshire 1988, Ray 1998, Rogers *et al.* 2006). Fluctuations and adjustments in wages, prices, markets, credit, interest rates, employment, technological change, international trade, and other variables affect growth rates, and thus are important in economic models. Nevertheless, more people consume more resources, make more products, create larger markets, require more housing units, and affect a greater area.

These simplified statements lead to two important attributes of urban regions (Fainstein and Campbell 1996, Ravetz 2000, LeGates and Stout 2003). Urban regions today are mushrooming in population, a trend much more due to immigration than to birth rate. Second, though housing goes upward in high-rises, development spreading outward at lower density has a much greater effect on natural resources and environmental conditions.

Ecological economics for resources and the environment

For some characteristics, times, and places in the urban region, the time-tested growth economics combined with regulatory economics work well (Aghion and Howitt 1998, Jones 2002). But the third component, ecological economics, is a key to effectively understanding and dealing with natural systems and their values for people.

Here we combine natural resources and environmental conditions in *ecological economics*, and differentiate this focus from the growth and regulatory approaches (Perman *et al.* 2003), although some economists use the core concepts in a broader, and some in a narrower, sense. Natural resources include both renewable and non-renewable resources. As might be considered in resource economics, they include both inputs into the economic system and resources undervalued or ignored by markets. At the other end are environmental conditions, for which environmental economics focuses on the by-products of production, wastes of consumption, and other human effects on natural systems. Ecological economics thus addresses the broad relationships between ecosystems and economics.

Not surprisingly, since ecological economics provides solutions for major societal issues poorly addressed by other economic systems, it brings somewhat different core attributes to the table for society. These are summarized as follows (Costanza 1991, Costanza *et al.* 1997a):

(1) Humans are one important component of, and dependent on, the overall system of natural processes and human activities.
(2) The core driving forces of preference, technology, social organization, and basic culture all evolve in response to ecological opportunities and constraints. This continuous adaptive evolution is not directional toward an equilibrium, but rather produces a fluctuating non-equilibrium.
(3) Humans have understanding and intelligence, and can manage for or against an economic goal.
(4) Individual resources are finite and the overall resource base is limited.
(5) The long-term future is given importance alongside the short-term.

(6)   Prudence based on uncertainty indicates that substitution and technology cannot indefinitely remove resource constraints.

The values of ecological economics and the contrast with growth economics are striking (Romer 1990, Nordhaus 1992, Aghion and Howitt 1998, Jones 2002). Growth economics highlights humans being central, consumption driving the system, intelligence being subservient to short-term market trends, and substitution and technology eliminating resource scarcity. In ecological economics, people understand, adapt and plan both long- and short-term for the broader natural-and-human system (or urban region) of which they are a part. Resources are finite and, in the face of uncertainty and expected surprises, prudence dictates conservation of natural resources for a more secure economic future.

Resources

Resource economics primarily addresses the many forms of natural resources that represent input capital for an economy. From a broad perspective, land and water and even space in an urban region are resources. Specific resources, of course, range from prime agricultural soil to a rare mineral, an aquifer, a forest, or a recreational greenway.

Resources may be scarce. They may approach depletion. They may be nearly all degraded. In each case no elegant economic model or "what if" exercise or agony about what the market indicates is needed to know that resource conservation is important (Dasgupta and Heal 1974, Perrings 1991, El Serafy 1991, Millennium Ecosystem Assessment 2005). With the resource lost the region is inherently poorer. Although the valuation of natural resources is a complex subject beyond the scope here (Costanza *et al.* 1997b, Patterson 2002, National Research Council 2005), the value lost is partly the short-term reduction in capital. But more importantly the long-term loss is represented by the constantly evolving physical/chemical/biological roles played by the resource. In this sense some of the loss is immeasurable and irreversible.

Planning and resource conservation are among the key solutions in resource economics. In market economics, usually a fair portion of the costs of resource use and pollution are shifted as *externalities* to the public, which pays taxes and fees to cover some of these costs. Since the full cost is rarely calculated or covered, resources and the environment degrade a little or a lot. In ecological economics, more of the costs are shifted to the resource users and the polluters. Consequently more resources are conserved and less of the environment is degraded.

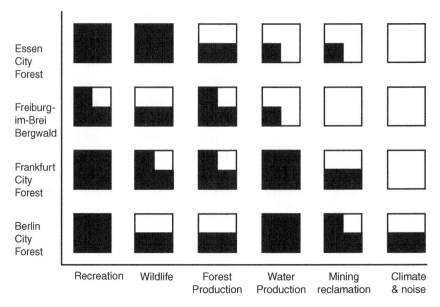

Figure 3.1 Urban forests providing valuable ecosystem services (nature's services) in Germany. Relative intensity of use: 100 % black, very high; 75 % high; 50 % medium; 25 % low; 0 % very low. Adapted from Osband (1984).

Unmentioned above and poorly represented in most economic models are *natural systems*. Yet these address major stated objectives of society and provide enormous value (Daily 1997, Atkinson *et al.* 1999, Daily and Ellison 2002, Ravetz 2000, White 2002, Millennium Ecosystem Assessment 2005). Most natural-systems values are quite familiar and taken for granted, until they are in short supply or run out. Clean water supply is a big and rapidly growing problem in numerous growing urban regions. Natural vegetation covering an aquifer is arguably the best way to sustain a clean-water supply (Figure 3.1). The natural vegetation provides this key "service" to society, and requires protection. Long-term natural-systems protection is a small cost compared with the service value provided to millions of people in the city.

Nature's services (natural-systems services, natural capital, nature's values) or *ecosystem services* (those where natural organisms play central roles) are widespread (Daily 1997, Millennium Ecosystem Assessment 2005, National Research Council 2005) and are especially provided by diverse types of greenspace. Nearby natural recreation areas, especially forested, provide a valuable service to city residents. Wetlands that absorb stormwater and reduce flooding provide a service to the city. Vegetated stream corridors reduce erosion and sedimentation. Natural soils absorb and break down chemical pollutants. The list of ecosystem services in urban regions goes on and on, each providing a key value to society

(Kremen and Ostfeld 2005, Robertson 2006). Except perhaps for protecting a water supply, ecosystem services and biodiversity do not appear to be correlated (Peter Kareiva, personal communication). Not surprisingly, markets in ecosystem services are emerging (Daily and Ellison 2002, Robertson 2006).

Formerly such values had a hard time fitting into economic models – variously referred to as non-market resources, externalities, market failure, the absence of a market, the underfunctioning of a market, the importance of informal markets, or an "e" term for environment added at the end to an equation. After protracted discussion among economists and others about steady-state economics (Boulding 1964, Daly and Cobb 1989, Daly 1990, Rogers *et al.* 2006), where resource conservation and the second law of thermodynamics are important foundations, natural resource economics and ecological economics (Costanza 1991, Jones 2002) have evolved and begun to reach front stage. However, this time the core community of economists has developed more robust models in which natural-systems values increasingly appear in some form.

Indeed, as economic growth came to the forefront in the 1940s, thirty years later Rachel Carson and big environmental challenges altered the public consciousness and public policy (McNeill 2000). *Environmentalism* has been rapidly maturing as one of history's "big ideas", and economic models are mutating to keep up with or effectively address the developing paradigm.

Conservation of resources, illustrated by the water supply and natural vegetation example above, is particularly important in urban regions, where land-to-people ratios are so limited. Traditional economic models have provided little motivation for resource conservation, for instance, because a scarce resource can be replaced by technological change or by changes in trade (Ray 1998, Pearce and Ulph 1999, Jones 2002). Still, societies that conserve their base retain a base to sustain them.

Today oil-rich Norway apparently puts its income in the bank for future generations, and spends the (growing) interest gained from this investment (Karl 1997, Listhaug 2005). With a finite annual budget, the people lead no Shangri-la life and have an incentive to conserve (Davis *et al.* 2001). Their important land and sea resources continue to look good, and even improve. Meanwhile Norwegian children and grandchildren can look to a bright economic future because their parents and grandparents invested in the future.

The environment

Environmental economics is here paired with, and overlaps, resource economics. Focusing on the by-products of production, wastes of consumption, and human overuse of natural systems, this half of ecological economics is perhaps more familiar as an approach poorly served by market forces (Costanza 1991,

Costanza *et al.* 1997b, Patterson 2002, Perman *et al.* 2003, Rogers *et al.* 2006). The heat, heavy metals, hydrocarbons, pesticides, nitrogen and phosphorus, noise, CO_2, and much more resulting from both production and consumption are often lumped as pollutants. Some call these (economic) resources out of place, which underlies, for example, recycling programs and industrial ecology that do operate within markets. Indeed wastes are an important and normal part of the production process and of society (Berger 2006). Such wastes are "everywhere," especially across the urban region, a huge challenge offering rich economic and design opportunities.

Nevertheless, most pollutants are unwanted and degrade environmental conditions. Widespread environmental impact analyses evaluate the levels of pollutant effects on natural resources and their value to people. Prevention, avoidance, and conservation efforts to reduce pollution effects on natural systems are common, and may or may not be costly. However, efforts to mediate, mitigate, and compensate pollutant effects are also made, often at considerable cost.

The giant in environmental economics is the *degradation* effect by these pollutants on natural and human resources. Aquifers, lakes, streams, estuaries, and seas suffer. Soil, diverse natural communities, wildlife, and rare species suffer. Recreation and aesthetics suffer. Human health suffers. The economic losses due to these production by-products and the wastes from human consumption are staggering and growing. With the failure of market and regulatory approaches, ecological economics has evolved to address them.

While pollutants are the core of environmental problems, a small but increasing degradation of resources is occurring from the presence of humans themselves. Population growth, urban sprawl, transportation, and technology are placing more people in more remote areas. The direct and indirect effects of this pattern are threatening and degrading more wildlife populations, rare species, natural communities, aquatic ecosystems, and fish populations. Ecological rather than market or regulatory economics is the key again here.

As mentioned in the preceding section, the degradation of resources by pollution and human overuse is a form of resource loss (Dasgupta and Heal 1974). Urbanization and associated pollution of the only good aquifer in an urban region is an obvious loss. Its short-term valuation based on costly substitution is readily calculated, whereas its long-term role as water supply, changing forest products source, soil-erosion protector, biodiversity protector, and changing recreation resource is difficult to estimate. In contrast, the diverse ramifications of increased urban heat and increased stormwater runoff, due to extensively replacing greenspace with impermeable paved areas cannot be estimated with any confidence. Consider the lost functions and the new costs for air-conditioning, industry, domestic water use, changed precipitation patterns, summer tourism, street trees, flood frequency, flood heights and damage, scouring of streams/rivers, fish populations and recreation, and stormwater pollutants carried into water bodies. The economic effects of an extensive loss of the greenspace resource are too widespread and uncertain at present, and into the future, to confidently estimate. Many other resources in the urban region fall into this category.

Whereas the preceding discussion focuses on resource protection and pollution avoidance and minimization, many of the same points apply to restoration, mitigation, and compensation for sites or areas already degraded by pollution and human overuse (Salvesen 1994, Cuperus *et al.* 2001, Forman *et al.* 2003). *Restoration* is a return to some previous state, which in the urban region typically means to semi-natural vegetation with little pollution. *Mitigation* means to minimize the effects of, which can be accomplished in many ways. *Compensation*, where mitigation seems impossible, means to provide benefits off-site equivalent to the losses on-site. A significant reduction in pollutant input and human usage, plus clean up of the accumulated pollutants, is usually assumed in restoration and mitigation projects.

Economics in time, space, and footprints

An important range of factors affecting urban region economics is presented in three categories: (1) time, stability, and sustainability; (2) spatial arrangement; and (3) ecological footprints.

Time, stability, and sustainability

Occasionally over pre-history and history cities have been wiped off the map by wars, "natural" disasters, or climate change. Now greenhouse-gas buildup with an expected 0.5 to 5+ m sea-level rise may eliminate more. Nevertheless most urban regions are nearly permanent, and are especially suitable for long-term thinking and action. The concepts of uncertainty, adaptability, and stability are all focused on the long term rather than short term. These are familiar to industries that expect to persist. The time period is decades or human generations, exactly the time period usually considered for sustainability.

Stability

Numerous solutions for issues such as uncertainty, surprises, flexibility, adaptability, and stability exist and have been proposed. Methods of analysis are equivalently diverse, from simple comparisons of options based on principles and distinctive features on the land to computer simulation modeling of complex systems (biological and/or social) with uncertainty levels, spatial remote-sensing/GIS data, and sensitivity analyses (El Serafy 1991, Perrings 1991). A few solutions are listed as illustrations, but governments and non-governmental organizations in each urban region will have their own issues and methods to achieve these goals for the long term.

The following approaches normally enhance flexibility, adaptability, and stability in an urban region: (1) conserve land; (2) protect specific resources; (3) develop buffers; (4) store resources; (5) diversify activities; (6) change technologies; (7) maintain diverse types of resources; (8) maintain redundancy, e.g., 3–5 examples of each structure or resource type; (9) save the rarest pieces; (10) have high connectivity in the transport system; (11) maintain high circuitry (lots of loops) in the transportation network; (12) provide widespread modal choice in transportation; and (13) channel development to a few satellite cities rather than indefinitely growing the central city. The reader's ideas can probably double this list. To achieve stability two long-term overriding challenges need solutions. Get through tough periods. And adapt to a changing world.

Consider the resident family on the outskirts of Kingston (Jamaica) or Jakarta who has a small fenced garden around the home. That garden may have dozens of species carefully planted and tended that provide food throughout the year: mangoes up top in a tree; an understory of coffee, bananas and papaya; manioc and other valuable shrubs; smaller plants below; vines in the right places; and chickens and other animals all around. The family is buffered from the some-times devastating economic fluctuations at the national level. Also, with a highly diverse (biodiverse) garden system, food plants may be seamlessly substituted to adapt to changing conditions. So, in addition to short-term economic activities, both the Norwegian nation and the Jamaican family have thought and invested in the long term. Economic game theory and "What if?" exercises can be done not only at these scales, but also for the urban region.

Sustainability

Sustainable development, sustainability, economic sustainability, sustainable environment, and so forth are much-used terms, yet little-used in this book, mainly because definitions vary all over the map and are usually tailored to the user's objective (Daly 1990, Rogers *et al.* 2006). All *sustainability* concepts focus on the long term, over decades or generations, rather than the usual planning and design horizon of years or a couple of decades. All normally imply a balance between natural and human conditions. Sustainable development emphasizes development, and how to do it. Sustainability suggests an overall condition, often global in perspective. *Sustainable environment* suggests a more specific spatial area, such as an urban region or portion thereof, amenable to planning.

Economic sustainability focuses on one portion of sustainability, for instance, one leg of a three-legged environment/social/economic stool (Barbier 1987, Campbell 1996, Perman *et al.* 2003, Rogers *et al.* 2006). This concept is quite different from sustained economic growth. Measures of economic sustainability have attracted particular attention (Pearce and Atkinson 1993, Pezzey *et al.* 2006). Also, sometimes sustainability economics is used to refer to ecological economics or environmental economics as described in the preceding section. Just as in those cases, a rich array of policies has been proposed based on sustainability (Howarth and Norgaard 1992, Pezzey 2004).

Finally, *urban sustainability* has a nice ring, but realistically is an oxymoron. Energy efficiency in buildings, public transport, growing food in window boxes, recycling of materials, self reliance, and such proposals are usually listed to describe urban sustainability, and typically are all positive goals. Given a city's huge concentration of people and massive inputs and outputs, the gain in energy, materials, food, etc. from such proposals is small or negligible. No pretense of a balance, where both people and nature thrive in a city, exists. People overwhelmingly dominate the area of a city or metropolitan area, and nature survives as shreds. However for a whole urban region, a nature-and-people balance is worth considering and evaluating (also see the ecological footprint section below, and especially Chapter 12, for alternative, more promising, ways to think about urban sustainability).

Spatial arrangement

A bare introduction to this often overlooked, but potentially large topic in economics highlights two dimensions: (1) urban-region patterns and economics; and (2) economic disparities: poor and rich.

Urban-region patterns and economics

As mentioned earlier, economic models are largely non-spatial (Costanza 1991, Jones 2002). In addition to the spatially explicit dimensions of ecological economics, two exceptions are noteworthy for the urban region. Urban economics is a mixture of market and regulatory approaches focused on the city, and to a lesser extent, the metropolitan area. *Economic geography*, on the other hand, evolved with an early spatial foundation (Christaller 1933, Losch 1954). It has a much broader focus on the land, and integrates, e.g., agricultural, forested, and urbanized lands with transportation. Economic geography considers the form of cities and spatial arrangement of land uses. This approach, thus, is consistent with the combined growth, regulatory, and ecological economics approaches elucidated here for the urban region (Braat and Steetskamp 1991).

Rural and urban interactions emphasize the importance of spatial pattern in an economic model (Hall 2002). Just as in development economics, they are another key to understanding urban region economics (Ray 1998). Two directional resource flows characterize the rural and urban relationship (though, as noted in Chapter 1, the term rural does not fit well in most urban regions). The traditional transfers are of rural agricultural products to the city, and of urban industrial and service products back to the rural area.

In today's urban region, resource transfers and people movements may be massive and spatially quite complex. Commuters jam highways and rail lines entering and leaving some cities, especially in industrialized nations. Air and water pollutants, diverse species, and much more move between rural and urban. Furthermore lateral movement and transfer is increasingly prevalent, especially around cities with external ring roads commonly present across Europe (Chapter 7). Therefore, in addition to agricultural economics, major transfers in the urban region emphasize transportation economics, solid-waste and sewage-treatment economics, public-health economics, and more.

These movements and transfers bring *land economics* into focus (Cheshire 1988, Fainstein and Campbell 1996, Ray 1998). Some people own land and some rent. Land value helps determine ownership, as well as ongoing housing cost for both owners and renters. But land prices or value also affect the size of area owned, an indicator of the amount of agricultural and other products from the land, at least for production that is proportional to land area. Analogously, land prices help determine whether outward urbanization is compact development or sprawl on large house lots.

In general, land value decreases with distance from city. However, the highly patchy pattern of diverse communities and greenspaces in the urban region creates a complex land-value patchwork, which is superimposed on the distance-from-city gradient. In a rapidly growing region, land values rise markedly in anticipation of growth. Also, corruption may act to artificially inflate land prices. Usually the patchwork is highly dynamic with rapid land-price changes in patches around the region appearing in a seemingly uncoordinated manner. Zoning by government and the public puts some limits to uncontrolled development, but often becomes eroded in effectiveness over time.

Municipalities in an urban region sometimes invest in transportation infrastructure to meet existing demand for resource or people transport (Gomez-Ibanez 1999, Forman *et al.* 2003). Alternatively, transportation investment is designed to stimulate development opportunities, which, without serious planning and protection, often have major environmental degradation and economic effects. Municipalities or private interests also invest in industrial and commercial development, which normally provides a significant boost in income to the municipality. And they invest in residential development, which is considered to provide a modest income, though service costs may exceed income.

Acquiring land for resource protection is just as sensitive to land value as are housing sales and ownership. Such land acquisition, as in the water supply case above, is difficult and expensive in an area near a city or a community expected to expand outward. Accomplishing land protection early, well before growth is expected, is much less expensive. It requires thinking and investing in the future, just as the Jamaican family and Norwegian nation do.

Economic disparities: poor and rich

Land value highlights another large, difficult economic issue. The benefits of growth fall unequally, creating or exacerbating the economic disparities between the rich and poor (Main and Williams 1994, Fainstein and Campbell 1996, LeGates and Stout 2003, World Bank 2006). Poverty is present in all cities and, relative to the land as a whole, often concentrated there. Wealth is analogously present and concentrated. Neither wealth nor poverty is good for the environment.

But the people and residences of the two groups are little mixed, indeed commonly quite segregated, and often separated by middle-income communities. Land values are stretched from high to low in nearby rich and poor neighborhoods, almost anywhere in and near cities. Market-driven investments in the neighborhoods tend to parallel the land values, helping to maintain both poor and rich areas. Consequently a patchwork of neighborhood land values is superimposed on the distance-from-city land value gradient.

One additional force makes this economic disparity a mammoth urban problem: immigration. Some of the highest population growth rates anywhere result from immigrants from rural to urban that "overnight" create *squatter settlements* (shantytowns, informal housing, favelas) (Perlman 1976, Main and Williams 1994, *State of the World's Cities* 2006). The people mainly arrive with no capital and do not pay for the land or its ongoing occupation, creating an externality cost for government and the city. Characteristically the arrivals are poor, unemployed, ill-prepared for urban life, and have little for shelter. Informal economic and social systems, mostly beyond the reach of economic institutions and government, control life in such communities. Government and NGOs sometimes help a bit with infrastructure or economic conditions. Since squatters illegally occupy a site, periodically the owner, whether government or private, expels them as land value changes. Thus many squatters move from location to location for various reasons, often in an outward direction from the city center.

Such squatter settlements mainly appear in specific types of locations: steep slopes, flood-prone areas, forgotten spaces around transportation corridors, and other spaces where normal land values are low and investments few. Often these are the disaster-prone areas affected by floods, earthquakes, and mudslides. They tend to be the worst for buildings and for constructing infrastructure, such as water supply and sewage treatment. In much of the world, squatter settlements on poor sites are an increasingly familiar sign of a rapidly growing urban region.

A quite different economic, social, and environmental pattern occurs with *sprawl* (Chapter 1), an urbanization process producing low-building-density built areas, particularly characteristic in North America (Handy 1992, Jenks *et al.* 1996, Gordon and Richardson 1997, Bullard *et al.* 2000, Lopez 2003, Frumkin *et al.* 2004). Sprawl thrive on market economics with weak government controls, where people with ample capital can own large house lots and travel largely by vehicle. Sprawl may correlate with population growth, though even no-growth areas often exhibit a net outward sprawl. The widespread nature of the process is illustrated regionally, where 95 % of the 74 Economic Areas in the US South are expected to experience some degree of sprawl, compared with 88 % of the 98 areas elsewhere in the USA (Burchell *et al.* 2005).

Modeling various compact-growth scenarios as alternatives to sprawl can be done using important spatial attributes in urban regions, such as adding an urban growth boundary (Ozawa 2004) or channeling development to the only county with public services available. Such models (Jenks *et al.* 1996, Gordon and Richardson 1997) can also compare the potential costs associated with each growth scenario. Even forest fragmentation, e.g., as a result of sprawl, correlates with urbanization area and may be a useful economic indicator (Wickham *et al.* 2000).

Ecological footprints

The final economic perspective for urban regions indirectly integrates many of the preceding dimensions. Instead of using monetary value or calories (Odum 1973, Odum and Odum 1981) as the universal currency, it uses area, such as hectares or acres, as the universal currency to which "everything" can be converted. The *ecological footprint* is the effect or "load" imposed on the biosphere by a population or person (Wackernagel and Rees 1996, Rees 2003, Mayor Farguell *et al.* 2005, Luck *et al.* 2001). Commonly it is measured as the total area of productive land and water surface required to support the population.

Consistent with ecological economics (Costanza 1991), the concept recognizes that: (1) irrespective of changes in trade and technology, humans remain tightly interlinked with natural systems, and the economic production and consumption process invariably uses an area of terrestrial and aquatic ecosystems; and (2) biophysical measures, rather than monetary, more effectively express the relationship between humans and ecosystems on Earth (Costanza 2000, van den Bergh and Verbruggen 1999, Lenzen and Murray 2001). Inverting the usual carrying-capacity question of how many people can live in a given area, "eco-footprinting" estimates how much area is needed to support a given population, irrespective of where it is.

What are the ecological footprints of people in urban regions? The average citizen of Europe, North America, Australia, and Japan, the most intensely urban regions worldwide, requires 5–10 ha (12.5–25 acres) of productive land and water per capita to support his or her current lifestyle. Residents of the Canadian cities, Toronto and Vancouver, have 7.7 ha footprints on average (Rees 2003), and residents of the megacity London about 11 ha (28 acres). In striking contrast the residents of developing nations on average have a footprint of about 1.0 ha (2.5 acres). The population of each of the Canadian cities has a footprint about 300 times greater than the area of the metropolitan area. The much-larger London requires a total productive area, not only larger than its urban region, but equal to the total productive land area in Britain.

These ecological footprints are based on the total equivalent productive area to provide the resource inputs to the city or person. However, as evident in the ecological economics discussion above, both incoming resources and outgoing effects on the environment are important. Thus, in an extensive study of the 29 largest European cities around the Baltic, the city-population footprints were 565–1130 times larger than the city areas (Rees 2003). That study added the area required for waste assimilation to the resource-consumption area.

Urban sustainability, as outlined in the previous section, is not promising, because the human imprint, inputs, and outputs have normally overwhelmed natural systems in a city or metropolitan area. However, a different bigger per-spective might provide a solution and also provide further insight into ecological footprints. Think of the primary inputs and outputs for a city. What areas or landscapes do they mainly come from and go to? If the group of landscapes, along with the city, were considered as a whole system, it could be planned and managed for a positive balance of both nature and people. That would be a sustainable system with the city a key part.

As an example, suppose the bulk of a city's food comes from a distant grain-growing area, a livestock area, and close-by market-gardening area. Much of the outside-manufactured products originate in a distant forested area and a mining–industrial area. Water supply mainly comes from a forested aquifer in the urban region. Also, most water pollutants and air pollutants end up in a large downwind lake, solid waste in a huge dumpsite, and recreation and its impacts in a nearby forested area. The city itself also provides industrial, commercial, and residential resources. So, to plan this city's whole system for nature and people, or urban sustainability, requires planning and managing the city with its nine primary linkage areas. In this example, four of the outside linkage landscapes are distant and five are in the urban region. The approach also highlights a next spatially explicit step in footprint analysis.

In public policy terms, ecological footprint analysis has mainly been used as shock treatment, to alert people to their consumption or over-consumption patterns in market-growth and other economies, and to highlight consumption disparities among different peoples and areas. It has provided an easily measurable quantitative method, which is useful to the public and somewhat rigorously confirms generally familiar patterns. Making footprint analysis spatial within a city or culture or region (Costanza 2000, Luck et al. 2001) offers promise for policy recommendations. For example, comparing the footprint of suburban residents with and without transit-oriented development could help highlight the relative value of TOD. Comparing the conversion of a community from oil- to coal-generated power, or vice versa, would be indicative. Multivariate comparisons of populations with different consumption habits and resource uses, analogous to public health studies, could lead to useful policy changes.

Finally, through whichever economic lenses one looks, the urban region is a rapidly changing powerhouse tightly entwining the big population with the finite land around. In the face of rapid outward urbanization, natural-resource loss, and widespread environmental degradation, an altered economic approach appears to be important and available. Combining growth and regulatory economics with ecological economics for resources and the environment offers a promising approach. Spatial arrangement also appears to be central to economic solutions for natural systems and their uses in urban regions.

Social patterns

While social and economic patterns broadly overlap, it is useful to separate the social dimensions, since they are also tightly linked to environmental patterns. For example, mudslide and flood-prone areas in cities attract squatter settlements and limit the types of social interactions and the quality of life therein. An expanding residential neighborhood of large houses and house lots focused around schools and shopping malls effectively destroys a productive agricultural or large natural area. It also consumes a huge amount of energy, produces greenhouse gases accordingly, and degrades water bodies, aquatic ecosystems, and fish populations. Based on such examples, both poverty and wealth may degrade nature, and therefore our future.

Social patterns focus on groups of people, their interactions, and their spatial and organizational arrangements, obviously key factors in understanding urban regions and uses of natural systems. Also important from a different perspective are the core components of culture, such as aesthetics, traditions, morals, learning, and language. Social patterns and culture are the two themes introduced in this and the next section.

Four somewhat dissimilar perspectives are used to explore social patterns relative to the topic of this book: (1) social linkages and spatial scale; (2) squatter settlements and the poor; (3) transportation in urbanization; and (4) land protection and social pattern.

Social linkages and spatial scale

Neighborhood, as a space of nearby residential buildings and people, and *community*, as a group of people with vibrant linkages, are useful places to begin considering social patterns and urban regions. What makes a neighborhood a community? People interacting and remaining in a neighborhood creates a community (Ravetz 2000, LeGates and Stout 2003, Forman *et al.* 2004, Handy 2005). For example, safe roads, sidewalks, paths/walkways/nature trails, bicycle routes, meeting places, playgrounds, ball fields, tiny parks, local-community-event sites, and town conservation lands help create neighborhoods with actively interacting people. These are spatial objects available for planning and designing. A high frequency of local trips by residents walking or bicycling, rather than driving, catalyzes interactions and a community. Safe, attractive walkways are especially significant as linkages. Appealing meeting places for children and neighbors are of paramount importance. Other types of neighborhoods can be described.

In a broader perspective, a sense of place creates and maintains neighborhoods. Small parks and greenspaces help define the character of a *place*, which elicits human responses such as understanding, coherence, welcome, danger, or mystery (Jacobs 1992, Kaplan *et al.* 1998). Nature and the built environment combine as the central components of a place for people.

At the urban-region scale we are all in a giant sandbox together. To establish and sustain the vibrant social linkages of a community, minimizing chaos and conflict is valuable. That requires collaboration of residents and at least some planning of space. The task is increasingly hard and urgent; so many people from afar keep entering the sandbox.

Regional social groups and organizations and institutions usually exist, but mostly in a scattered, intermittent, and peripheral way. Regional cyclists recreate together on weekends and work to create a regional trail network. Analogous groups may exist for hiking, canoeing, off-road-vehicle riding, bird-watching, sailing, and so forth. Service organizations may have regional activities. Public-health institutions may work regionally with the public. Solid-waste disposal and recycling efforts may link the public regionally. Some conservation organizations protect land around a region. Nevertheless, thinking about, working together, and planning an urban region is a challenge because social linkages are mainly at the scale of smaller units within the region.

Squatter settlements and the poor

In cities, social interactions within and among communities reflect the overall concentration of people plus the spatial arrangement of neighborhoods and other land uses (Fainstein and Campbell 1996, Bullard *et al.* 2000, Macionis and Parillo 2001, Hall 2002, Ozawa 2004). As indicated above, the economic and environmental implications of poverty are far-reaching, so we start with certain social dimensions of low-income (or slum) communities (Jacobs 1992, Main and Williams 1994, Vigier 1997, Hall 2002, *State of the World's Cities* 2006).

First it should be emphasized that stable poor communities are probably present in all cities. A sense of place is strong. Neighbors not only protect their place against internal degradation, but also against outside threats. Threats range from encroachment by politically more-powerful wealthy or middle-class communities to environmental justice issues such as building a noisy highway or a polluting factory nearby. Social linkages in such a community are kept vibrant and strong.

Much more challenging are the *squatter settlements*, shantytowns, informal housing, and favelas born of unrelated immigrants from afar (Perlman 1976, Main and Williams 1994). Normally the sites are the environmentally worst in the area, such as steep mountain slopes (e.g., Caracas, Rio de Janeiro), nearby hillslopes (Tegucigalpa), oft-flooded river floodplains (Bangkok), around railways and highways, and by polluted streams and drainage ditches where food can be grown. Greenspaces and water bodies are combed for food and other resources (Figure 3.2). Most informal communities are located close to potential jobs, such as by factories or to help serve nearby wealthy neighborhoods (Sao Paulo). Temporary squatter settlements also often appear just inside the edge of the metro area. Later, as land prices rise, they are replaced, while new squatters colonize the now-further-out metro area edge. Few informal communities appear much beyond the metropolitan area.

Squatter settlements are widely plagued by poor water supply, no sewage treatment, no public transport, few public-health services, crime, little policing, undependable electricity, few jobs, uneven food supply, makeshift shelter, and a very high ground-floor population density. Yet in the face of adversity, people, even from far unrelated rural areas, may create vibrant communities. Probably all urbanists have seen this, but let me illustrate with my story.

One glorious day, an ecologist in a motorboat took a colleague and me through the mangrove-swamp islets in the mouth of a river in Rio de Janeiro. We

Figure 3.2 Fishing for local resources in a polluted channel by an adjoining low-income neighborhood. Guatemala. Photo courtesy of Elizabeth and Bryant Harrell.

first zipped along an attractive estuarine channel behind a barrier beach lined with ten-story hotels facing away to the sea. Behind each hotel, our view was highlighted by a large blackish 1.5 m-diameter concrete pipe ending right at the edge of our channel. Flushing a toilet in any hotel room efficiently sent black-water rushing down and out. The aroma was strong, the estuary over-enriched.

Soon we stopped on a nearby degraded mudflat, collected mangrove "seeds," and stuck them in the mud a meter apart. This was at least a symbolic contribution to the ecologist's impressive mangrove restoration project funded by the hotels. That introduction widened my thinking about urban pipe systems and water bodies, a useful lesson when later working on a plan for an urban region (Forman 2004a).

More informative were the discoveries when slowly motoring among the mangrove islets. The place was filled with shantytown houses of discarded wood, plastic, tin, and anything else found in or near the river. Down the middle of wider channels were rows of crooked wooden poles and tree trunks holding a maze of drooping electric wires, which we were careful to avoid. The immigrant residents had created their own electric infrastructure. Occasionally we passed a small, flat wooden boat with facing seats for four people which was poled or paddled past us. This was the "water bus," the immigrants' own transport system. Once we passed a higher spot with a tiny school, a playfield, and a "water school-bus" with six seats. Children's education was a priority for the squatters in this community.

Finally, an amazing sight appeared on the main river channel, a line of per-haps a thousand large plastic soda-bottles strung together. From a point on the far shore the bottle line projected diagonally across and upriver, leaving only a narrow space for boats to get by. Day and night, month after month, the line of bottles caught other plastic bottles floating down the river, the discarded jet-sam and flotsam from upriver communities in the urban region. I could see two men at the far-shore point gathering in the valuable floating plastic bottles and tossing them into a truck. None would be wasted; urban resources provide jobs and income. Adversity, basic values, and ingenuity had created an active social community among apparently unrelated immigrants, right in Rio.

Still, this site and others with informal communities are environmentally among the worst for buildings and human settlements (Perlman 1976, Main and Williams 1994, *State of the World's Cities* 2006). Normally caring organizations and individuals, along with government agencies, attempt to provide limited social services for the people. Occasionally governments forcibly remove or try to eliminate such communities, sometimes with social or political repercussions. Occasionally better housing, or incentives for it, is created on a better site near jobs. However, pressure builds to recolonize the original difficult site, unless a widely recognized and policed land use of value to society is established on the site. Especially in the rapidly growing cities of developing nations worldwide, informal communities are covering large areas. Effective answers remain elusive.

Transportation in urbanization

Outward urbanization responding to population growth and opportu-nity is the norm in history. The social implications of *sprawl* (Figure 3.3) have gen-erated much discussion (Jenks *et al.* 1996, Gordon and Richardson 1997, Daniels 1999, Bullard *et al.* 2000, Benfield *et al.* 2001, Getting to Smart Growth 2002, 2003, Frumkin *et al.* 2004, Burchell *et al.* 2005). Not knowing your neighbors, alienation, scarce meeting places, little walking, mostly vehicle driving, no place to play, inconvenient for the elderly, unsafe, no time to volunteer, and commu-nity organizations that wither, are familiar refrains for life in sprawl areas. A societal priority for planning and design that directly addresses sprawl and such issues remains embryonic, though global climate change might become a cata-lyst (McCarthy *et al.* 2001, Gore 2006).

Several key dimensions of transportation, one piece of the big puzzle, are helpful here. Providing alternative modes of transportation, such as rail transit, buses, car driving, walking, bicycling, etc., is widely recognized as a key way

Figure 3.3 Residential developments relative to highway and informal walking paths. The highway and shallow ponds isolate and affect neighborhoods. Single-family homes on left are accessed by curved local streets and dead-end cul-de-sacs. Multi-unit housing at bottom has a small central common area and several informal walking paths radiating outward. Denver, USA. Photo courtesy of US Federal Highway Administration.

to provide flexibility to those traveling, and to reduce clogged routes (Handy 1992, 2005, Warren 1998, Benfield *et al.* 1999, Warner 2001, Forman *et al.* 2003, Ozawa 2004). If it takes half a day to drive across a megacity such as Sao Paulo, one may use an alternative transportation mode or simply not take the trip. A second approach, traffic calming, provides a set of techniques to slow traffic and improve the safety and quality of life within neighborhoods.

Third, communities that have been bisected by a highway commonly have both reduced social interactions and greater environmental justice issues (Jacobs 1992). Creating walkways and bikeways under or over the bisecting highway can help mitigate the degradation and reestablish a sense of community. However, designing the structures to be well vegetated and wide for good lateral vision also facilitates crossing by wildlife, thus revitalizing surrounding animal populations and nature.

A fourth approach, *transit-oriented development* (TOD), refers to urbanization centered around nodes with public transportation (Cervero 1993, 1998, Calthorpe 1993, Dittmar and Ohland 2004). Such development is typically moderate- to high-density mixed-use within an easy walk of a major transit stop. Mixed-use especially refers to residential and shopping areas in proximity, though industry and employment may also be included. Planners usually use 800 m (1/2 mile) as the radius for TOD. However, an easy walk also means a network of attractive safe paths or sidewalks, particularly radiating outward from the transit node (Handy 1992, 2005).

Transit-oriented development should reduce vehicular travel somewhat and also enhance walking and shopping close to home. Those in turn should translate into greater social interactions and sense of community, compared with a sprawl area. Also, if a sense-of-place results from relating to and caring about the combination of nature and human structures somewhere, then generating a sense of place for people will doubtless depend on a serious protection of, and provision for, natural areas in a TOD nodal area.

The San Diego trolley, Los Angeles commuter rail, and Portland (Oregon) light rail systems all have transit-oriented development around some stations along radial commuter routes (Cervero 1993, Ozawa 2004; Hollie M. Lund, 2006 website, www.csupomona.edu/-rwwillson/tod). In California, TOD residents have higher rates of transit use than people in adjacent areas, the city as a whole, and other cities and regions. Residents are about five times as likely to commute by transit as workers across the city. Transit-oriented development offers promise as an alternative to sprawl in urban regions, even though improvement in rates of local shopping trips, walking, environmental sensitivity, and social interactions remains unclear. Nevertheless, at this early stage, TOD, like "new urbanism," is still focused on development. Pairing it effectively with natural systems is an obvious giant missing step in urban regions.

Land protection and social pattern

Finally, land protection bears highlighting as a social dimension in urban regions. The social interactions involved in the use of a common resource are highlighted in the so-called "tragedy of the commons," where land owned in common is degraded by overuse. No-one feels responsible for it, and no cooperative planning and management develops (Hardin and Baden 1977, Botterton 2001). Government land intensively used by the public would have the same fate if government did not plan, manage, and police it.

However, consider parcels of land protected for nature and natural-systems-related uses by individuals, conservation organizations, and government at many levels. When the land parcels are small, isolated, and close to a city, their future is in doubt, no matter what their legal status is. As the city grows and rolls outward these small isolated nature reserves and natural parks are under increasing threat, even if the local community remains strong. Under urban pressures, gradually many become, for example, ball fields, grass–tree–bench parks, infrastructure sites, or squatter settlements.

If the protected areas are large, they tend to persist longer in a semi-natural state, but still become eroded. The huge somewhat-natural Ajusco area next to Mexico City may support water supply, rich biodiversity, wood products, game, squatter settlements, weekend houses for the wealthy, illegal drug gangs, police, armed anti-government groups, and the military (Pezzoli 1998). Imagine the land-use conflicts. Other cities have close-by large natural areas, such as Barcelona's Collserola and Brasilia's two large adjacent parks, with their own issues, including encroachment (Acebillo and Folch 2000, Forman 2004a).

The best most-*sustainable natural areas* around cities are probably those that clearly fit into a larger context, which is widely understood by government and the public. Thus a parcel in the middle of a green wedge projecting into Stockholm or in London's greenbelt or in a key greenway of Minneapolis/St. Paul's (USA) greenway network is likely to long persist (Munton 1983, Parsons and Schuyler 2000, Hall 2002, Elmqvist *et al.* 2004). Widespread recognition that the broader greenspace pattern would be disrupted by the degradation or loss of an essential piece protects the site.

Frederick Law Olmsted's Emerald Necklace in Boston has been threatened by development countless times, seriously damaged in spots occasionally, and persists today because government and the public understand and cherish it as a connected system in their midst (Zaitzevsky 1982). A city is an aggregation of sometimes-mapped social neighborhoods, as in London, but it is also a social entity as a whole (Bartuska 1994, Warner 2001). In Boston, the valued Emerald Necklace is protected by social interactions and caring by both the local adjoining communities and the city as a whole.

Culture

Culture has some advantages over economics and social patterns in planning urban regions. Economic conditions, like politics and government, may fluctuate rapidly and widely. Social patterns may also fluctuate widely, but usually somewhat more slowly, since social interactions provide a network for stability, helping a community more easily get past a difficult time. Culture, normally, is still more stable, often gradually changing and adapting over generations (Forman 1995).

Although sometimes broadly used to include social, economic, political and other dimensions, I use culture in its traditional core sense (Seddon 1997, Nassauer 1997, Buell 2005). *Culture* refers to the traditions, aestheics, arts, language, morals and learning of a group, that are passed through generations. Thus cultural cohesion is a bonding force, a long-term linking of people by common aesthetic, intellectual, and moral traditions. Culture provides stability.

Culture may be rich in either rural or urban settings, but tends to be concentrated, organized, and institutionalized in many cities (LeGates and Stout 2003). Museums, art associations, universities, language schools, concert halls, theatres, mechanical/technical institutes, and major libraries are manifestations. Architecture, art, and music around a city provide daily reminders of culture.

The urban region normally has considerable *cultural diversity*, the different cultures coexisting. Normally cultural groups are somewhat separate but spatially overlapping (Hall 2002, LeGates and Stout 2003). Where people are packed together, some conflicts are inevitable, though mutual respect for different cultures often makes things work, i.e., supports cultural diversity, in a region. A central value of cultural diversity is the richness of art, music, dance, celebrations, traditions, dress, language, and much more, provided for an urban region. These are deep enduring values which lend welcome and vitality and meaning to a place (Eaton 1997, Buell 2005, Nassauer 2005). Think of culturally diverse and vibrant Buenos Aires, Paris, New York, and San Francisco.

Human culture relative to natural systems is now explored from three perspectives: (1) nature in culture; (2) biophilia and the building; and (3) urban agriculture. The third topic also helps to integrate economic dimensions, social patterns, and culture.

Nature in culture

Nature permeates and is of central importance in human culture, just as ecosystem services provide major economic values to society. Art, songs, celebrations, stories, and traditions are rife with nature. For some people, nature has a for-its-own-sake intrinsic value, irrespective of human attitudes. Thus we have no inherent right to destroy or degrade nature. Other people treasure the "existence value" of nature. I value the migrating herds of caribou in the arctic and the existence of aardvarks and tomb bats, even though I have never seen, and might never see, them in the wild. Still others gain inspiration from nature – the reflection of a still pond and the "cathedral" of an Australian mountain-ash (*Eucalyptus grandis*) or Chilean monkey-puzzle (*Auracaria*) or US redwood forest.

Aesthetics is another widely treasured cultural value of nature – the beauty of a rich tropical rainforest edge or the glory of a single golden grass waving endlessly across a plain (Yaro *et al.* 1990, Nassauer 1997, Eaton 1997, Seddon 1997). Symbolism is especially valued by some cultures, such as the gnarled mountain pine in China and Japan and a ring of redwoods in California. Or a large nearby primeval forest may symbolize or represent danger and evil, thereby evoking fear. Even nature as a rich source of metaphors that enhance human understanding provides value to some.

Two brief stories bring culture and nature alive. A shopping mall in Minneapolis/St. Paul (USA) was built on a former wetland and years later essentially went bankrupt, leaving inexpensive temporary stores mixed with broken-window spaces (Joan Nassauer, personal communication). Conservationists convinced the city to remove the mall and recreate a wetland there. An attractive trail with long curving walkway-bridge was designed to serve neighboring communities as well as a wider array of nature walkers and birdwatchers. One local recent-immigrant community was aghast. Not only were the convenient affordable stores being closed, a wetland was beginning to appear while the safe familiar parking lot was disappearing. A wetland, no less! Wetlands are sources of evil, places where bad things happen to people, as generations of children well know in that culture.

The second story is of a forestry expert explaining how a local wooded area in a developing country could have triple the production and income from a gradual replanting with pine or eucalypt (Forman 1995). The village leader ponders the opportunity and invites the expert for a stroll through the woods. The host points out a tree that provides nuts in the dry season, a moist spot that protects their drinking water, a tree where he was married, a vine for the annual religious celebration, some unburnable trees protecting the woods on the windward side, some decrepit trees that provide flutes for the children, and tall arching trees for reflection and inspiration. The forester is warmly thanked, and then returns home to look more thoughtfully at the conifer plantations near his own community.

Nature is a centerpiece of culture and runs deeply in people. Being packed together in a city cannot extinguish that essence of humanity. Urban executives line their offices with stunning mountain scenes and seascapes which may have deep meaning to them. Urban immigrants may keep a treasured plant growing or talk about special places or show a faded photograph of relatives next to familiar trees and shrubs. Almost always these are from their home village or town, their roots. Nature is a central component of place, and of a sense of place.

These values and patterns lead to two key spatial principles useful for planning. Nature attracts or repels, or is appreciated at a distance. And different cultures coexist somewhat separately in an urban region. A wise spatial arrangement of nature and people can sustain the diverse urban-region population.

Biophilia and the building

This book on urban regions only lightly touches fine-scale patterns such as house lots and towns. Yet one very-fine-scale pattern, the building, illustrates an especially important linkage between nature and culture. People in buildings can be effectively cut off from nature. Especially for people working day after day, or during long periods of recovery from illness, that disconnect appears to be significant (Kellert and Wilson 1993, Kellert 2005). Research studies indicate that linkages with nature improve human health and recovery from illness, improve mental well-being and "quality of life," enhance worker satisfaction and productivity, and reduce stress (Ulrich 1984, Orr 2002, Frumkin *et al.* 2004, Stephen Kellert, personal communication).

These patterns seem to be associated with *biophilia*, the inherent human affinity for nature, whereby people evolved with, fundamentally depend on, and are inspired by nature (Wilson 1984). All this has spawned *biophilic design* thinking (Kellert *et al.* 2007). Buildings not only can minimize adverse environmental and human health effects (e.g., the so-called LEED design approach in architecture), but equally important, buildings and landscapes foster human health, performance, and productivity by enhancing connections to the natural environment.

Yet biophilic design is not simply anthropocentric. "Bringing buildings to life" offers significant benefit to nature itself. For example, structures can be designed to: provide habitat for targeted rare species; enhance surrounding natural systems; serve as stepping stones for species movement across a built area; attract a richness of fine-scale nature or small species on the texture of building surfaces; and even educate people for nature protection elsewhere. Since buildings may exist by the hundreds of thousands or more in urban regions, the cumulative positive effect of biophilic design could be quite remarkable.

Urban agriculture

Urban agriculture refers to the growing of food in and close to cities, though commercial flower growing is sometimes included in the concept (Ponting 1991, Smit and Nasr 1992, Jacobi *et al.* 2000). Urban agriculture may occur on any suitable site, including window boxes, balconies, rooftops, temporarily vacant spaces, community gardens (allotments) in designated public spaces, market-gardening areas (truck farms), remnant suburban or peri-urban farm fields, and fields unsuitable for building. Greenhouses of various sorts and indoor spaces for hydroponics and other high-technology biomass production are also used. Small spaces sometimes grow livestock, poultry, and other domestic animals, though normally land values are too high for animal production. Aquaculture is important in some urban regions (e.g., Calcutta/Kolkata, Bangkok) (Costa-Pierce *et al.* 2005). Family food growing is usually most successful close to one's residence. However *market-gardening* (truck farming), i.e., intensive commercial vegetable-and-fruit production in an area of small fields close to a city, seems optimum to provide local food for city markets and restaurants. Unusual examples of market-gardening are an "agricultural park" next to Barcelona and locations in a greenbelt around London (Howe 2002, Forman 2004a).

Several hundred million people apparently are involved in urban agriculture, with fresh vegetables being the most important product. In Antananarivo (Madagascar) and Bissau (Guinea-Bissau) 90 % of the city's vegetables are grown in urban agriculture (Mougeot 2005). Sofia (Bulgaria) and Addis Ababa (Ethiopia) receive ≥80% of their milk, Hanoi 50% of its meat, and London 10% of its honey from urban agriculture. Mexico City, Moscow, English cities, Australian cities, Havana, Rosario (Argentina), Vancouver (Canada), and many other cities share in this self-made bounty (Losada *et al.* 1998, Howe 2002, Mougeot 2005, Houston 2005). With a "100 mile breakfast" campaign, Vancouver has farmers and consumers increasingly producing and buying local foods. Not surprisingly, worldwide, with food production some 5 to 15 times greater per unit area around cities than in rural areas, the number of farms and farmers as well as the value of products is increasing (Jacobi *et al.* 2000, Smit 2006).

Historically food-growing was thoroughly integrated with communities and cities (Losada *et al.* 1998). But especially in some cultures, city planners and developers largely covered suitable small growing spaces with concrete and with imitations of nature in the form of city parks. In cities across the USA an average food item in a supermarket has traveled 2100 km (1300 miles) to be there (Smit and Nasr 1992). In contrast, for Accra (Ghana) government provides many incentives for local food growing, and 90% of the city's vegetables consumed are local (Asoniani-Boateng 2002, Mougeot 2005).

What are the goals and advantages of urban agriculture? (A) Economically, it: provides fresh food for markets, restaurants, and families; supplements income, especially for the poor; eliminates most transportation costs and associated road/vehicle/fuel use and pollution/greenhouse-gas emissions; and recycles organic wastes. (B) Socially, local food-growing: enhances interactions among neighbors; provides outdoor gardening opportunities for diverse social groups to work together toward a tangible constructive goal; and reduces hunger and malnutrition in poor areas. (C) Culturally, it: enhances aesthetics; discourages dumping debris on vacant spots; provides flowers and other bits of nature to enrich living spaces; and teaches urban children about soil, plant growth, animals, natural pest controls (Mougeot 2005), and food. (D) Environmentally, urban agriculture: provides greenspace; reduces air temperature and pollutant buildup; absorbs rainwater that reduces flooding and stormwater pollutant runoff; and recycles wastes such as garbage from food markets and restaurants to fertilize crop fields or to grow pigs.

Problems with urban agriculture, of course, also exist. Pesticides and excess nitrogen fertilizer seep down and pollute the groundwater. Water used for irrigation may be rich in pathogenic bacteria from human wastewater. In dry climates much scarce water is used. Insect and other vectors carrying malaria, chagas disease, and many other public health menaces are enhanced, especially by poor-drainage water in tropical cities (Robinson 1996, Asomani-Boateng 2002, Asare Afrane *et al.* 2004; Burton Singer, personal communication). Plants grown on chemically polluted vacant lots or brownfield areas may absorb high levels of heavy metals and other toxins (Kirkwood 2001, Berger 2006). Aquaculture normally produces prodigious amounts of food, yet, where human wastewater containing pathogenic bacteria is used to support the production, a cultural aversion to its use has to be balanced against the prevalence of poverty and food shortage (Costa-Pierce *et al.* 2005).

Remnant *farmland areas* in suburban or peri-urban areas also provide many values to their communities (Forman *et al.* 2004). These include diverse farm products, and the potential for production on prime agricultural soils in the future. The historical symbolism of farmland in town, the active roles of farm families, the educational dimensions of farms, and the availability and convenience of fresh produce in town farm-stands are important values. Agricultural areas near roads, railroads, paths, and scenic points contribute significantly to preserving the open and rural character of a town. They enhance game populations, and increase the town's wildlife biodiversity by providing habitat for species requiring large open areas. Perhaps most important is the ethics of protecting prime food-producing areas in a world of extensive and growing hunger.

Or maybe *teaching children* lies at the core of culture. Urban agriculture teaches where food comes from, how one's own plants grow and change, what happens after flowers open, where seeds are hiding, what grubs look like, how fast weeds grow, what pests appear, which flowers will help control pests, what birds do in gardens, how nice soil is compared with dirt, and how slimy earthworms feel. Collective knowledge about food and nature is passed from generation to generation in urban gardens.

Finally, following Chapter 1 on regions and land mosaics and Chapter 2 on planning land, this chapter on economic dimensions and social patterns completes the presentation of underlying principles and concepts focused on people. We now turn to Chapter 4 on natural systems and greenspaces, the final essential foundation provided as a springboard to understand and plan urban regions from a natural systems and human-uses perspective.

4

Natural systems and greenspaces

Ask a child with paints to make a large picture of a city look ecological. Splotches of greenery will be added around the buildings, perhaps with some birds in spots. Similarly, have the child make a remote natural valley look lived in, and some charming awkward houses will be drawn in the picture, along with people and a street. Of course no one pretends that real ecology is represented in the first image, or serious design and planning in the second image. Rather this is art, which sometimes even appears on huge highway billboards or the sides of trucks as green marketing.

After introducing regions, land planning, and socioeconomic dimensions in previous chapters, we are now ready to focus on natural systems, especially ecology. This challenging central topic for understanding and planning urban regions is introduced along with greenspaces, where natural systems have the potential of thriving long term.

Five major topics, which progressively build on each other, are presented: (1) ecosystem, community, and population ecology; (2) freshwater and marine coast ecology; (3) earth and soil; (4) microclimate and air pollutants; (5) greenspaces. Important themes and overlaps among the topics will become evident. The first four, even the fifth, are key foundations and motifs throughout the book.

Natural systems are effectively a scientific way of saying nature (Chapter 1). Rather than presenting the science of natural systems in their separate disciplines of soil science, hydrology, meteorology/atmospheric science, ecology, etc., the salient principles are nicely integrated through the lens of ecology.

Only major ecological principles of particular value in urban regions are introduced. In most cases an example illustrating the importance of a principle to society is given. No hot-off-the-press hypotheses appear. Rather, we focus on widely known much-tested principles and concepts that can be used with considerable confidence in analysis and planning (Karr 2002). To understand their development, the underlying evidence and variability, other ecological principles, and fuller definitions of terms, see basic ecology texts such as by Barbour *et al.* (1987), Krebs (1994), Smith (1996), Morin (1999), Ricklefs and Miller (2000), Townsend *et al.* (2000), and Odum and Barrett (2005). To further sense the scientific questions, methods, analyses, results, discussion, and excitement of ecological discovery, peruse recent issues of journals such as *Ecology, Ecosystems, Biological Conservation, Conservation Biology, Journal of Animal Ecology, Journal of Applied Ecology, Journal of Ecology, Landscape Ecology, Limnology and Oceanography, Oecologia, Oikos,* and scattered articles in *Nature* and *Science.*

To help understand why this particular set of natural systems principles is important, visualize the following distinctive characteristics of urban regions. Hardly any other place on Earth has these attributes. Furthermore, virtually all the ecological principles were developed from research mainly done elsewhere, so this chapter also emphasizes the applicability of ecological principles to these unusual urban places.

Distinctive characteristics of urban regions include: (1) packed people, increasing in density toward the center of the region; (2) extensive impermeable surface also increasing toward the center; (3) large aggregations of rectangular residential lots; (4) major transportation routes, sometimes with strip development, radiating from the central portion, and at times jammed with commuters; (5) central heat-island effect; (6) agricultural and/or wooded land relatively continuous in the outer portion, fragmented in the middle, and nearly absent in the central portion; (7) many busy irregular fine-scale road networks; (8) extensive areas with a diversity of prominent non-native species; (9) highly diverse air pollution in the central and downwind portion; (10) groundwater levels lowered and polluted in large areas; (11) extensive surfacewater pollution, commonly a mix of agricultural runoff, stormwater runoff, human wastewater effluent, and industrial waste; (12) stream systems widely disrupted; and (13) wetlands extensively eliminated. Other distinctive characteristics of urban regions of course could be added, but these illustrate the kinds of issues that ecological principles will be called on to address for society.

Before turning to ecosystem and population principles, a range of exceptionally important ecological principles at broader scales are briefly mentioned. Most important are the *landscape and regional ecology* principles (Forman 1995, Burel and Baudry 1999, Farina 2005, Perlman and Milder 2005, Turner 2005, Odum and Barrett 2005). These were briefly introduced in Chapter 1 and will be progressively described and widely used throughout this book on urban regions.

Continental ecology and global ecology principles include a heterogeneous array related to paleoecology, biogeography, biomes/major vegetation types, winds and ocean currents, pollutant formation and transport, and long-distance migratory birds. Rather than including a separate section here, these principles are occasionally integrated into the upcoming section on ecosystem, community, and population ecology, as well as the following sections on water, air, and soil ecology.

Ecosystem, community, and population ecology

The main focus is terrestrial, i.e., the ecological patterns and processes on land. Three broadly overlapping topics, progressively narrowing in scale, are introduced: (1) ecosystems; (2) natural communities; and (3) natural populations.

Ecosystems

Local ecosystems, from natural to degraded, appear in aerial photos or satellite images as the basic spatial units of landscapes in an urban region (though the ecosystem concept can apply, e.g., from a tiny acorn to the globe). *Energy flows one way* through an ecosystem, that is, from sunlight to producer, through the food chain, and ends up as heat dissipated in the atmosphere. In contrast, materials and mineral nutrients *either flow one way through or cycle within* an ecosystem. Ecologists consider ecosystems to be basic units of ecology and of the Earth's surface.

An *ecosystem* is a space where species interact with the physical environment (Ricklefs and Miller 2000, Odum and Barrett 2005). A city pond, a meadow, and a patch of forest are ecosystems. *Ecosystem structure* refers to the distribution of energy, materials, and organisms, while *ecosystem functioning* refers to the flows of energy and materials in food chains and cycles. The term *materials* (somewhat analogous to matter, elements, or biogeochemicals) is used as a contrast to energy, and primarily refers to water, *mineral nutrients* (chemical elements including nitrogen, phosphorus, potassium, sulfur, calcium, zinc, etc. required by organisms), and chemical pollutants.

Biomass is the amount of living tissue present, and is typically in five forms related by feeding or *trophic levels*: (1) *producers*, the green photosynthesizing plants; (2) *herbivores*, which eat plants; (3) *predators*, which consume herbivores; (4) *top predators*, which eat predators; and (5) *decomposers*, which break down and gain nutriment from dead tissue of all five groups. In an area of land the total biomass of the first four feeding levels progressively decreases and is appropriately called a *pyramid of biomass*. For example in a large suburban park, the total biomass of producers (e.g., trees and shrubs) is typically huge, herbivore biomass (e.g., caterpillars, mice, and deer) moderate to small, predator biomass very small, and the total biomass of top predators (e.g., coyotes, large wildcats, eagles) is tiny (Figure 4.1).

Figure 4.1 A keystone top predator, Florida panther (*Puma concolor coryi*). A globally threatened subspecies with fewer than 100 animals believed to exist; endemic to South Florida. Hollingsworth photo courtesy of US Fish and Wildlife Service.

Using light energy and chlorophyll, green plants photosynthesize, which effectively absorbs carbon dioxide and water to make carbohydrate and give off some oxygen. Thus plant cover, from green roofs to coastal marshes and farmland, helps somewhat in reducing greenhouse gases by absorbing CO_2, though forests and woodlands have the added advantage of holding the carbon for decades. The carbohydrate made by plants is the basis of biomass and these food chains.

Although much variation exists, the food chain may have an *energy efficiency* of about 10 %. Each time biomass energy changes in form much heat is given off (as described by the second law of thermodynamics), and in this case about 90 % of the biomass energy in one level never makes it to the next level. The food chain or pyramid has several important implications in urban regions. First, rarely are there more than four or five trophic levels in any terrestrial ecosystem. Most animals, and normally most animal species, present are herbivores. Second, not much energy remains for top predators, which are often scarce.

Third, *biological magnification*, the progressive increase and concentration of a particular material through the food chain, means that predators (and especially top predators) get a concentrated dose. If the material is toxic, predators and top predators therefore are impacted the worst. Some heavy metals and long-lived pesticides in urban regions biomagnify to toxic levels, just as some radioisotopes from the periodic "tiny" releases of radioactivity from nuclear power plants accumulate through the food chain to lethal levels in predators.

A *food web* combines the food chains in an ecosystem and indicates where each species fits, i.e., what it eats and who eats it (Morin 1999). That permits evaluation of the vulnerability of a species, as well as the stability of the ecosystem. A bird species that only eats seeds of one tree species disappears if the tree does not flower or is killed by a pest, whereas a bird that feeds on three types of seeds may readily persist in a park. Food webs also often have many feeding links lower and fewer links higher in the web, which, overall, results in less stability for higher trophic levels.

In view of the several preceding threats to predators and top predators, a final principle is especially poignant and important. Some species at higher trophic levels are called *keystone predators* because they exert an influence or control (far in excess of their limited biomass or abundance) over many species lower in the food web, as well as the ecosystem as a whole (Figure 4.1). For example, in San Diego (California) parks, the keystone predator, coyote (*Canis latrans*), controls the populations of mid-size mammals (house cats, skunks, opossums) which feed on the young of, and tend to eliminate, many native bird and mammal species (Soule 1991). Thus the presence of the top predator, even at a low density, helps maintain a diverse native fauna and a stable natural community or ecosystem.

The decomposers in the ecosystem primarily break down dead leaves and wood into mineral nutrients and heat. A whole *decomposer* or *detritus food web* operates in dead wood and particularly in the leaf litter and humus on the ground. The species include bacteria, fungi, and numerous types of tiny animals, which greatly affect urban soil conditions, from pH to water and humus availability. The urban heat-island effect, as well as global warming, accelerates decomposition, thus decreasing the leaf litter and humus cover that protects and enriches soil.

The final big subject of ecosystems is the material or *mineral nutrient (biogeochemical) cycles*, especially of sulfur, carbon, phosphorus, and nitrogen, all major components of organisms (Barbour *et al.* 1987, Odum and Barrett 2005). The *sulfur cycle* effectively links fossil fuel, especially coal burning in factories and power facilities, to nearby high sulfur dioxide levels in the air that damage plant and human health. The associated acid precipitation acidifies lakes, dissolves mortar and concrete, and degrades built structures, from gravestones and sculptures to large buildings across the region.

The "atmosphere–organism" *carbon cycle* has rapid exchanges of CO_2 between organisms and the air in ecosystems. Plants absorb CO_2 from the air, and all organisms in the food web metabolize organic (carbon-containing) compounds in cellular respiration, which liberates CO_2 back to the air. Carbon is stored in forest vegetation, cool soils, the sea, and the atmosphere. Carbon is stored long term in limestone as well as in gas–oil–coal deposits which are rapidly being combusted by concentrations of people in urban regions and elsewhere. This process increasingly pumps carbon, i.e., CO_2 and CH_4 (methane), into the atmosphere as greenhouse gas. Associated higher global temperature and sea-level rise promise big problems for many urban regions.

In the "organism–soil" *phosphorus cycle*, phosphorus is absorbed by roots, moved through the food chain, and returned to the soil in decomposition. Rocks and ocean sediments are long-term storage reservoirs for phosphorus. The more complex "atmosphere–organism–soil" *nitrogen cycle* includes the basic phosphorus-type model. In addition though, a series of bacteria with specialized functions converts ammonia (NH_4) to nitrite (NO_2), nitrite to nitrate (NO_3), nitrate to nitrogen gas (N_2) in the air, and nitrogen gas to ammonia.

Urban regions are bathed in nitrogen and phosphorus (Smith 1981, Gilbert 1991, Craul 1999, Nowak 1994, Santamouris 2001, Sieghardt *et al.* 2005). A concentration of high combustion engines used in transportation and industry covers the land with nitrogen oxides (NOX). These air pollutants are also major components of acid precipitation. Fertilizers rich in nitrogen and phosphorus are poured onto farmland that remains the matrix cover in most urban regions, so agricultural runoff into water bodies is rich in these mineral nutrients. Market-gardening areas near cities are typically nitrogen and phosphorus hotspots. Furthermore, human wastewater is rich in nitrogen and phosphorus, so downstream of sewage treatment plants and across residential areas with septic systems these mineral nutrients tend to be abundant.

An excess of nitrogen or phosphorus causes *eutrophication* (algal blooms due to nutrient enrichment), so lakes, ponds, estuaries, bays, and near-shore sea areas tend to be unnaturally green. Recreational swimming areas are degraded. Recreational and commercial fishing may be improved or worsened, but the species caught are different due to eutrophication. Heavily eutrophicated ponds and lakes may lose their oxygen, and consequently many fish and other species, at lower levels.

Natural communities

A *natural community* is the assemblage of interacting species in an ecosystem (Krebs 1994, Morin 1999). Communities dominated by different plant species are *vegetation types* (or ecosystem types), which are determined primarily by climatic and secondarily by soil conditions. Across the globe we recognize broad vegetation types or *biomes*, such as tundra, boreal forest (taiga), temperate deciduous forest, grassland, and tropical rain forest. Within each of these, vegetation types are often associated with microclimatic conditions and differ less dramatically, such as on the north vs. south sides of mountains and coastal vs. inland areas. Also soil conditions frequently differentiate fine-scale vegetation differences, such as calcareous vegetation on limestone and acidity-dependent vegetation of a former peat bog. Mapping microclimatic variations and soil types within an urban greenspace is a key step in planning and designing for habitat heterogeneity or a diversity of vegetation types.

Urban areas are extremely rich in vegetation or habitat types (Gilbert 1991, Sukopp *et al.* 1995, Godde *et al.* 1995, Wheater 1999, Boada and Capdevila 2000). One of the many ways to classify urban vegetation is a simple division into three groups (Hough 2004): (1) native plant community, such as remnant woods, dominated by native plants; (2) cultivated plant group, such as a garden area dominated by plants developed by horticulture for urban conditions; and (3) naturalized plant community, dominated by plants which acclimated or adapted to urban conditions without human action (e.g., breeding or planting). Many in the last group are from warmer climes or other countries, and are favoured by disturbance. Indeed the number of natural communities or vegetation types is an important component of nature's richness (biodiversity), and is also a basis for the richness of experience available to the public.

Vertical, horizontal, and species dimensions describe three types of community structure. *Vertical community structure* typically refers to the distribution or stratification of layers of vegetation that are primarily determined by light intensity. Consider a forest with canopy, subcanopy, understory, shrub, and herb layers. Each layer might receive about 10 % of the sunlight received by the layer above it, so plants of the herb or ground layer are adapted to thrive in very low light conditions. Animals in the forest have different food and cover requirements and are relatively different in each vertical layer. Also, dead wood in standing trunks and branches is a major habitat for many animal groups. In maintaining high faunal biodiversity, e.g., in a semi-natural park area, the number of vegetation layers is generally considered primary and the diversity of plants secondary. The relative loss of vertical community structure is a useful measure of habitat degradation.

Figure 4.2 Herbivores in heavily browsed woods, white-tailed deer (*Odocoileus virginiana*). Note exposed mineral soil with near-absence of leaf litter and humus cover. Michigan, USA. Photo courtesy of US Department of Agriculture.

Often one or two forest layers appear to be absent because of fire, livestock grazing/browsing, overpopulation of a vertebrate herbivore, logging, or other human activity sometime in the past. Removal of a shrub layer, for example, by deer overbrowsing emphasizes that animals do not simply respond to vegetation, but affect and in some cases mold the vegetation form we see, especially at edges (Figure 4.2). The loss of a shrub layer is particularly significant because it serves as cover for so many ground animals. Also it is a primary determinant of light, temperature, and moisture conditions on the soil, where concentrated seeds and invertebrates are often important food sources for animals. Urban park management may need to balance the value of vegetation cover and biodiversity against shrub-removal security concerns in spots.

Horizontal community structure may be described as vegetation along a gradient or in patches. A *gradient* represents a species assemblage gradually, rather than abruptly, changing across an area, and characteristic of certain natural conditions such as in a tropical rain forest. However many ecological patterns, including lichen diversity, soil attributes, and rare species, have been analyzed along a generalized urban-to-rural gradient (McDonnell *et al.* 1997). Usually, however, rock, soil, and water conditions are patchy with fairly abrupt boundaries, so the vegetation is *patchy* with relatively distinct edges. Human activities normally accentuate the sharpness of edges. Indeed, in greenspaces, abrupt and relatively straight *hard edges* between vegetation types tend to be depauperate in species (Forman 1995, Fortin 1999). *Soft edges* may appear as a squeezed-together gradient, as a curvy or convoluted boundary, or as a narrow strip of fine-scale patchiness or mosaic pattern. Soft edges, typically rich in biodiversity and much used by wildlife, offer many design opportunities in the urban region.

The *species structure of a community* refers to the richness, relative abundance, and composition of species present (Morin 1999, Perlman and Milder 2005). *Species richness*, the number of species present, is the core of the biodiversity concept. *Relative abundance* ranges from strong dominance or abundance of a single species to a high degree of evenness, where several species are relatively abundant and no species dominates. *Species composition* refers to the particular species present, in contrast to their richness and relative abundance. While species richness is a key to biodiversity, a strong dominant species in a community may be quite natural, or may be eliminating other species and warrant evaluation for management. Species composition emphasizes that species are not created equal, but that certain species are rare, dominant, interior, keystone, non-native, pest, and so forth. Planning, designing, and managing greenspaces is not just about species richness, but especially about species composition (Zipperer and Foresman 1997). Diversifying street trees increases bird diversity, with avian richness and composition also depending on which tree species are used. Replacing native trees with non-native trees can be expected to decrease bird diversity.

Also, communities change naturally in the process of *ecological succession*, a directional sequence of natural communities replacing one another over time. An abandoned field that is progressively dominated by herbaceous plants, shrubs, small trees, and large trees illustrates a successional sequence in which the relative abundances of numerous species change over time. Plant and animal species may differ at each stage and gradually replace one another through various mechanisms. At some stage called an *old-growth community*, the large trees may become *self-reproducing*, so another stage does not replace them. However, individual trees or groups of trees die from time to time, e.g., from blowdowns and lightning strikes, and the resulting small space or gap is often filled by species of an earlier successional stage. This process of forming and filling holes in the vegetation is called *gap dynamics*, with large gaps, especially, mimicking the original abandoned-field sequence. Protecting old-growth in an urban region is a high priority, because of its scarcity and the presence of old-growth-related rare species. Maintaining successional species, those characteristic of the early and mid stages, is a well-known goal in management of certain natural vegetation. *Retrogressive succession*, where human effects progressively degrade natural vegetation is particularly prevalent in urban regions, and can be reversed by removing the cause and in some cases planting species of the next successional stage.

For more than a decade ecologists have basically dropped "balance of nature" and equilibrium community from their vocabulary. Instead they emphasize the *non-equilibrium* nature of nature, since the scientific evidence overwhelmingly highlights change as the norm. *Disturbance* as something causing a rapid major change in the community may come from inside, e.g., a pest outbreak or fire, or from outside, e.g., overhunting or tornado blowdown. Thus, like the gap dynamics at a fine scale, disturbance-induced *patch dynamics* means that the vegetation at a broad scale resembles a mosaic of successional stages (Pickett and White 1985). Patch dynamics maintains high biodiversity. Indeed the prevention of disturbance, rather than disturbance itself, is the threat. Furthermore some *disturbance-adapted species*, such as fire-adapted species, evolved with and essentially depend on periodic disturbance. Without disturbance these species are outcompeted and gradually disappear. Patch dynamics is particularly important in natural landscapes in the outer portion of urban regions.

Natural populations

Forming the basic unit of a natural community, a *population* refers to all the individuals of a single species in an area (Smith 1996, Ricklefs and Miller 2000, Townsend *et al.* 2000). The pandas of a mountain range and "polkadot" palms of a large swamp are populations. *Exponential growth* refers to a population that increases in proportion to its size. Populations differ in their intrinsic genetically determined rate of growth, basically the difference between birth rate (natality) and death rate (mortality). Sex ratios and age structure (e.g., the proportion of pre-reproductive, reproductive, and post-reproductive individuals) are internal attributes affecting growth rate. *Doubling time*, how long it takes to double the population size, is often a useful integrator of these variables.

External factors also affect population growth rate. The balance between immigration and emigration rates is important for some populations. *Carrying capacity* refers to the maximum number of individuals an area can support. As population size gradually approaches carrying capacity, *environmental resistance* becomes stronger and growth rate decreases, eventually reaching zero when population size reaches carrying capacity. Environmental resistance thus gradually changes the exponential J-shaped curve into an S-shaped curve. Environmental resistance includes weather and many human effects, as well as inherent *density-dependent factors* such as increased mortality and hormonal limitation on birth rate, which tend to regulate or dampen fluctuations in population size. Sometimes a population like rabbits, deer, and kangaroos exceeds the carrying capacity, causing severe vegetation damage and soil erosion, so the population crashes and subsequently rebuilds slowly. An unusually high immigration rate or proportion of pre-reproductive individuals is an early indicator of a population that may overshoot the carrying capacity and damage the environment.

Consider one large population of your favorite species in a large area; then consider four small populations on four separate patches with individuals never moving between patches (McCullough 1996, Hanski and Gilpin 1997). A *metapopulation* is the intermediate case: its individuals are distributed in separate patches, but organisms move from patch to patch. This pattern is particularly important in urban regions where the land is fragmented by highways and residential areas, which convert large natural populations into interacting small separate subpopulations. Small populations are more likely to disappear (go locally extinct) due to demographic fluctuations and inbreeding genetic deficiencies. *Metapopulation dynamics* refers to the rate of loss and recolonization of subpopulations on the patches. Characteristics of the patches largely determine species loss, whereas characteristics of the matrix and corridors between patches largely determine species recolonization. Metapopulation formation, species loss from patches, and species recolonization are all readily affected by land-use pattern and planning. For example, maintaining at least one large patch is particularly effective at reducing metapopulation dynamics, because individuals are likely to continually disperse outward from it to the small patches.

Natural selection is the genetically based change in a population over time, a process with four key components: (1) overpopulation (more individuals than can survive in the next generation); (2) variation (individuals that differ genetically and in their use of resources and environmental responses); (3) competition (individuals compete for limited resources); (4) and survival of the fittest (the best adapted or most genetically fit individuals survive and pass their genes to the next generation). So-called *K-selected species* have large individuals and reproduce slowly, whereas *r-selected species* have small individuals and reproduce rapidly. The former tend to dominate a site for a long time, while the latter rapidly colonize disturbed sites. Natural selection is a central process in evolution and in forming new species, including *endemic species* which only naturally exist in one area. The Capetown (South Africa), Perth (Australia), and Concepcion (Chile) areas are rich in endemic species.

Adaptations are attributes genetically determined over generations that provide an advantage or increased fitness to an individual or population. For instance, when industrial pollution blackened tree trunks in Britain, light-colored pepper moths on the bark were increasingly visible and subject to bird predators; the moth species adapted over generations and became darker and less visible. In contrast, *acclimation* (or acclimatization) is an adjustment of a single individual to gradually changing conditions. Also an animal may become *habituated* or accustomed to a continuous or frequent human activity so that the animal is no longer disturbed by the activity.

Several other types of species are especially important in urban regions. *Rare species* have small populations, and are inherently in jeopardy near urbanization. Conservation biologists recognize different types of native rare species, some decreasing and some stable in population size. *Non-native species* (sometimes called *exotic*, introduced, alien, or non-indigenous species) originated elsewhere. Residential areas are major centers of non-native species for the urban-region ring. *Invasive species* are non-natives that successfully colonize and reproduce in a natural community. *Naturalized species* are effectively former invasives which have become well integrated into, rather than unnaturally dominant in, native food webs and ecosystems (Sorrie 2005, Muehlenbach 1979). *Interior species* live in a large patch only or mainly distant from its boundary. In contrast, *edge species* are only or mainly near the boundary of a patch of any size. *Specialist species* have a narrow genetic tolerance and are typically limited to a specific type of habitat. *Generalist species*, on the other hand, have a wide genetic tolerance and thrive in a variety of habitats and edge conditions. *Keystone species* have an influence on the natural community far in excess of their biomass or abundance, and may be of major importance for land-use planning (Figures 4.1 and 4.2). Finally, *species of conservation importance* is a general concept mostly referring to rare species or keystone species recognized to be of particular ecological or societal value in an area.

An organism increases growth rate as resources increase, typically growing until a limiting resource is no longer sufficient to permit further growth (Smith 1996, Ricklefs and Miller 2000, Townsend *et al.* 2000). *Competition* for limiting resources benefits one species and inhibits the other. Environmental conditions, such as pH and temperature, influence an organism's use of resources, but are not depleted by the process. *Diffuse competition* where a species competes with many species for many resources, each resource being a small portion of the total used, is widespread. *Species coexistence* in nature is also the norm, and largely results from: environmental heterogeneity, such as hiding places and varied-size food patches; the use of many rather than one resource; having different food preferences; and switching diets as food availability fluctuates.

Unlike competition between species at one trophic level, predators and prey represent two levels in the food chain (Morin 1999). *Predators* (including top predators) are animals that consume other organisms, while *prey* are organisms consumed by a predator. When prey density changes, the predator changes its rate of food consumption. *Predator–prey cycles* illustrate a *negative feedback system*, whereby one component stimulates a second which inhibits the first, resulting in some stability for both components. In this case, more prey leads to more predators, which leads to fewer prey, which leads to fewer predators, which leads to more prey, which leads to more predators, and on and on. This predator–prey cycle does not produce constancy in either component, but rather each cycles up and down in numbers over time, and neither goes extinct. The persistence and regular cycling is a form of *regulation* or *stability* for both populations. Many other factors in an ecosystem affect predator and prey populations, so the relative importance of predator–prey cycles varies widely by species and location.

Herbivores and plants are somewhat analogous to a predator–prey cycle. In addition natural selection plays a key role. Over many generations a plant species adapts to the chewing and sucking of herbivores by changing the chemistry of leaves and stems. Chemicals either unpalatable or toxic to the herbivore species, i.e., *defensive chemicals*, accumulate, which effectively protects the plants from being completely defoliated. Indeed leaf chemistry varies greatly from species to species. Tree species planted along streets help determine what and how many herbivore insects (and seed-eating birds) are present, and thus what and how many insect-eating birds are present. When periodic explosions of an insect population occur (Robinson 1996), certain plant species are extensively defoliated, while others with the appropriate chemical defenses are barely touched.

Public health issues in urban regions often depend on the growth rate and dispersal of certain *species vectors* which rapidly spread disease. In medieval Europe, there were rats that carried the fleas that carried the plague bacteria that killed the people walled up in cities. Urban wetlands and community gardens (allotments) in the tropics may support mosquito populations that carry malaria protozoa that kill people (e.g., Harare, Zimbabwe). Understanding life cycles of the different species, and especially the spatial patterns affecting dispersal or transmission rates, are keys to public health solutions.

Finally, species movement from site to site by walking, flying, water transport, or wind transport is a key ecological characteristic. It is usually difficult for seedlings to grow to maturity close to their parent, and most *plant dispersal* is by seeds carried and dropped elsewhere by animals or wind. Many vertebrates have a *territory* around the den or nest that is defended, mainly against other individuals of the same species. The *home range* is a larger area used in daily movements, especially foraging for food. When young reach subadult stage, *animal dispersal* normally occurs, where the individual looks for a mate and establishes its own home range at some distance. Also, *migration*, a cyclic movement between locations, helps some animals avoid difficult environmental conditions and access beneficial conditions. Large semi-natural areas in an extensive built area tend to be valuable feeding and resting spots for migrating birds. However, home range foraging and animal dispersal are most important in urban-region planning, since these strongly depend on the sizes and arrangements of habitats.

Freshwater and marine coast ecology

We now turn to diverse types of water bodies, their aquatic ecosystems, and their immediately surrounding land. The first big topic, *freshwater*, includes hydrology, groundwater, wetlands, streams, rivers, lakes, and urban-region effects. The other complex topic, *marine coasts*, includes rocky shores, sandy beach strips, coastal wetlands, estuarine bays, various saltwater ecosystems, dynamic forces, and urban-region effects.

Freshwater

The *hydrologic cycle*, describing the accumulations and flows of water on Earth, is a simple way to begin (Ahrens 1991, Moran and Morgan 1994, Smith 1996, Wetzel 2001, Odum and Barrett 2005). Water accumulates as water vapor in the atmosphere, as liquid in *waterbodies* (i.e., groundwater, lakes, streams/rivers, and seas), and as ice in mountain and polar glaciers. Cooling the water vapor produces precipitation (rain and snow) which falls on land and water. Some rainwater soaks into the ground and some moves across the land in streams and rivers to the sea. *Evapotranspiration* from the land and plants, plus evaporation from water bodies, sends water vapor back to the atmosphere, completing the cycle.

Deep groundwater is normally in an *aquifer*, i.e., a porous rock or sandy area full of water, suggestive of an underground lake (Gibert *et al.* 1994). Except in limestone areas, water at the upper surface of an aquifer moves very slowly, often only tens of meters or a few hundred meters a year, so pollutants reaching an aquifer tend to accumulate. *Shallow groundwater* within meters of the ground surface saturates earth and soil spaces, with the *watertable* being the top of the saturated zone. Shallow groundwater flows through the ground into, and commonly helps maintain, surface water bodies such as streams and lakes. It also sustains plant roots and vegetation.

Wetlands have water at or above the ground surface for prolonged periods most years, with marshes (dominated by herbaceous plants, such as grasses and sedges) having the longest inundation period, peatlands (often dominated by shrubs and peat moss) commonly intermediate, and swamps (dominated by trees) shorter inundation seasons (Salvesen 1994, Smith 1996, Keddy 2000, Mitsch and Gosselink 2000, Parsons *et al.* 2002). The seasonal rises and falls of water level, plus microhabitat heterogeneity, are important wetland characteristics. In most urban regions, wetlands have been mainly drained or filled, so the remaining ones are often centers of rare species.

Lakes and ponds usually have sufficient horizontal water flow to remain oxygenated, and hence support ample populations of fish and other species (Wetzel and Likens 2000, Wetzel 2001, Kalff 2002). Internal habitat heterogeneity is especially important. *Littoral zones*, the shallow water edges of lakes, tend to support rooted vegetation and numerous microhabitats for a diverse fauna, from microorganisms to fish. Lakeshores with natural vegetation, which are often scarce in urban regions, provide valuable inputs and protection for aquatic ecosystems in small lakes and ponds. Lake bottoms are typically covered by fine inorganic and organic sediments, which support bottom-dwelling animals. Water layers in the lake, largely characterized by temperature and oxygen differences, contain somewhat different fish and other species. Some lakes have *anaerobic conditions*, i.e., no oxygen, in the bottom layer permanently or seasonally. *Ephemeral ponds* (such as vernal pools) contain surface water for only a portion of the year, which essentially eliminates fish, but provides habitat for many unusual species of conservation importance (Colburn 2004).

A stream/river system receives precipitation water from a *drainage basin* (catchment or watershed area) (Figure 4.3) (Decamps and Decamps 2001, Wetzel 2001, Kalff 2002, Wiens 2002). In streams and rivers, water velocity is a major determinant of both their habitat structure and aquatic ecosystems. In moist climates, streams and rivers are differentiated by a *stream-order system*. In this the smallest perennially flowing streams, called first-order streams, receive groundwater flows, but only receive surfacewater in tiny intermittent (ephemeral) channels. Two first-order streams combine to form a second-order stream, two second-orders combine to form a third-order stream, and so forth. Second- to about fourth-order streams typically are straight to somewhat curvy, flow rather fast, slowly erode downward, maintain heterogeneous mainly rocky or sandy bottoms, and have narrow floodplains alongside. About fifth-order-and-up rivers commonly are meandering or convoluted, flow slowly, gradually accumulate sediment, have river bottoms predominantly of silt (relatively fine material), and flow through wide floodplains.

The *river continuum concept* highlights the sequence from small first-order stream to large river, with gradual changes in water velocity, inputs of important dead leaves, light availability, presence of rooted vegetation, floating-algae production, stream-bottom micro-heterogeneity, curvilinearity of streamsides, fish populations, and much more. Just above where rivers enter the sea, freshwater is flowing downriver while tidal seawater pushes upriver, producing a *freshwater tidal zone* sometimes extending for several kilometers. This unusual habitat

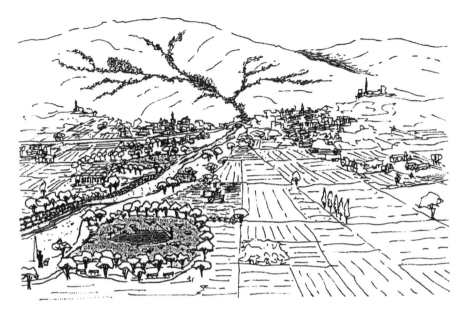

Figure 4.3 Upper streams and lower river with associated wetlands. All three types
of water bodies are typically connected to the underlying groundwater, and have
quite different structure, functions, and protection requirements. The river has been
straightened (channelized) and rocks line the riverbank. Also a bridge and nearby
built structures have fixed the channel in position, reducing its natural capacity to
migrate back and forth in its floodplain. First- to fourth-order streams in the hills,
recreation in a wetland and river floodplain, and agricultural runoff issues in
cropfields are illustrated. Adapted from Taco I. Matthews' drawing (Forman 2004a).

supports many rare species of conservation importance, yet is often degraded by
urban activities.

These river patterns are modified by a cascade of typical urban-region
attributes, including: widespread built areas; impermeable surfaces on
slopes; high peak flows from piped/channelized stormwater; extensive squeez-
ing/straightening/piping of tributaries (and river); major infrastructure conduits
along a river; small maintainance roads crisscrossing floodplains; elimination
of most wetlands; limited natural vegetation remaining across the land; very little stream–corridor vegetation; lowered water tables; human wastewater, some
or most untreated, from a concentrated population; numerous old and new
industries; very high road density; huge traffic flows; and much more. These
commonly result in: normal low-water flows; periodic big floods with structural
damage; waste removal and cleaner cities; heavily polluted downriver flows;

blockage of fish migration; and severely degraded aquatic ecosystems and fish populations. Yet urban rivers can be the people's central focus and joy.

Water pollutants cause physical changes (e.g., covering fish-spawning gravel beds with fine sediment, or adding sun-heated roadside ditchwater to a cool stream); chemical changes (e.g., adding nitrogen and phosphorus from fertilizer runoff, organic matter from sewage effluent, or toxic substances from an industry); and biological changes (e.g., an explosion of blue-green algae, loss of fish due to loss of oxygen, or disappearance of mussel populations due to sediment-laden muddy water). Urban regions are especially characterized by four types of water pollution: (1) *agricultural runoff* from the urban-region ring adds nitrogen and phosphorus from fertilizers plus sediment from surface erosion; (2) *stormwater runoff* from roads, buildings, and other surfaces carries hydrocarbons and nitrogen oxides from vehicles, heavy metals from many sources, and an array of atmospheric pollutants that have settled on surfaces, only to be rain-washed into stormwater pipes and water bodies (it also picks up heat and dissolved chemicals from the surface materials themselves); (3) *septic and sewage effluent*, the former from dispersed residential locations and the latter from sewage treatment facilities, adds organic matter, nitrogen, and phosphorus to water bodies; (4) *industrial wastes* are extremely diverse and sometimes little known, but include inorganic materials such as heavy metals, plus numerous organic substances associated with the manufacture of plastics, paints, and other products common in markets, hardware stores, and automotive operations.

Sediment runoff from cropfield erosion and construction sites produces muddy water, smoothes the bottom (a loss of important microhabitats), and clogs up fish gills and other filter-feeding animals. Toxic substances kill aquatic organisms of many sorts. Organic matter from sewage and some industrial pollution causes an exponential growth of bacteria decomposing it, thus creating anaerobic conditions, which kill fish and other aquatic organisms. An excess of nitrogen or phosphorus normally produces *eutrophication*, a nutrient-enrichment-caused explosion of algae near the surface, and sometimes anaerobic conditions lower down as bacteria decompose dead algal cells falling to the bottom. Continuous, relatively clean, water along a river system is important for passage of stream-to-sea-to-stream *migratory fish*, such as salmon and eels, a special challenge where the river passes a city.

Urban-region streams are usually heavily impacted by the removal of woody *riparian vegetation* covering the floodplain, channel straightening (channelization), adding rocks or concrete along streamsides, and directing the stream into a large underground pipe. *Water quantity* or hydrology, rather than water quality, is the main issue for flooding and low flows. Impermeable surface cover in the drainage basin produces rapid and high water flows (Arnold and Gibbons 1996,

Spivey 2002, Forman *et al.* 2003, Jared 2004, Frazer 2005). *Peak flow*, the maximum height reached, is the primary determinant of flood damage. Riparian vegetation, plus vegetation depressions and other stormwater detention basins, are primary solutions for reducing peak flows. These solutions also permit more water to soak into the ground so streams do not dry out as readily in dry seasons. Stream corridors (blue-green ribbons) are often prominent in urban regions.

Marine coasts

The abundance of cities on or near a coast means that extensive areas of coastline are within urban regions and subject to the manifold effects of concentrated people. Coastal areas themselves are as complex as the built areas, so identifying and addressing the key problems is not simple (Beatley *et al.* 1994, Smith 1996, Breen and Rigby 1996). Nevertheless, it is clear that coastal areas are especially rich in biodiversity. The combination of freshwater, brackish, and saltwater habitats, the highly heterogeneous coastline produced by gradual land-surface variations relative to water surface, the dynamic nature of the zone affected by daily winds, waves, and water flows, plus extreme weather events, provides countless microhabitats for rare species and natural communities. Also coastlines are used by many migratory fish, turtles, and birds that move out to sea, up rivers, across the land, and along coastlines.

Four major terrestrial coastline types are easily recognized. *Rocky coastlines* experience high wave energy, have a rather distinct elevational zonation of plants and animals, and often an abundance of tidepools full of rather unusual species. Second, *sandy coastlines*, including barrier islands, under natural conditions commonly have dunes, grass-covered areas, woody vegetation, lots of feeding shorebirds, sometimes nesting sea turtles, and high wave energy. But beaches are magnets for urban people, who largely eliminate the dunes, grass, woody cover, shorebirds, and nesting turtles. Linear hard structures such as jetties and walls are often built and rebuilt and rebuilt again after periodic big storms, and significantly alter, not only the beach area, but also the marine area (Pilkey and Dixon 1996). Barrier beach islands, river deltas, lower floodplains of rivers, and other sandy areas near cities often experience *saltwater intrusion*, where seawater moves inland a distance, under surface freshwater. This results either from excessive pumping out of freshwater, or the upslope diversion of freshwater that would normally recharge the groundwater. Saltwater intrusion causes wells to produce undrinkable water.

The third major coastline type, *coastal wetlands*, includes marshes (grassy), mangrove swamps in the tropics, and mudflats at the mouth of rivers. These wetlands have gradual elevational changes and gradual salinity changes from freshwater to brackish to saltwater. Also, daily, monthly, and storm tides repeatedly cover parts or all of these wetlands. Coastal wetlands, such as those formerly on the seaward side of New Orleans, are important wave energy absorbers against major storms (Farber 1987, Danielsen *et al.* 2005, Costanza *et al.* 2006). Yet coastal wetlands are particularly subject to loss in urban regions. The wetlands are drained and filled, sometimes extending the coastline outward to add high-value city-center real estate for offices, condominiums, parks, and shipping docks (e.g., Boston; Chicago; Kagoshima, Japan). The fourth terrestrial coastline type, *bays, coves, and harbors*, are indentations along the seacoast with concentrated uses because of their relatively protected low-wave-energy condition. Docks, boat anchorages, and accompanying functions are located there.

Several of the coastal land types are, in effect, *estuaries*, the coastal water-bodies and aquatic ecosystems where river freshwater carrying sediment and nutrients mixes with saltwater from the sea. Rivers commonly empty into estuarine bays, coves, and harbors, which are bordered by coastal wetlands. The combination of environmental resources makes an estuary exceptionally productive, so normally food chains are long, food webs complex, and fisheries and shellfisheries quite productive. Furthermore, the rich resource base combined with considerable horizontal and vertical heterogeneity usually results in extremely high biodiversity in estuaries.

Underwater habitats, including seagrass beds, coral reefs, submerged rocky areas, clear sandy areas, mud bottoms, and, in urban regions, plenty of sunken ships, boats, and debris used by fish, are equally diverse and important. Seagrass beds are rich areas for fish and are good indicators of unpolluted water. Coral reefs, among the Earth's most species-rich habitats, also require clean water. Estuarine bays and coves support both dense and diverse fish and shellfish populations, as do near-shore areas along the seacoast.

These coastal areas are all highly dynamic. Storms on land produce floodwaters, and storms from the sea bring strong winds and large waves. Big storms such as hurricanes (cyclones) periodically occur, as well as tsunamis in some regions. Sand is eroded, transported, and deposited by sea currents along the coast (Pilkey and Dixon 1996). Groundwater and wind arrive from the land. Thus coasts are in the path of, and shaped by, these forces. Yet coastal areas are also highly resilient, returning quickly from a disturbance, though often in different form. Adding to the effect is gradual sea-level rise (Chapter 12).

Many other urban effects on these coastal spatial patterns and processes are evident. A river passing a city receives masses of pollutants which form elongated plumes in coastal areas. Such pollutants may block light penetration, cause eutrophication, cover the bottom with sediment, damage tide pools, degrade estuarine shellfish beds, and diminish near-shore sea fish populations. Sewage from the city may enter the coastal ecosystems, causing problems associated with excessive organic matter and mineral nutrients. Major industrial centers with docks and shipping are usually present by coastal cities (Breen and Rigby 1996), resulting in highly polluted deep water that spreads by currents along the near-shore coast. The dispersed discarding of solid waste, or dumping in recognized locations, produces accumulations of debris on the sea bottom. Concentrated recreational boating and commercial fish/shellfish boating widely distribute organic wastes in the water, and significantly affect both fish and bird populations.

Houses, condominiums, hotels, and other commercial structures often line the coasts of urban regions (e.g., Australia's Gold Coast, Southern Spain, Miami Beach), especially sandy coastlines. A paucity of protected coastal strips plus masses of migratory beachgoers and boaters normally results in a coastline with but shreds of nature. It seems likely that intense human impacts expanding from cities along coastlines, which are among the world's most ecologically valuable habitats, will continue, unless interrupted by rapid sea-level rise and more severe weather events associated with global climate change.

Earth and soil

Two key ecological topics are introduced in the context of urban regions: (1) earth and soil in this section; and (2) microclimate and air pollutants in the following section.

Earth (or *earthen material*), resulting from the breakdown of rock to smaller particles, provides many important functions in urban regions (Costa and Baker 1981, Gilbert 1991, Craul 1999, Bartels 2000). These include aquifer water supply, a stormwater drainage system, and a patchy deep structure underlying different soil types on the surface. Porous sand and gravel is extensively used as fill around buildings and transportation infrastructure, so that water can readily drain through rather than accumulate in it. However, earthen fill is inherently unstable on slopes, such as the downslope side of a highway where the surface of fill is often highly erodible, and thus a sedimentation threat to a water body below (Lal 1994, Forman *et al.* 2003). Overall, however, *erosion* and *sedimentation* rates in commercial, public, and low-density residential areas are low (Craul 1999). Indeed downstream areas are often scoured out leaving streams short on sediment. Earth is easily moved, mixed, and smoothed, so all cities have important sand-and-gravel sources. Usually coastal plains, eskers, river floodplains, or more-expensive rock-crushing operations provide urban fill.

The proportions of *sand* (relatively coarse material of large particles), *silt* (intermediate-sized particles), and *clay* (fine particles) determine the *texture* of earth or soil. For example, sandy soils have more porosity, better drainage, more oxygen, and less compaction susceptibility. In contrast, clay-dominated material tends to drain poorly, form surface puddles, erode, compact readily, and hold mineral nutrients better.

Soil, on the other hand, as the upper portion of earth altered by organisms, is a rich dynamic mixture of mineral particles, water, air, roots, organic matter (blackish dead material), fungi, bacteria, and tiny soil animals. (This is not dirt which refers to unclean or filthy material). Soil types differ in organic matter, texture, chemistry, and other attributes, and are patchily distributed across the ground surface, as represented in soil maps. Vertically a typical soil profile is usually composed of somewhat distinct soil horizons. The topsoil or A-horizon has an organic layer of leaf litter and humus over a layer of mineral particles and organic matter, that in turn is on a leached layer with most mineral nutrients washed out by percolating rainwater. Beneath the A-horizon layers is a subsoil or B-horizon and beneath it a C-horizon mainly of decomposed rock materials. Organic matter and available nutrients are scarce in these two horizons.

Soil profiles and soil types partially reflect underlying rock types, but mainly reflect differences in climate and microclimate. Planners and designers have used soil maps showing the water-holding capacity (primarily determined by soil texture) of each soil type, together with the juxtaposition of types, to absorb stormwater and thus minimize flooding in communities (Woodlands New Community 1973–74, Morgan and King 1987, Galatas 2004).

Urban soils are significantly modified by human activities, especially related to construction history (Craul 1999). Indeed, where impermeable surface cover due to roads, parking lots, and buildings is extensive, very little soil of any sort exists. Some urban parks have little or no natural soil remaining; New York's 340 ha Central Park was entirely covered by 10 000 wagon-loads of sand from nearby Long Island (Phillip Craul, personal communication). Near metropolitan areas remnant woodland/forest is normally on poor agricultural soil and poor building sites, while farmland is on good agricultural soil.

The preceding structural attributes of soils have major effects on tiny soil animals, water, and chemistry, which largely determine how soil works (Cothrel et al. 1997, Steinberg et al. 1997). Soil animals in prolific numbers, such as earthworms, snails, slugs, and beetles, move up and down in the soil (except where it is highly acid). This increases porosity, drainage, and oxygen conditions, and mixes organic matter and mineral nutrients, all of which generally benefit plant growth. On the other hand, high soil temperature from the urban heat-island effect accelerates decomposition and disappearance of soil organic matter and physiologically stresses plants and animals.

Silt-dominated soils and loamy soils (with a fairly even mixture of sand, silt, and clay) are particularly good for growing crops and other plants. This is because excess water tends to drain away, the soil holds adequate water for plant growth (except in droughts when wilting occurs), oxygen reaches relatively far down, and soil particles (especially the fine clay particles present) hold ample mineral nutrients to support plant growth. Loamy and silt soils are also best for septic systems in residential areas, because water neither puddles nor flows too rapidly to a water body. Also, oxygen helps in decomposition of the organic matter, and the wastewater mineral nutrients may be held in the soil and absorbed by roots.

Soil acidity (pH) is strongly affected by rock type, yet rainwater running over the widespread mortar and concrete surfaces in built areas typically makes urban soils less acid (higher pH), which affects the plant species that can thrive. Soil acidity affects mineral nutrient availability and root absorption. Not surprisingly urban soil chemistry is extremely diverse and particularly important from mineral nutrient and pollutant perspectives (Sieghardt 2005). *Mineral nutrients* such as nitrogen, phosphorus, sulfur, calcium, magnesium, potassium, iron, and so forth are required in modest amounts for plant growth. If in short supply, nitrogen–phosphorus–potassium (N–P–K) fertilizer is typically added. Some elements, such as heavy metals, zinc, cadmium, nickel, and copper, are required by plants in tiny or trace amounts.

However, at high concentrations, resulting from pollution, these trace metals or elements are often toxic (Sieghardt 2005). Also, many polluting organic substances, e.g., from the plastics, paint, and automotive industries, are directly toxic to plants, soil animals, and microorganisms in soil. Hydrocarbons, particularly from fuel combustion and vehicle use, tend to coat urban soils, creating a "hydrophobic" film or crust (Gingrich and Diamond 2001). In consequence, water from light rains does not readily soak into the soil and root area, but remains on the surface and is evaporated. Nitrogen oxides (NOX) from vehicles and machinery also coat the soil, which often results in one or two plant species becoming dominant, and many others therefore becoming scarce or absent (effectively a terrestrial eutrophication process).

Several other attributes of urban soils are rather distinctive (Craul 1999). The horizontal and vertical variability is typically greater, patchier, at a finer scale, and different than in rural areas. Fallen leaves may be removed from under the scarce vegetation, so the organic matter that holds mineral nutrients and facilitates root growth tends to be limited in urban areas. Human materials and contaminants – metals, plastics, glass, asphalt, masonry, pesticides, road salt, and much more – are patchily distributed, break down slowly if at all, and leave accumulations of many pollutants and toxins in the soil. Also residential septic systems and cesspools add concentrated water, organic matter, and nutrients in spots.

Digging down to see an urban soil profile often reveals layers of sand or rubble (e.g., pieces of bricks, mortar, concrete) with sharp boundaries, sometimes with a buried layer of mixed organic matter and mineral particles, which reflect the history of building activities on the site. The sand and rubble layers are porous, and the layers with mortar or concrete have a high pH.

Soil compaction and associated poor plant growth is widespread in urban areas, from the 50–100 cm thick compacted material under ballfields to regularly walked trails. Construction equipment often significantly compacts the subsoil, so when a site is later covered with topsoil, water cannot drain well and accumulates in the topsoil inhibiting plant growth. Even soil compaction due to vibrations from nearby vehicle traffic, trains, and diverse equipment is significant and widespread in urban regions.

Overall, in metropolitan areas the history of human construction is a major determinant of soil conditions. In the urban-region ring, agricultural and other land-use history plus geographic/topographic location are progressively more important determinants of soil conditions.

Microclimate and air pollutants

Microclimate and air pollution, the two subjects introduced here, are of major ecological importance in urban regions (Landsberg 1981, Oke 1987, Ahrens 1991, Schmandt and Clarkson 1992, Moran and Morgan 1994, Forman 1995, Smith 1996, Santamouris 2001, Arnfield 2003). *Microclimate*, the history of weather conditions in small spaces, differs on north vs. south slopes, upslope vs. downslope, near vs. far from a coast or other water body, and on different sides of buildings. Wind, solar angle, and source of water vapor are the major reasons.

Solar radiation composed of short wavelengths (the visible spectrum) and long wavelengths (infrared radiation) is absorbed by soil, vegetation, and especially dark impermeable surfaces, which then reradiate energy to the sky (especially at night) in the form of infrared radiation. Infrared radiation is effectively heat, so the air is heated. The abundance of dark impermeable surfaces around cities helps produce a *urban heat-island effect* (Landsberg 1981, Moran and Morgan 1994, Arnfield 2003, Hough 2004). Upward-moving city air at night carries pollutants out and draws in air from the surroundings. However, if a *temperature inversion* (a warm air layer, e.g., over the metro area) forms, the air is blocked from moving upward, so heat, particles, and gaseous pollutants in the air accumulate.

For an environmental condition such as temperature, an individual organism has a somewhat narrow optimum range in which it can survive, grow, and reproduce (Barbour *et al.* 1987, Gilbert 1991, Odum and Barrett 2005). Above and below that optimum the individual grows, but does not successfully reproduce. Above and below the growth ranges the organism survives, and beyond the survival ranges the organism dies. *High temperature*, for example associated with the heat-island effect or global climate change, increases an organism's metabolic rate which in turn may exceed the organism's ability to get resources. High temperature also accelerates development, such as leaf-bud opening and flowering, which consequently may be damaged by spring frosts. Still higher temperatures desiccate plants, and ultimately inactivate enzymes causing death.

Species composition changes with higher temperature, so, for example, on bird feeders in residential areas, cool-region species are gradually replaced by warm-region species. Higher temperature may lead to more air pollution and more severe effects on organisms. A major way to reduce air temperature is to maintain or restore vegetation cover, particularly of trees (Zipperer and Foresman 1997). Trees provide cool shade and also pump considerable water to the air in evapotranspiration, a process that cools the surrounding air. For example, in Berlin, a small greenspace of about 5–30 ha (12–75 acres) may cool the air 0.5– 2 °C (1–4 °F), a medium 30–300 ha greenspace some 2–3.5 °C, and a large >300 ha (750 acre) greenspace may reduce summer air temperature 3.5–5.5 °C (6–10 °F) (von Stulpnagel *et al.* 1990). Also a large greenspace such as Tiergarten in Berlin, Mont Royal in Montreal, or Golden Gate Park in San Francisco, cools summer temperatures in the city for several hundred meters or over a kilometer downwind (Schmid 1975, von Stulpnagel *et al.* 1990).

The *coastal effect* of a major water body such as a large lake or the sea, often cools air temperature inland for several kilometers in spring and early summer, and similarly warms the air in autumn. The coastal effect also increases relative humidity and fog or cloud conditions.

Air-borne particles (particulate matter) in urban regions typically serve as condensation nuclei around which water droplets form, producing aerosols. When cooled, rain falls so that cities and areas downwind of cities often have somewhat elevated rainfall amounts. So, while high temperature tends to desiccate, greenspaces, street trees, green roofs, and other urban vegetation help to maintain moist air.

Wind, another major urban microclimatic factor, comes in three forms (Brandle *et al.* 1988, Forman 1995, Arnfield 2003). *Streamline airflow* occurs in parallel layers over relatively smooth surfaces such as large fields and smooth rounded hills. *Turbulence*, composed of eddies with (usually) up-and-down air movements, forms where streamline airflow is disrupted by abrupt boundaries, such as buildings and cliffs. Vortices are strong helical winds with a vertical axis. Particles are lifted off surfaces to become air-borne, most by vortices, much by turbulence, and least by streamlines. Streamline airflow is common in cropland areas of the urban-region ring. However, turbulent airflows generally predominate around urban areas where the upper surfaces of buildings and vegetation are so uneven in height. Turbulence not only lifts particulate matter into the air, it increases heat loss and desiccates plants, so many species in natural areas are at a competitive disadvantage in turbulent wind conditions.

On windless nights in hilly or mountainous urban regions, *cool air drainage*, the downslope flow of cool air, pushes a city's warm air and pollutants upward and out. This serves as a free cleaning and ventilation system (e.g., Stuttgart, Germany; Spirn 1984, Christina von Haaren, personal communication). The primary requirement is to keep nearby hillslopes in unbuilt condition, with woodland rather than grass cover apparently preferable (Gross 2002). Secondarily high-rise buildings are kept out of the main valley-bottom air-drainage channels.

Noise is a particular issue in urban areas where it affects people and animals (Forman *et al.* 2003, Miller 2005). At modest noise levels, some species and people become habituated (accustomed) to noise so that it no longer disturbs them. However, traffic noise from roads and highways with more than about 10 000 vehicles passing per day appears to significantly degrade bird communities nearby, with the width of the degradation zone increasing with traffic volume (Reijnen *et al.* 1995, 1996, Forman *et al.* 2002). In addition to traffic volume and distance from road, the proportion of truck traffic is significant in determining noise levels. Not surprisingly, busy commuter rail lines also have significant, though less-known, ecological effects. Also artificial night lighting has a range of ecological effects around cities (Rich and Longcore 2006).

Several types of air pollutants are of major ecological importance in urban regions, and also cause human health problems (Smith 1981, Gilbert 1991, Craul 1999, Nowak 1994, Santamouris 2001, Forman *et al.* 2003):

(1) *Ozone* (O_3), which forms photochemical smog (by combining with other chemicals in sunlight); both can damage many plants.
(2) *Hydrocarbons*, which are liberated in fossil fuel combustion often cover and alter soils, and are also part of photochemical smog.
(3) *Carbon monoxide* (CO), which presumably can kill animals in local spots.
(4) *Carbon dioxide* (CO_2) and methane (CH_4), which are global greenhouse gases leading to higher temperature, sea-level rise, and more extreme weather events.

(5) *Sulfur oxides* (SOX), which are especially produced by fossil-fuel burning industries and power plants, have significant local effects on some plant species, cause acid precipitation that acidifies certain water bodies, and forms particles/aerosols in the upper atmosphere.

(6) *Nitrogen oxides* (NOX), which are produced by vehicles and other high-compression engines, cause acid precipitation that acidifies certain water bodies, blanket the region with nitrogen (noticeably changing plant species dominance in water, on land, and by highways), and forms particles/aerosols in the upper atmosphere.

(7) *Heavy metals* (e.g., zinc, cadmium, copper, nickel), which originate from wear and chemical breakdown of surfaces such as in vehicles, bridges, and machinery, inhibit various metabolic processes and therefore at sufficient levels are toxic to numerous organisms.

(8) *Particulate matter* (particles), originating from fuel combustion, vehicle and machinery wear, road dust and wear, fire, and wind erosion from cropland and construction sites. Particulate matter is commonly classi-fied as PM10 (small particles) or PM2.5 (very small particles), with the latter being especially damaging to lung health of people and presum-ably wildlife. Particulate matter and *aerosols* (particles or gases combined with water droplets) also reduce incoming solar radiation.

Species, of course, respond differently to different pollutants (Smith 1981). Thus for ozone, pine (*Pinus*) and sycamore (*Platanus*) are very sensitive, but maple (*Acer*) and fir (*Abies*) are not. Yet for sulfur dioxide, pine and elm (*Ulmus*) are very sensitive, while maple and sycamore are not. Since air pollutants blanket an urban region, varying from place-to-place in type and concentration, the distri-bution of urban plants and vegetation is significantly molded by air pollution.

Greenspaces

Greenspaces, as unbuilt areas in an urban region, contain and may sus-tain natural systems where ecological patterns, processes, and changes are in most-natural or least-degraded condition. Yet greenspaces, like built areas, are exceedingly diverse and significant to society. Therefore this section highlights important greenspace types and illustrates key functions and ecosystem services provided for the benefit of society.

A greenspace may be covered by a single natural system or by many, such as evergreen forest, deciduous woodland, shrubby hilltop, rock outcrop, meadow, hedgerow, aquifer, pond, stream, wetland, vernal pool, and soil. Periodically the central ecological and natural-systems values elucidated will be supplemented by pointing out other issues, such as economics, social meeting places, aesthetics, and public health.

A glimpse of the urban region as a whole is useful before focusing in on specific greenspace types. Many ecological patterns and processes have been compared between urban and rural or suburban areas (Gilbert 1991, Bird *et al.* 1996). Added insight is often gained by analyzing *urban-to-rural gradients*, from city center to the outer urban-region ring dominated by natural vegetation and/or agriculture. For example, studies of over 200 urban areas show that lichens decrease sharply from rural to urban sites, presumably largely due to the combination of desiccation and air pollution (Schmid 1975, Gilbert 1991). Spatial patterns and plant species change along the gradient (Steinberg *et al.* 1997, Williams *et al.* 2005, Hahs and McDonnell 2007). Soil characteristics such as hydrophobic conditions, fungi, heavy metals, organic matter, and mineral nutrients also change markedly from rural to urban greenspaces (McDonnell *et al.* 1997, Pouyat *et al.* 1997). The presence of rare species and many wildlife characteristics also sharply decrease along this gradient (Gilbert 1991, Bird *et al.* 1996, van der Ree and McCarthy 2005). The number of native species decreases, but the number of non-native species often increases more, so semi-natural city-center sites may have the greatest number of total species (Kowarik and Langer 2005). Landscape ecology patterns also change along an urban-to-rural gradient (Luck and Wu 2002, McGarigal and Cushman 2005). Discontinuities in response curves may be expected along urban-to-rural gradients, such as a drop in temperature at the metro-area border (Spirn 1984). Ecological measurements along numerous radii of an urban region might be expected to have the highest variability in the vicinity of the metro-area border or inner urban-region ring.

In Northern and Central Europe, plant species richness has been correlated with a city's area and population (Klotz 1990). Species number increased steeply up to a city of about 130 km^2 and 100 000 inhabitants. Plant species number remained essentially constant from about 130 to 420 km^2 (50 to 160 mi^2) and population 100 000 to 1 300 000, beyond which the curves noticeably rose. The smallest cities measured (Ballensted and Schmalkalden, Germany), both in area and population (1.5 and 2.5 km^2; 10 000 and 17 000 inhabitants), had about 350 plant species and the largest city, West Berlin (481 km^2; 1 900 000 inhabitants), had approximately 1400 species.

Where would you go in your city to find the lowest, and highest, species richness? Ignore the zoo. In Dusseldorf, Germany, five groups of species (plants, butterflies, grasshoppers, landsnails, and woodlice) were measured in 38 habitat types (Godde *et al.* 1995). Overall, six habitat types had the fewest species, though considerable variation from group to group existed (in order, beginning with fewest species): sealed (asphalt) parking place; parking places (lot); river bank; intensive grassland (mowed); market garden; and avenue. The highest species richness for these groups was found in the following six habitats (in order, beginning with most species): wasteland; swamp woodland; moist meadow; gravel pit; parkland; and railway site. Thus, in general, the most species-impoverished locations are the most intensively designed, built, and managed, whereas several of the richest biodiversity sites are the least designed, least managed, and most overlooked locations in a city.

A still-broader view of greenspaces recognizes patterns in three somewhat-distinct major areas of an urban region (Table 4.1) (Spirn 1984, Gilbert 1991, Godde *et al.* 1995, Lynch and Hack 1996, Houck and Cody 2000, Lagro 2001, Ishikawa 2001, Greenberg 2002, Hough 2004, Wein 2006). First, the city and metropolitan area are primarily characterized by a broad-scale pattern of large greenspaces, within which a fine-scale pattern of small greenspaces is nestled. Distinct corridors and patches dominate both spatial scales. Second, the metro-area border and inner urban-region ring have an exceedingly complex greenspace pattern, created by wide and medium-width corridors combined with large and medium-size patches. Finally, the outer urban-region ring manifests a broad-scale pattern of large greenspaces composed of corridors, networks, patches, and matrix, enhanced by a fine-scale pattern of small corridors, networks, and patches.

The 75 greenspace types recognized are familiar in cities worldwide (Table 4.1). In a few cases the same names appear in different areas where they appear and function quite differently. Numerous other less-common greenspace types could be listed (Chapter 5), including central-city palace grounds (Tokyo), lava bed (Portland, USA), barrier across river (London), large mine-waste areas (Johannesburg), and cave sites (Kuala Lumpur, Malaysia) that are rarely present in cities.

Instead of attempting to list the ecological functions and values to society of all these greenspace types, representative functions and values for planning are selected. Although these generic patterns apply widely, each greenspace in each urban region also has distinctive functions and values which can and should be delineated in planning. Furthermore, an array of spatially separate or connected greenspaces is always present that functions as a system connected by flows and movements. Consequently both the individual greenspace and the broader system are important to planning and society. Greenspace functions and values are illustrated for the three portions of an urban region: (1) city and metropolitan area; (2) metro-area border and inner urban-region ring; and (3) outer urban-region ring. For each portion, broad-scale patterns are introduced and then fine-scale patterns.

Table 4.1 *Greenspaces forming broad- and fine-scale patterns in three major portions of an urban region*

City and Metropolitan Area

Broad-scale pattern of large greenspaces

Corridors: (a) river corridor with floodplain; (b) highway corridor; (c) railway corridor; (d) coastline; (e) steep slope

Patches: (a) large urban wood-lawn park; (b) large semi-natural park; (c) lower floodplain/delta area; (d) railway yard; (e) zoo

Fine-scale pattern of small greenspaces

Corridors: (a) tree line; (b) shrub line; (c) row of greenspace stepping stones

Patches: (a) small urban wood-lawn park; (b) small semi-natural area; (c) historic/cultural site; (d) vacant lot; (e) school yard; (f) cemetery; (g) street-side planted spot; (h) green roof

Metro-area Border and Inner Urban-region Ring

Corridors

Wide: (a) greenbelt (or urban growth boundary); (b) ring of large parks; (c) greenway; (d) green wedge

Medium to wide: (a) stream or canal corridor; (b) pipeline corridor; (c) electric powerline corridor; (d) roadside; (e) tree/shrub line; (f) row of greenspace stepping stones; (g) string of pearls

Patches

Large to medium: (a) heterogeneous suburban park; (b) large natural or semi-natural area; (c) golf course; (d) botanical garden; (e) market-gardening (truck-farming) area; (f) nursery-plants area; (g) lake; (h) wetland; (i) office-park area; (j) shopping center; (k) industrial area; (l) brownfield (with chemical pollution); (m) race track site; (n) municipal-use space

Medium: (a) small semi-natural area; (b) crop field; (c) meadow or fallow field; (d) community garden (allotment garden, leisure garden, war garden); (e) waterworks/supply facility; (f) sewage-treatment facility; (g) solid-waste/recycling site; (h) cemetery; (i) sand-and-gravel site; (j) large shrubby patch

Outer Urban-region Ring

Broad-scale pattern of large greenspaces

Corridors and networks: (a) river corridor; (b) highway corridor and rectilinear network; (c) coastline; (d) pipeline; (e) electric powerline; (f) emerald network

Patches and background matrix: (a) cropland landscape or matrix; (b) forest/woodland landscape or matrix; (c) grassland landscape or matrix; (d) desert landscape or matrix; (e) low-density residential area

Fine-scale pattern of small greenspaces

Corridors and networks: (a) stream corridor and dendritic network; (b) canal corridor and dendritic/rectilinear network; (c) hedgerow

Patches: (a) lake; (b) reservoir; (c) wetland (marsh, peatland, swamp); (d) rock quarry; (e) sand-and-gravel site

City and metropolitan area

River corridors typically serve as a major infrastructure conduit for the city and are squeezed by numerous engineered hard structures. Small gravel maintenance roads on the floodplain are common. Normal water levels are low, in part due to groundwater pumping, while peak flows and flood hazard are higher (Groffman *et al.* 2003). Pipes carry stormwater, human wastewater, and industrial pollutants rapidly to the river which becomes severely polluted. Stagnant pools and wetlands seasonally breed clouds of mosquitoes and midges (Robinson 1996). Natural vegetation on the floodplain absorbs some floodwater and supports rich biodiversity. Recreation may be important in river corridors.

Highway corridors are sources of vehicular and road pollutants, which are washed by stormwater into water bodies. Traffic noise degrades nearby habitat for many wildlife species. Railway corridors suffer loud intermittent noise, but still may be rather effective conduits for wildlife movement. Steep slopes covered by residential developments or squatter settlements have elevated hydrologic, erosion, and sedimentation problems, and are more susceptible to periodic environmental disasters. Coastlines have very little protected area so remnant natural beach, dune, vegetation, and wetlands along the coast are rare habitats with concentrations of rare species. Also, the paucity of natural coastal area means that nearby aquatic ecosystems are severely degraded.

In the metro area a large semi-natural park, as the best facsimile of nature, commonly has educational and inspirational value, but also may have security problems. Commonly a few relatively rare species are present. Furthermore a large semi-natural greenspace (Konijnendijk *et al.* 2005): (1) serves as a major source of species dispersing to other city greenspaces; (2) reduces flood hazard; (3) cools surrounding downwind built areas for a considerable distance; and (4) attracts birds migrating across the metro area. A large urban wood-lawn park also provides the last three benefits. A railway yard tends to be rich in non-native species mainly transported by trains.

At a finer scale, tree lines provide continuous cover and cool shade that enhance movement by people and birds. A shrub line provides cover and movement enhancement for terrestrial animals, and serves as a visual barrier for people, e.g., between house lots. A row of greenspace stepping stones enhances directional movement of some species.

Vacant lots often teem with non-native species, and illustrate the process of succession plus the regeneration power of nature. School yards have deeply compacted soil, and may serve as roosts for gulls, geese, or ducks on windy nights. The urban cemetery is usually biologically impoverished, especially due to the scarcity of shrubs. Finally, green roofs capture and evaporate precipitation water, reduce stormwater runoff, cool the building and air, and provide a bit of biodiversity (Hien *et al.* 2007).

Metro-area border and inner urban-region ring

A wide greenbelt outside the metro area constrains outward urbanization, while providing a trail system, cool clean air for the metro area, and convenient market-gardening areas (Munton 1983, Pandell *et al.* 2002, Whitehand and Morton 2004, Bengston and Youn 2006). A ring-of-parks, analogous to a greenbelt sliced by radial transportation corridors, loses the greenbelt connectivity, but has more neighboring residents who can use and care for the parklands (e.g., Budapest). Greenways or trails could connect the parks. Greenways particularly facilitate walking and biking recreation, but also provide water and habitat protection (Briffett *et al.* 2000, Briffitt 2001, Fernandez-Juricic 2000, Erickson 2006, Binford and Karty 2006). A green wedge projecting into a metro area provides proximity to greenspace for many residents, but especially facilitates recreational movement between city and countryside. Green wedges also enhance clean air flows in the city, and the movement of species from countryside to urban greenspaces.

Narrow stream corridors in this metro-area border area are mostly straightened, even discontinuous with water flows in underground pipes. A pipeline corridor may be especially effective for wildlife movement, and maintenance activities usually result in creating a strip of common edge species. A *string of pearls*, as a tree-lined trail connecting small semi-natural patches (Forman 2004a), facilitates walking recreation and some species movement between the small patches or parks, and is easier to establish and protect than a greenway.

Patches around the metro-area border are extremely diverse. A golf course is an intensive water, fertilizer, and pesticide user, and usually is biologically impoverished because in construction the natural habitat heterogeneity was largely homogenized, and now shrub cover, dead trees, and logs are scarce. A nursery-plants area also pours on the water, fertilizer, and pesticide, while serving as a major source of non-native and invasive species which are widely dispersed across residential and commercial areas. A botanical garden, also using water, fertilizer, and pesticide, grows an exceptionally rich collection of native and non-native plants, and presumably is also a source of dispersing exotic and invasive species. A market-gardening area, perhaps using still more water, fertilizer, and pesticide, provides convenient fresh food and sometimes family and social benefits, while often polluting the groundwater beneath. A wetland commonly absorbs stormwater which reduces flood hazard, supports rich biodiversity, and provides swarms of seasonal mosquitoes. Finally a lake here usually is ringed by development for the relatively wealthy, is polluted, and offers boating and fishing recreation.

At a finer scale, a community garden provides family food, social benefits, and rich food for wildlife. A sewage-treatment facility cleans human waste, but is often a source of pollution overflows in heavy storm events and air-borne pathogenic microorganisms. A solid-waste/recycling site also addresses the problem of societal wastes, and is a magnet for concentrations of rats, gulls, or other scavengers, often including people who recognize wastes as economic value.

Outer urban-region ring

In the outer portion of an urban region, a broad-scale pattern of corridors, often connected into networks, is superimposed on the matrix and large patches. A river corridor upriver of the metro area may be quite clean or be heavily polluted by agricultural runoff, whereas downriver of the city, river water is polluted by urban and industrial processes. Highway corridors have traffic-noise-created degradation zones on each side, and serve as major barriers to walkers and wildlife movement (Cuperus *et al.* 2001). The emerald network of large natural-vegetation patches connected by green corridors is an effective flexible solution to sustain wildlife in the face of urbanization and climate-change processes.

Forest/woodland, grassland, desert, and especially cropland are the background matrix and landscapes in urban-region (urb-region) rings. As such, they are predominant sources of species and form the framework for water conditions. Forest/woodland is particularly valuable for recreation in addition to wood production and wildlife cover, though flooding, erosion, and sedimentation are chronic problems (Theobold *et al.* 1997, Konijnendijk *et al.* 2005). Grassland and desert are both subject to erosion, sedimentation, and loss of wildlife. For low-density residential areas, medium or relatively small house-lot sizes overwhelmingly degrade biodiversity and water conditions, whereas very large house lots may maintain only moderately degraded ecological conditions. The latter, therefore, may be suitable as buffers around large protected natural areas.

At the fine scale, a dendritic stream-corridor network is especially important in the outer urban-region ring. The stream network within large natural areas has a connected integrity which facilitates fish movements. On the other hand, in an agricultural area the stream network usually has narrow strips of riparian vegetation that provide shade and fallen wood as fish habitat, but that do relatively little to prevent stream pollution from runoff of eroded sediment and excess fertilizer. Trees and other woody vegetation along stream/river channels is largely removed by seasonal floods in some mountain areas (e.g., Santiago, Chile), and by human action in parts of England where channels are managed somewhat like large stormwater drains. Hedgerows often facilitate species movement across the landscape and reduce agricultural runoff. Lakes and especially reservoirs may have little development ringing them so that semi-natural shorelines help maintain clean water, viable aquatic ecosystems, fish populations, and fishermen.

In short, greenspaces are as rich in benefits to society as are built spaces, such as commercial, industrial, high-rise residential, single-unit residential, and highways. Many greenspaces, however, are inextricably tied to location. Buildings and roads and parking lots can be built in many locations. In contrast, greenspaces containing lakes, wetlands, rivers, floodplains, mountain slopes, and so forth are solidly fixed on the land. Wisely arranging built spaces around greenspaces is the planner's and society's challenge.

5

Thirty-eight urban regions

Selecting cities, determining boundaries, mapping regions

What would you do if you wanted to understand urban regions? So many cities, so variable in size and geography – the task seems daunting. You might enjoy traveling to and studying a good batch of them, but these are large complex objects and the enterprise would take years. Or you might devour books and articles on the subject, also a protracted process, which would provide a skewed picture dominated by a limited number of much-studied cities. Or simply talk to the experts (who wrote those books and articles). Here is the story of how I learned.

Selecting cities worldwide

To get the big picture at the outset I pulled out maps, all sorts, and sketched the shapes and sizes of cities and especially how they are arranged relative to water bodies and mountains. Since I have lived in parts of North America, Europe, Latin America and Australia, initially the focus was on other areas, then gradually becoming worldwide. Quickly I was able to group the sketches into three big categories characterized by: (1) continent or geographic area; (2) location relative to rivers, bays, seacoasts, etc.; and (3) city size, as indicated by area. I decided that my mix of urban regions to be studied should include the typical range of variation within each of these groups, and that probably other useful categories or subcategories exist which should at least be represented. Also I estimated that 25 urban regions would be the minimum needed (which might or might not provide clear results), and that 50 regions would be the maximum possible based on time and resources available.

When I am stuck indoors, the opportunity to study maps, read books and articles, and talk with experts is a pleasure, so these continued intensively for

Figure 5.1 The 38 urban regions worldwide selected for analysis. City population sizes (>250 000) range from high to low in each of six geographic areas: Europe; North America; Latin America; Africa; West to East Asia; and South Asia to Australia.

many moons. Over 350 cities worldwide were at least briefly considered, some in considerable detail. This process produced a set of primary criteria and secondary criteria, which formed the basis for ultimately selecting 38 urban regions (Figure 5.1).

Primary criteria

(1) World distribution (to represent a breadth of geographic areas/continents and cultures).

(2) Wide range of predominant land covers around cities (e.g., forest/woodland, cropland).

(3) Wide range of city population sizes (in six categories, 0.25–0.5, 0.5–1, 1–2, 2–4, 4–8, and 8–16 million inhabitants in the city).

(4) Wide range of forms of metropolitan areas (e.g., rounded, lobed, elongated).

(5) Wide range of city locations relative to major water bodies (e.g., by river, lake, sea).

(6) Barcelona (Spain) was included (since it seemed to be the best-analyzed case study available).

Secondary criteria
(Generally used in selecting between two good candidate cities.)

(1) Not near another city larger than itself or >250 000 population (thus Vienna, Washington, and Lahore were not considered).

(2)   Good- or adequate-quality satellite image of urban region available at time of city selection (thus Bogota, Lagos, Xi'an not considered).

(3)   Form of metropolitan area seemingly can be extrapolated to many other cities (thus Johannesburg, Naples, New York not considered).

(4)   Considerable useful ecological and planning information on the region readily available (thus Berlin, Chicago, Ottawa included).

(5)   Author familiarity with the urban region through residency, work, or travel.

Population of a city rather than of the metropolitan area or urban region was used as the indicator of size, because somewhat consistent and reliable data are available for all of the cities (Turner 2005). Population data for the cities chosen were for 1999–2002, with four exceptions: Iquitos (Peru) was 334 000 in 1998, Abeche (Chad) 188 000 in 1993, Samarinda (Indonesia) 335 000 in 1990, Rahimyar Khan (Pakistan) 234 000 in 1998. All were adjusted to the year 2000 using a 5% annual growth rate. For convenience in comparing major worldwide cities, "small" cities have 0.25 to 1 million, "medium" cities 1 to 4 million, and "large" cities 4 to 16 million inhabitants. Satellite cities located around a major city are smaller, with <250 000 inhabitants.

Geographic area was used instead of continent to better capture the combination of location and culture, and so that each geographic area included a similar number (five to eight) of urban regions studied (Table 5.1). The six geographic areas were: Europe, North America, Latin America, Africa, West to East Asia [Moscow and Erzurum (Turkey) to Sapporo], and South Asia to Australia [Rahimyar Khan (Pakistan) to Samarinda (Indonesia) and Canberra].

I have lived in or near nine of the urban regions (Atlanta, Barcelona, Canberra, Chicago, London, Mexico City, Ottawa, Philadelphia, Tegucigalpa), and visited another ten (Bangkok, Kuala Lumpur, Beijing, Portland (Oregon), Edmonton, San Diego, Stockholm, Bucharest, Rome, Berlin). This general familiarity helped especially in understanding the urban region beyond the city. However, I have never seen half of the 38 regions analyzed, which highlights the importance of the three primary analytic approaches used: measuring remotely sensed images, literature survey, and consultations with knowledgeable persons.

An initial set of cities was selected because they are well known in the literature for illustrating key planning and ecological patterns, or I had accumulated especially useful ecological information on them (Barcelona, Berlin, Stockholm, San Diego, Portland, Chicago, Atlanta, Ottawa, Mexico City, Brasilia, Canberra, Beijing). Large satellite images were printed for 120 urban regions under consideration. From these, several additional distinctive, but widely applicable, patterns stood out, and those cities were selected accordingly (Santiago,

Bucharest, Edmonton, Iquitos [Peru], Moscow, Bangkok). The remaining 20 cities were selected to provide a balanced set of urban regions that accomplished the primary and secondary criteria listed (Table 5.1).

Determining urban-region and metro-area boundaries

A major city, as well as its boundaries, normally cannot be seen on a satellite or other aerial image. Instead, a large mass of built area stands out, often relatively compact, but sometimes diffusely grading into greenspace (unbuilt area) at its edge. The city is inside the large built area, typically with the city center centrally located. This large central clearly visible object, the essentially continuous built area, is herein called the *metropolitan area* or simply *metro area* (see Figure 1.1). In general, its boundary is easily determined and marked on the large images. In spots one of two mapping decisions has to be made. How wide does a strip of greenspace need to be to exclude a peripheral built area as a separate suburb or town, or to cut off a greenspace wedge projecting into the metro area? A greenspace width of 1 km (0.6 mi) was used as a general guideline. The other mapping decision related to low-density housing development, mainly an issue around portions of US cities. The general guideline cutoff used was a large group of 2 ha (5 acre) house lots; areas with smaller house lots were considered built and larger lots unbuilt.

More interesting and significant to this book is how to determine the *urban-region boundary*. Arguably the central importance of an urban region relates to interactions – flows and movements – back and forth between the city and its surroundings. In this sense the urban region is a functional region. One can ask and learn, "How does it work? What are the types, directions, and intensities of interactions?" Delineating the urban-region boundary thus simply requires an estimate of the intensity of flows and movements along radial lines extending out from the city center. The boundary is where inward flows noticeably increase and outward interactions noticeably decrease.

Numerous attributes affect movements and flows inward and outward, and data on as many as possible were mapped on the large aerial images. Several attributes initially thought to be important turned out not to be so, because they usually did not extend very far beyond the metropolitan area (ends of commuter rail lines, communities with substantial commuter populations, airports, sewage-treatment facilities, solid-waste disposal sites, reduced air-quality areas). Rivers and highways hardly ever delineated urban-region boundaries. Wetlands were too scarce around urban regions to be important determinants.

In contrast, mountain ranges often delineated urban-region boundaries. Major political/administrative borders often determined boundaries of a region. Another nearby major city (>250 000 population) with its own urban region frequently helped delineate a region boundary. However, boundary location seemed dependent on the relative size of the core city vs. the other city. If the other city was larger, a point 40 % of the distance from core city to other city was arbitrarily marked as a preliminary estimate of boundary location. If the other city was smaller and >250 000 population, a 60 % point was marked. If the outer city was smaller and also <250 000 population, the 70 % point was marked. One-day recreation or tourism sites were also significant boundary delineators, the idea being that both the core city and the sites are important economically and culturally to each other, and cannot be too far apart relative to the transportation system. Major biodiversity areas, either impacted by or dependent on protection by people of the city, sometimes helped delineate boundaries. Outlines of the drainage area around major water supplies were often important in determining urban-region boundaries.

Finally, where none of the preceding seemed important, a radius of approximately 100 km was chosen. Partly this was because almost all the other delineated urban-region boundaries were in the 70–100 km range. And partly it was selected as a typical maximum distance on a paved highway with traffic that large numbers of people would travel back and forth in one day for shopping, recreation, and so forth. Even around remote cities with unpaved radial roads, 100 km may be a reasonable distance, for example, for once-a-week shopping or business trips.

Mapping urban regions

While literature, consultations, and direct observations helped understand urban regions, the core analysis were measurements of spatial patterns on large remotely sensed images (c. 70 × 100 cm and 1:200 000 scale). Consistent base maps for urban regions using Landsat satellite geospatial data, with a 30 m cell or pixel size, were generated from the Earth Science Data Interface (2006 website) of the University of Maryland's Global Land Cover Facility, College Park, Maryland, USA. The data were organized in color spectra that provided the ability to separate red, green, and blue, as well as more advanced bands that could separate urban areas from forest, meadow areas, and the like. The data were finally manipulated in the ESRI ArcGIS 9.1 geographic information system. The Arc Tool "Composite Bands" allowed for an image to be created, and point scaling was used to create a number of different images. All flights for the remotely sensed images of the 38 urban regions were in 2001 ± 1 year (except Santiago, Tegucigalpa, and Sapporo, flown in late 1999).

The mapping process was a series of steps that created an aerial image as part of a comprehensive database (see Appendix I). The steps began by extracting the raw band of color from the Earth Science Data Interface and ended with image processing in Arcview. The resulting image was printed in color with the city in the approximate center. Preliminary determination of the boundaries for many urban regions found that boundaries rarely extend in any direction more than 100 km from center city, so images were printed with a diameter of approximately 200 km.

After considerable consultation with GIS (geographic information systems) specialists, I decided that hand measurements using planimeter and ruler on clear plastic sheets over the satellite images was the optimal procedure. Sufficient quality, accuracy, and consistency to accomplish the objectives seemed unavailable by GIS analysis without considerably more time and resources than were available. Especially worrisome was the uncertain ability to correctly differentiate by computer the hundreds of land covers in urban regions worldwide. Doubtless I would make some errors in image interpretation, but based on 35 years of landscape ecology work and analyzing/ground-truthing maps and GIS images in many nations, the decision to use manual rather than GIS measurements ended up an easy one.

With the 30 m cell-size resolution on the satellite image, two-lane highways were often invisible or hard to follow, whereas most of a multilane highway length was clearly evident. Streams and small rivers disappeared in agricultural and built landscapes, the predominant land covers in most urban regions, except where wide stream-corridor vegetation was present. In forested landscapes, rivers and streams less than about fifth-order (Wetzel 2001, Kalff 2002) could seldom be followed, because the open strip in the tree canopy was too narrow. The rare forested wetland generally was not differentiable from forest. Hedgerows, individual houses, and narrow two-lane roads normally were invisible.

In contrast, major airports, shipping/ferry ports, and large mines or quarries were quite prominent. Normally I could differentiate: marsh from other open areas; mangrove swamp from other woodland/forest; dammed reservoirs from lakes; mountain ridges (but not hills); villages and hamlets; and direction of surface-water drainage and stream/river flow.

Urban-region boundaries, metropolitan areas, and numerous specific objects were marked on the large images (c. 70 × 100 cm). Spatial measurements (area, shape, distance, and number) were then made for both marked attributes and unmarked patterns. When new information and corrections became infrequent for an urban region, the markings were copied with black pen onto a small image (43 × 43 cm) of the region.

An illustrator then converted the major land covers marked into brightly colored maps and computer-reduced the images for printing in this book. The colors represent a balance between standard urban-planning practices and the ability to differentiate land covers in a black-and-white photocopy of the image.

Finally, key sites and features marked were converted to pictograms for easier recognition.

Key spatial attributes

Having selected the 38 cities around the globe and determined the boundaries of their urban regions, a still more daunting task lay in wait. What is important in a region? Somehow the most important areas and sites needed to be identified and mapped. Some are visible on the aerial images, but many are too small and had to be located from published maps, literature, and consultations with knowledgeable persons. "Important" here refers to attributes needed to understand the natural systems present, their human uses, and more broadly, how the region works.

Land cover types

For most analyses, the numerous land covers present in the 38 urban regions worldwide were distilled to 14 major ones, which were mapped – nine area-cover types and five linear-cover types. All but one (region boundary) were visible on the satellite images. Because some linear features are narrow and less distinct, published maps were occasionally used for clarification in the mapping process. A few analyses used other subtle or less common cover types which were not mapped (e.g., railroads, different crop types, different building-density areas). The 14 major land covers were:

(A) Salt water (sea, coastal bay).
(B) Freshwater (lake, reservoir, river, major stream).
(C) Forest/woodland (forest of tall trees and relatively continuous closed canopy, woodland of smaller trees and relatively open canopy; in some climates the two types are intermixed and difficult to separate on aerial images).
(D) Small-tree farming (especially coffee, tea, and oil palm plantations).
(E) Cropland (cultivated/tilled fields covering at least 75 % of the area, the non-cultivated portion usually being forest/woodland and/or built area; includes non-irrigated and irrigated land).
(F) Grassland/pastureland (includes ranchland, paddockland, and grass-dominated savanna with scattered trees or tree clusters; cropland fields are much smaller than the relatively distinct pastures/paddocks [rarely evident in desert/desertified areas] which characterize most pasture-land).

(G)   Desert/desertified area (desert with <25 cm annual precipitation, some-times sand-covered, but more typically stony/rocky with sparse low shrubs and other plants; desertified area overgrazed, eroded, salinized, and/or water-table-lowered by human activity).

(H)   Metropolitan area (the essentially continuous built area including a cen-tral major city).

(I)   Small/medium built area (satellite city or a town; based on area, not population; villages, hamlets normally not included).

(J)   River/major stream (in moist climates typically at least a fourth-order stream [Wetzel 2001]; in dry climates with seasonal surface-water flows, a relatively wide [typically >30 m] floodwater channel).

(K)   Multilane highway (usually a four-lane divided highway; sometimes more traffic lanes).

(L)   Paved two-lane main road (asphalt/tarmac/concrete surface; mapped in regions with few or no multilane highways outside the metro area).

(M)   Unpaved main road (mapped around cities with few or no paved main roads outside the metro area).

(N)   Urban-region boundary (determined as described in the preceding section; not visible on aerial image).

Site types

From several dozen possible spatial attributes considered and partially mapped, the following 26 types of major sites were considered especially impor-tant and found to be widespread. The absence of a mapped area or site on an image usually means it does not exist. In some cases it may exist, but no infor-mation was located. Although often many small sites exist, only major sites were mapped.

(1)   Dock area for shipping and/or ferry.

(2)   Industrial area (usually a group of industries; occasionally a single large one).

(3)   Sewage treatment facility (often only many small ones are present).

(4)   Solid-waste disposal (sometimes combined with a recycling facility).

(5)   Water-quality poor (data not easily obtained or interpreted, so typically a water body downslope of a city, industrial area, sewage treatment facility, or mine site is marked).

(6)   Air-quality poor (data not easily obtained or interpreted, so typically locations in center city and downwind of the city or an industrial area are marked).

(7)   Flood-hazard area (data not easily obtained, so typically metro-area locations up to a few meters higher than adjoining rivers or seacoasts are
marked).

(8)   Fire-hazard area (data not easily obtained, so typically locations in dry
climates where forest/woodland adjoins built areas are marked).

(9)   Airport (includes the primary busy airport for city, other passenger airport, military airport, and, in at least one case [Edmonton], a large
decommissioned airport).

(10)   Noise of flying aircraft (data not easily obtained, so locations extending
c. 3–5 km (2–3 mi) out from major runways [Miller 2005] are marked).

(11)   Nearby slopes facing city (covered by natural vegetation, agriculture, or
built area [wealthy to slum], these slopes are of special visual, recreation, erosion, flooding, cooling, air-cleaning, and biodiversity importance associated with proximity to city).

(12)   Market-gardening area for vegetables/fruits near city (sometimes called
truck farming, this area provides low-transportation-cost fresh produce
for city markets and restaurants [Chapter 3], and is often threatened by
development).

(13)   End of commuter rail (most radial commuter-rail lines serve the metro
area, but some extend outward to satellite cities and towns [e.g., London,
Philadelphia]).

(14)   Commuter residential area (data not easily obtained, so locations are
mostly based on consultations with knowledgeable persons).

(15)   Biodiversity area (data not easily obtained, so, as elaborated below, locations mostly based on parks, natural areas, etc. on published maps, and
on distinctive topographic and landscape ecological patterns).

(16)   Wetland (generally rare in urban regions due to drainage and filling).

(17)   Salt flat or intermittent lake (intermittent lake seasonally or periodically
dries up, often leaving a salt flat).

(18)   Volcano (active or apparently recently so).

(19)   Mine site or quarry (for valuable minerals, coal, or sand/gravel; data not
easily obtained; since quarries and most mine sites are usually small,
only large mine sites prominent on images are marked).

(20)   Water-supply source (location where water is extracted from a water
body; a few cities depend on groundwater wells or streams, so general
areas are marked).

(21)   Drainage area around water supply (without topographic maps in many
cases, the general area seemingly most important for water-supply protection is marked).

(22)   Recreation/tourist area for one-day trip (based mainly on parks, natural areas, cultural features, etc. from published maps and travel guides, plus consultations with knowledgeable persons).

(23)   Mountain range (approximate ridgeline marked).

(24)   Political/administrative boundary (boundary of another major political/ administrative unit, e.g., nation [San Diego/Tijuana example], state or province [Philadelphia], or county [Edmonton], is marked when some-what perpendicular to a radius of the region to help determine the region's boundary).

(25)   Sixty percent of distance to a smaller, but >250 000 population, city (recognizing that the outside city has its own urban region).

(26)   Seventy percent of distance to <250 000 population city outside region (recognizing that the outside city has its own urban region, and that it is a smaller city than in the preceding case; 70 % [and/or 60 %] occasionally appears twice in about the same direction on an image because of two cities in that direction).

The last six site types (numbers 21–26) were particularly important in determining the boundaries of urban regions, as highlighted in the preceding section.

Biodiversity areas (number 15 above) have a special importance in urban regions because of the high concentration of people and their manifold effects. These sites usually have importance far beyond maintaining rare species and natural communities. Many biodiversity areas maintain water resources, provide diverse recreation, protect cultural heritage resources, provide high visual quality, minimize soil erosion and stream sedimentation, provide wood products, and accomplish other key objectives of society.

Since rigorous data on biodiversity areas in urban regions are rarely available or easily obtainable, a protocol for identifying these was developed. In a few cases (e.g., Edmonton, Chicago, London) ecologists and other experts have identified many of the most important biodiversity areas, which provides a good basis for mapping. In many urban regions published maps, travel guides, and other literature identified biodiversity areas in the form of parks, natural areas, wildlife conservation areas, and the like, which were mapped. Then for some urban regions, knowledgeable consultants were able to pinpoint a few key biodiversity areas.

However, in addition the aerial images and other maps were further examined through the lens of landscape ecology (Forman 1995, Farina 2005, Groom *et al.* 2006) to identify and mark probable key biodiversity areas. This highlighted five important types of biodiversity areas:

(a) *Rare features* in the urban region were marked: topographic (e.g., volcano [Portland], high point next to the sea where migrating birds congregate [Barcelona]); geologic (limestone area [Barcelona], island [Bangkok]); vegetational (mangrove swamp [Kagoshima], palm oasis [Rahimyar Khan]); water-body (lake [Nairobi], freshwater wetland [Ottawa]); and human-created (reservoir [Beijing], large long-established military base rich in biodiversity [San Diego] [Goodman 1996, Leslie *et al.* 1996]).

(b) *Large areas of natural vegetation*, differentiated by type where possible (e.g., marsh, forest), were marked.

(c) *Isolated areas* of potentially viable natural vegetation were marked (palm oasis [Rahimyar Khan], salt flat [Tehran]). These are likely to support many uncommon species and to be key sources of species for populating surrounding areas.

(d) *The nearest lobes* of natural vegetation projecting toward a metropolitan area (Rome, Bucharest) were marked. These are key sources of species that colonize parks and other greenspaces across the metro area, as well as nearby access points to nature for recreationists from the city.

(e) *Strategically located sites for regional connectivity*, i.e., species movement and water flow, were marked. These include gaps or narrows connecting large natural areas and a row of natural "stepping stones" (Ulaanbaatar, Nantes).

Even using these diverse approaches for pinpointing major biodiversity areas, the maps are incomplete, often woefully so. Nevertheless, because of the many other major benefits provided for society, as listed above, these sites are among the most important anywhere in the urban region.

In addition to the 26 widespread site types above, 16 others were infrequent in the urban regions, but were deemed important and mapped:

(1) Ice and snow (atop high mountains; Santiago, Ulaanbaatar, Portland, Erzurum, Sapporo, Tehran).

(2) Lava bed (Mexico City, Kagoshima, Portland).

(3) Canal (Beijing, Cairo, Bangkok, Berlin, Moscow).

(4) Forty percent of distance to a neighboring larger city (Ottawa, Philadelphia, San Diego, Rahimyar Khan; this recognizes that a larger nearby city probably has a larger urban region).

(5) Barrier across river (Seoul, London).

(6) Greenbelt (London, Ottawa, Seoul).

(7) Urban-growth boundary (Portland).

(8)   Saltwater intrusion into aquifer (Barcelona).

(9)   Aquaculture area (Bangkok).

(10)   Demilitarized zone (Seoul).

(11)   Oasis and fort in use (Rahimyar Khan).

(12)   Native People's land (Edmonton, San Diego/Tijuana).

(13)   Transportation tunnel entrance/exit (London).

(14)   Concentrated greenhouses (Barcelona).

(15)   Large recently logged clearcuts (Canberra).

(16)   Intermixture of cropland, small-tree farming, and bits of natural vegetation (Cairo, Rahimyar Khan).

Nuclear power plants, though not mapped, operate in 17 of the 31 nations represented in this analysis (John P. Holdren, personal communication), and are present in several of the 38 urban regions. Recently the nuclear power plants in the London and Portland urban regions have been decommissioned with fuelcells removed. In the Portland case the facility was then dynamited into oblivion.

Numerous published maps were used to aid in mapping areas and sites within urban regions, the most common being: national and regional road maps; world atlas maps; DeLorme topographic maps (USA); National Imagery and Mapping Agency (declassified) maps (USA); keyhole website maps. When in doubt, the pattern visible on the 30 m-pixel Landsat aerial image was used.

Built areas mapped are mainly intermixed residential and commercial areas. Since residential and commercial areas have very different effects, both as areas and as sources and sinks, future work might usefully separate them. Similarly, future work might differentiate areas by population density, socioeconomic status, and so forth, since the effects on the areas and on surrounding areas, are so different. Future work might also usefully include power sources, such as hydro, nuclear, coal, biomass, wind and solar, since the effects of production and transport on the urban region markedly differ. Also mapping other hazard areas, such as earthquake (Sapporo, Mexico City, San Diego), cyclone/hurricane (Bangkok, Kagoshima), and volcanic eruption probabilities, is desirable. Intercity rail lines were omitted since their trains only stop in the city, but mapping rail stations for commuters outside the metropolitan area would be valuable. Adding such common urban history and planning parameters to the major patterns explored in these analyses should provide valuable added insight into the big picture.

Finally, people knowledgeable in specific urban regions and who have experience with mapping will know that all boundaries, land-cover types, and sites marked should be considered approximate, and that all the maps are incomplete and probably contain, hopefully minor, errors. More complete maps with these

Plate section

Cover Types

■ Salt water

■ Freshwater

■ Forest/Woodland [1]

■ Small-tree farming [2]

■ Cropland [3]

■ Grassland/Pastureland [4]

■ Desert/Desertified area [5]

■ Metropolitan area [6]

■ Small/Medium built area

▬ River/Major stream

▬ Multilane highway

– – – – Paved 2-lane main road

·········· Unpaved main road

🔘🔘🔘 Urban region boundary

1 Natural/semi-natural growth or plantation forestry

2 Fruit orchard, coffee, tea, date palm, oil palm, or agroforestry

3 Cultivated/tilled area (e.g., rice, wheat, maize, beans, vegetables)

4 Grass-dominated, with or without livestock (paddocks/ranchland), or wooded savanna

5 Bare earth surface with or without separated shrubs/other arid plants

6 Major city plus adjoining nearly continuous built area

Site Labels

✈ Airport

✵ Air quality poor

🦅 Biodiversity area

🏘 Commuter residential area

⚓ Docks for shipping or ferry

∨ Drainage area around water supply

🚊 End of commuter rail

🔥 Fire hazard area

🌊 Flood hazard area

🏭 Industrial area

🌷 Market gardening area for vegetables/ fruits near city

⚒ Mining site

⋀ Mountain range

·)) Noise of flying aircraft

P Political/administrative boundary

🌲 Recreation/tourist area for one-day trip

🏝 Salt flat or intermittent lake

♒ Sewage treatment

🔻 Slope facing city

⛰ Solid waste dump/tip/disposal/recycling

▲ Volcano

▦ Water quality poor

🚰 Water supply

〜 Wetland

60 60% of distance to a smaller, but > 250 000-pop., city

70 70% of distance to a < 250 000-pop. city outside region

✳ See Color Figure 1 caption

Color Figure 1 Key to cover types and site labels for maps of 38 urban regions (Color Figures 2–39). Asterisks (*) on certain maps refer to the following: *Bangkok* (aquaculture along bayshore; canal inland); *Beijing* (canal); *Cairo* (canal); *Chicago* (canal); *Edmonton* (Native People's land); *Kagoshima* (lava flow); *London* (barrier across river within metro area; greenbelt around metro area; transportation tunnel entrance/exit to the southeast); *Mexico City* (ice, snow, and alpine tundra atop volcanoes; lava flow to the southeast); *Nairobi* (ice, snow, and alpine tundra); *Philadelphia* (saltwater intrusion to the southeast; 40 % of distance to a larger city, i.e., New York to the northeast); *Ottawa* (greenbelt around metro area; 40 % of distance to a larger city, i.e., Montreal to the east); *Portland* (urban growth boundary around metro area; ice, snow, and alpine tundra atop volcano; lava flows to the northeast); *Rahimyar Khan* (oasis and fort in use in desert; 40 % of distance to a larger city, i.e., Bahawalpur to northeast, is marked by main road; double-asterisk area is cropland with scattered small-tree farming and natural vegetation); *Rome* (ice, snow, and alpine tundra); *Santiago* (ice, snow, and alpine tundra); *Seoul* (greenbelt around metro area; demilitarized zone to the north; barriers across mouths of rivers to the southwest); *Sapporo* (ice, snow, and alpine tundra); *Tehran* (ice, snow, and alpine tundra atop volcanoes; lava flows on volcano slopes to the northeast); *Ulaanbaatar* (canal by metro area; ice, snow, and alpine tundra to the northeast).

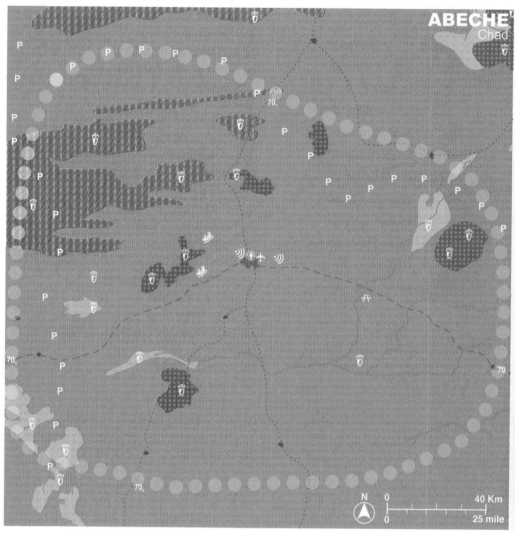

Color Figure 2 Abeche, Chad urban region. See Color Figure 1 for key to cover types and site labels.

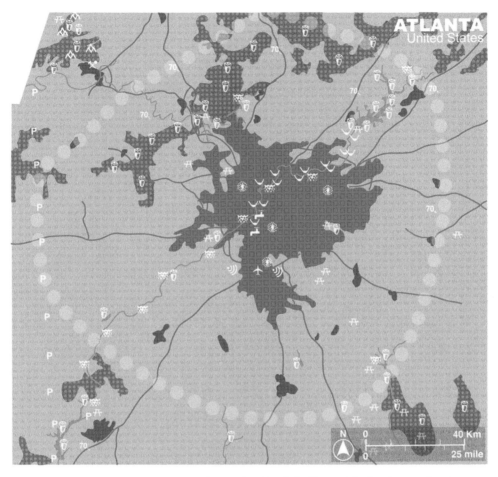

Color Figure 3 Atlanta, USA urban region. See Color Figure 1 for key to cover types and site labels.

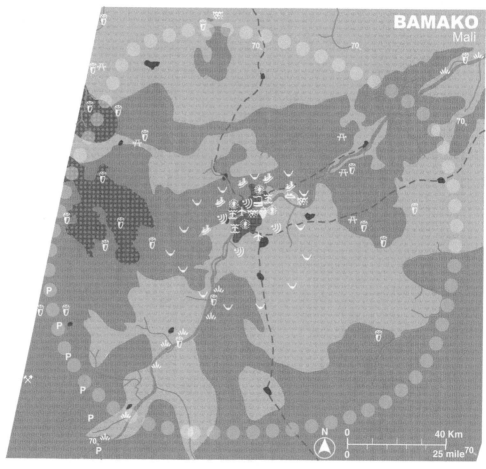

Color Figure 4 Bamako, Mali urban region. See Color Figure 1 for key to cover types and site labels.

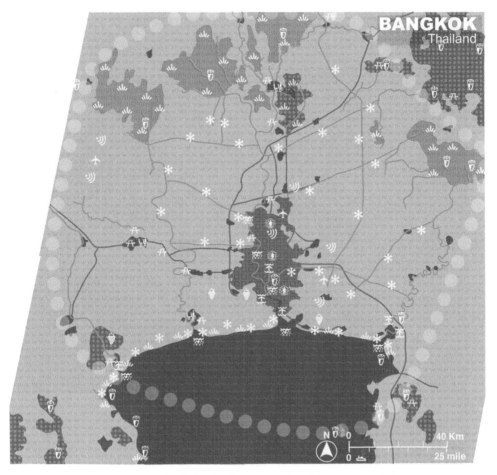

Color Figure 5 Bangkok, Thailand urban region. See Color Figure 1 for key to cover types and site labels.

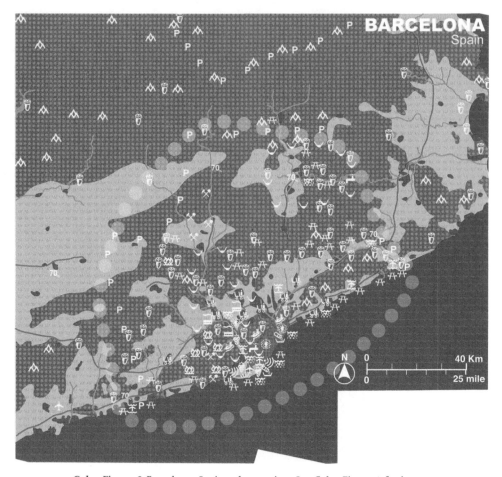

Color Figure 6 Barcelona, Spain urban region. See Color Figure 1 for key to cover types and site labels.

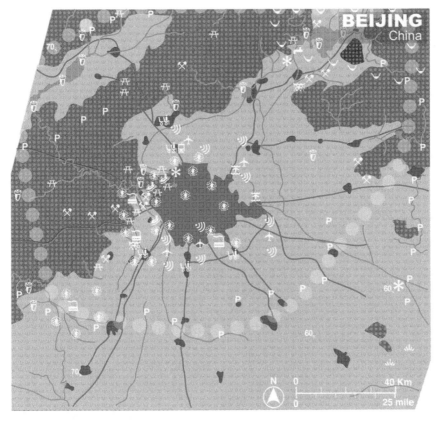

Color Figure 7 Beijing, China urban region. See Color Figure 1 for key to cover types and site labels.

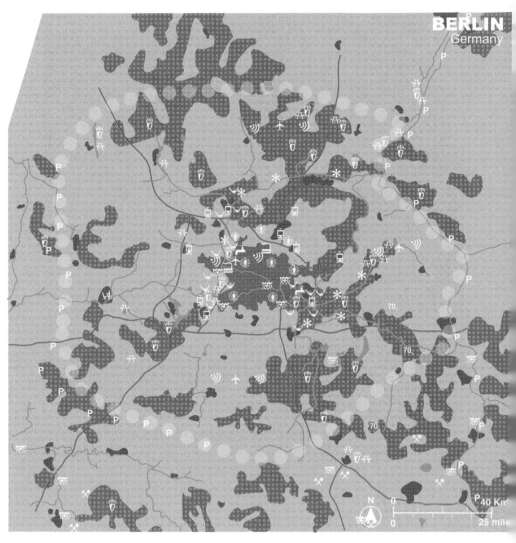

Color Figure 8 Berlin, Germany urban region. See Color Figure 1 for key to cover types and site labels.

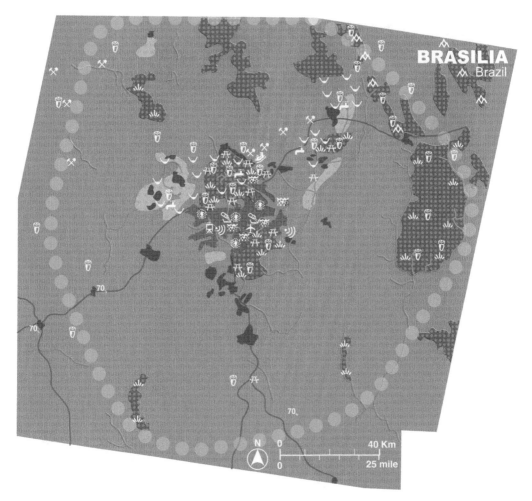

Color Figure 9 Brasilia, Brazil urban region. See Color Figure 1 for key to cover types and site labels.

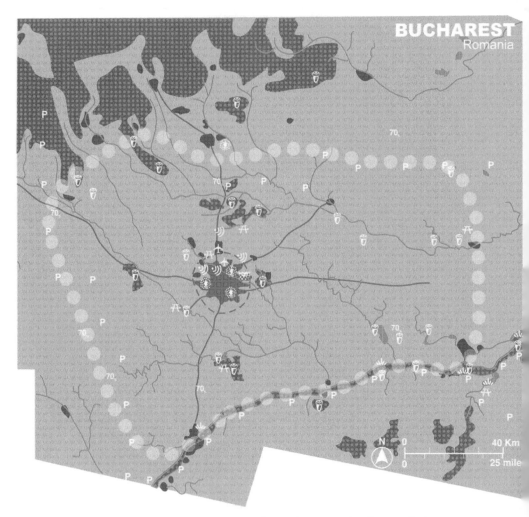

Color Figure 10 Bucharest, Romania urban region. See Color Figure 1 for key to cover types and site labels.

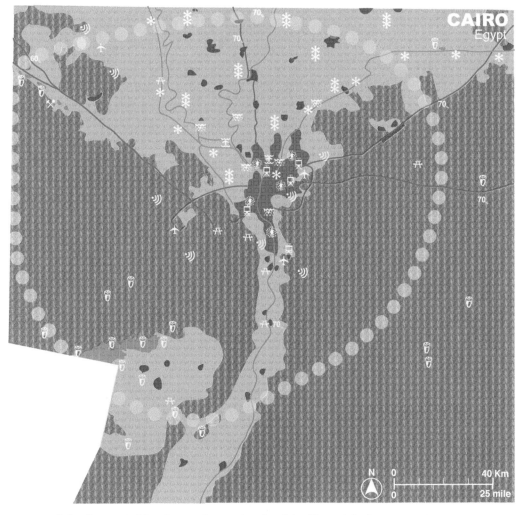

Color Figure 11 Cairo, Egypt urban region. See Color Figure 1 for key to cover types and site labels.

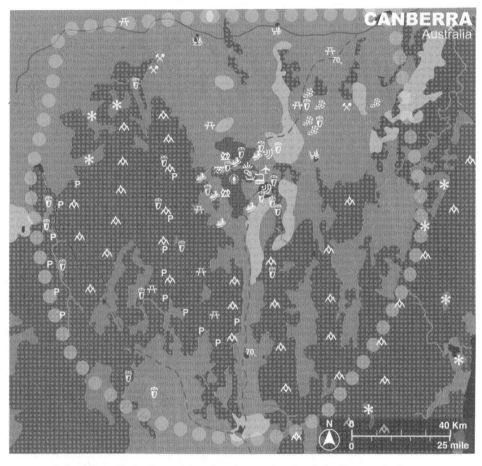

Color Figure 12 Canberra, Australia urban region. See Color Figure 1 for key to cover types and site labels.

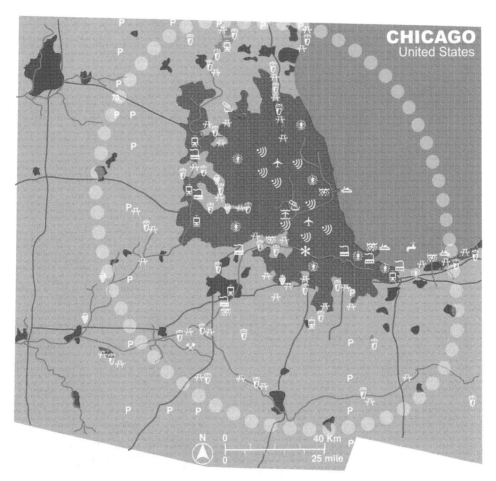

Color Figure 13 Chicago, USA urban region. See Color Figure 1 for key to cover types and site labels.

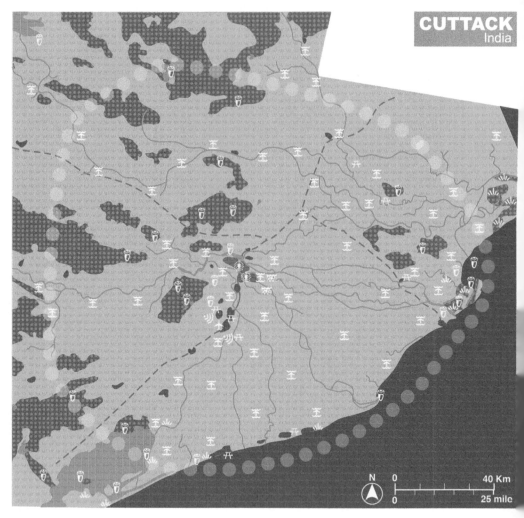

Color Figure 14 Cuttack, India urban region. See Color Figure 1 for key to cover types and site labels.

Color Figure 15 East London, South Africa urban region. See Color Figure 1 for key to cover types and site labels.

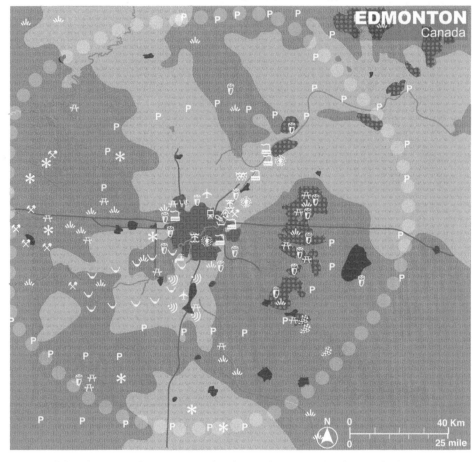

Color Figure 16 Edmonton, Canada urban region. See Color Figure 1 for key to cover types and site labels.

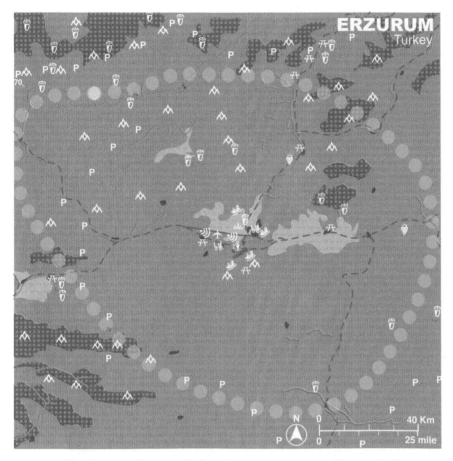

Color Figure 17 Erzurum, Turkey urban region. See Color Figure 1 for key to cover types and site labels.

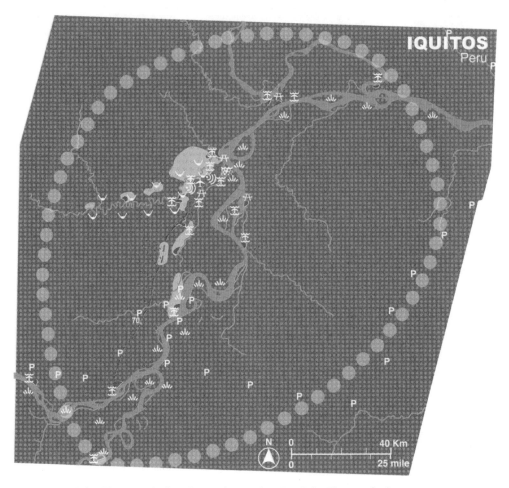

Color Figure 18 Iquitos, Peru urban region. See Color Figure 1 for key to cover types and site labels.

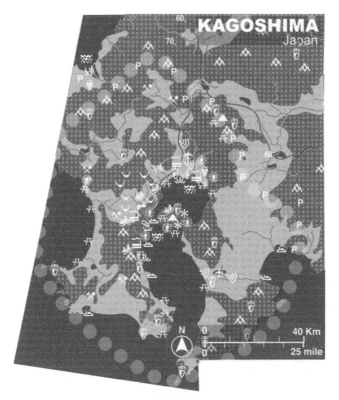

Color Figure 19 Kagoshima, Japan urban region. See Color Figure 1 for key to cover types and site labels.

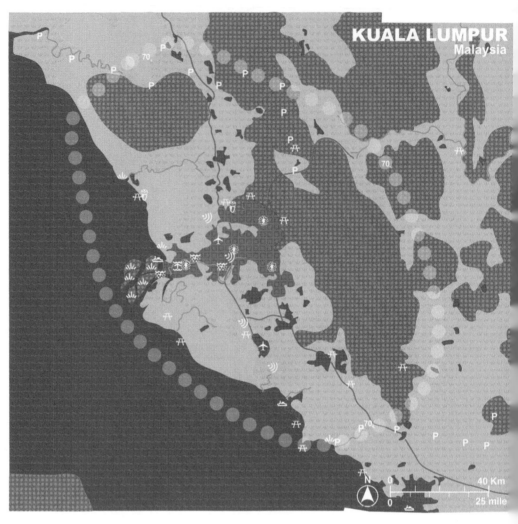

Color Figure 20 Kuala Lumpur, Malaysia urban region. See Color Figure 1 for key to cover types and site labels.

Color Figure 21 London, United Kingdom urban region. See Color Figure 1 for key to cover types and site labels.

Color Figure 22 Mexico City, Mexico urban region. See Color Figure 1 for key to cover types and site labels.

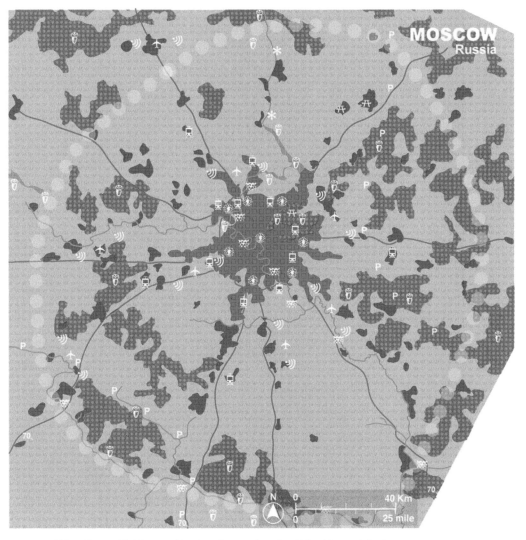

Color Figure 23 Moscow, Russia urban region. See Color Figure 1 for key to cover types and site labels.

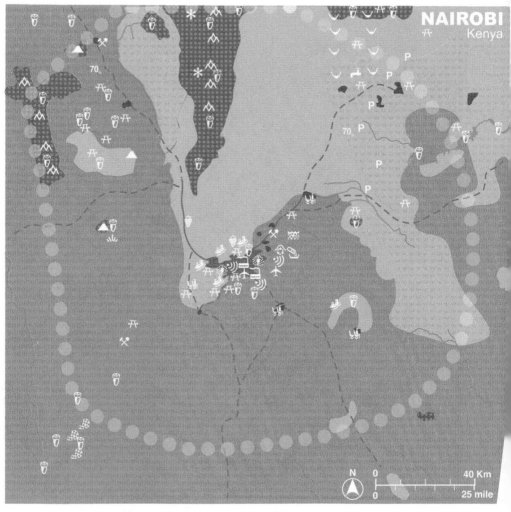

Color Figure 24 Nairobi, Kenya urban region. See Color Figure 1 for key to cover types and site labels.

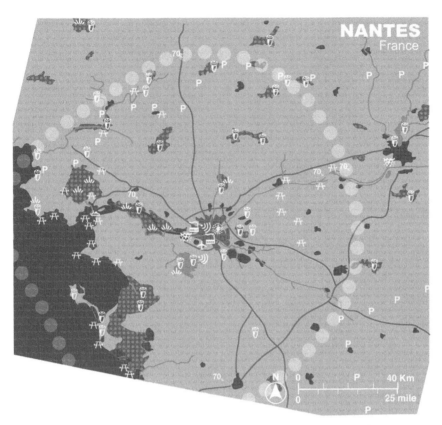

Color Figure 25 Nantes, France urban region. See Color Figure 1 for key to cover types and site labels.

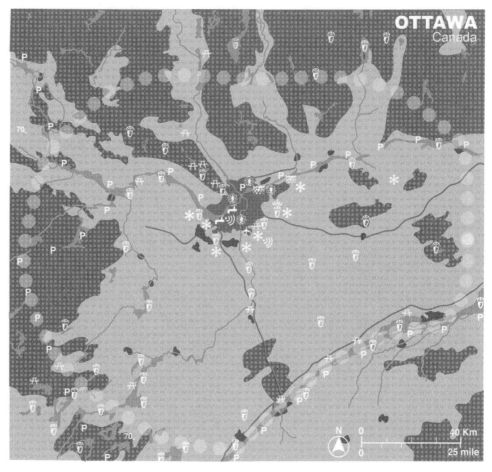

Color Figure 26 Ottawa, Canada urban region. See Color Figure 1 for key to cover types and site labels.

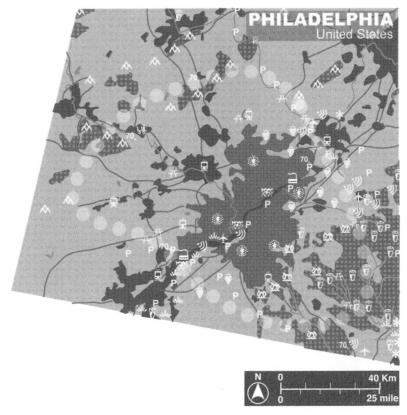

Color Figure 27 Philadelphia, USA urban region. See Color Figure 1 for key to cover types and site labels.

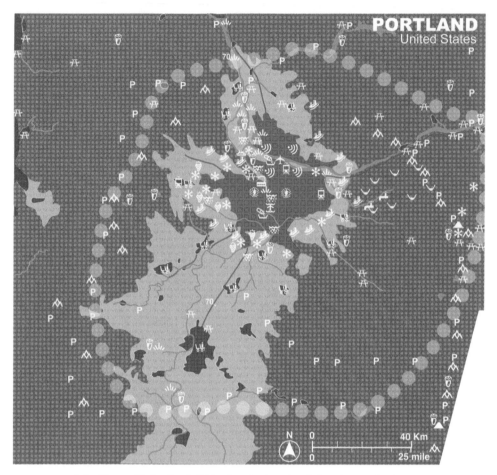

Color Figure 28 Portland, USA urban region. See Color Figure 1 for key to cover types and site labels.

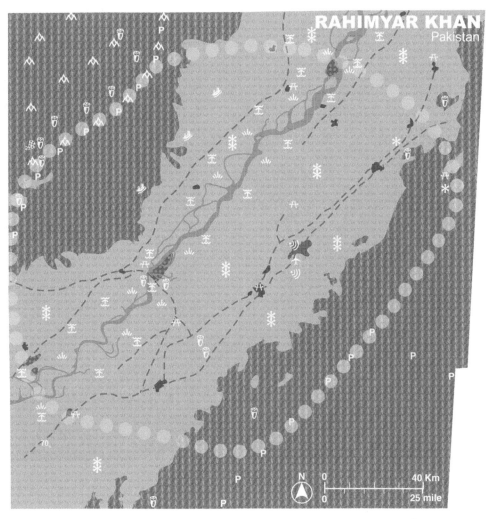

Color Figure 29 Rahimyar Khan, Pakistan urban region. See Color Figure 1 for key to cover types and site labels.

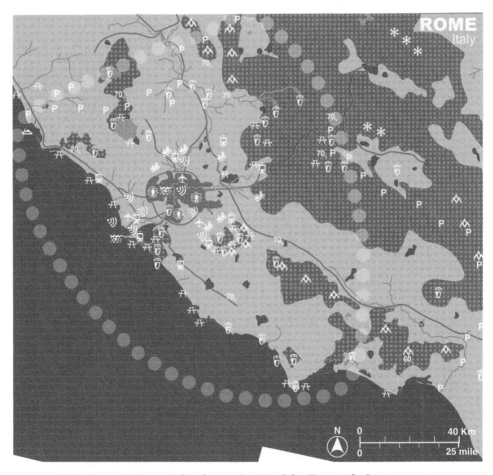

Color Figure 30 Rome, Italy urban region. See Color Figure 1 for key to cover types and site labels.

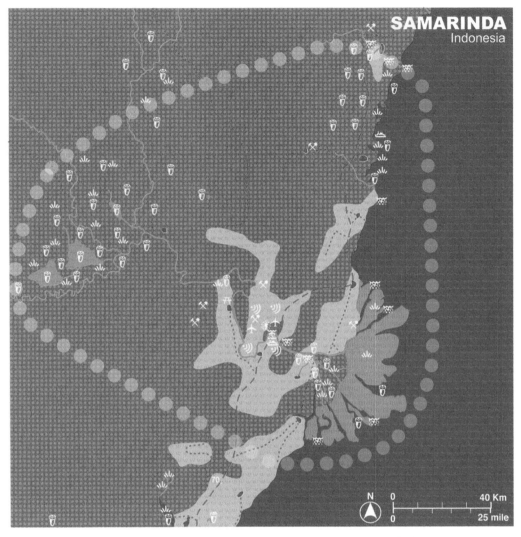

Color Figure 31 Samarinda, Indonesia urban region. See Color Figure 1 for key to cover types and site labels.

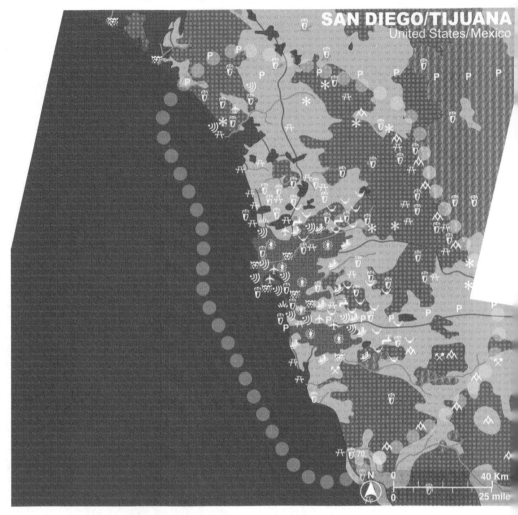

Color Figure 32 San Diego/Tijuana, USA/Mexico urban region. See Color Figure 1 for key to cover types and site labels.

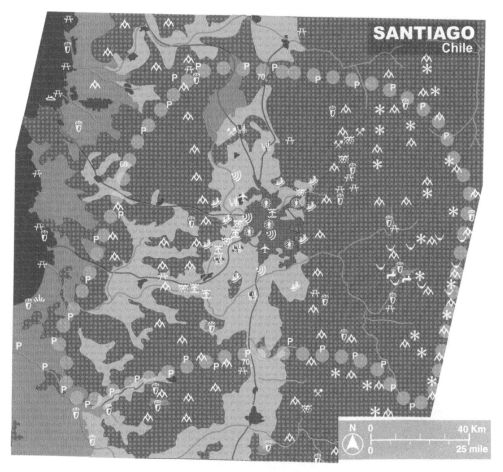

Color Figure 33 Santiago, Chile urban region. See Color Figure 1 for key to cover types and site labels.

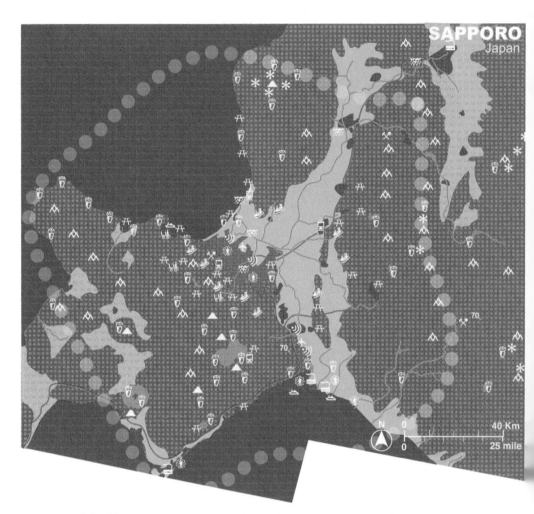

Color Figure 34 Sapporo, Japan urban region. See Color Figure 1 for key to cover types and site labels.

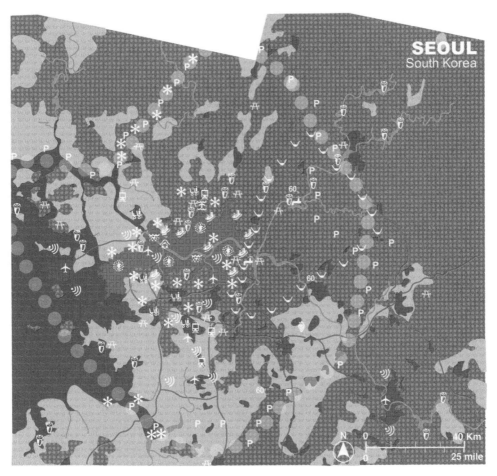

Color Figure 35 Seoul, South Korea urban region. See Color Figure 1 for key to cover types and site labels.

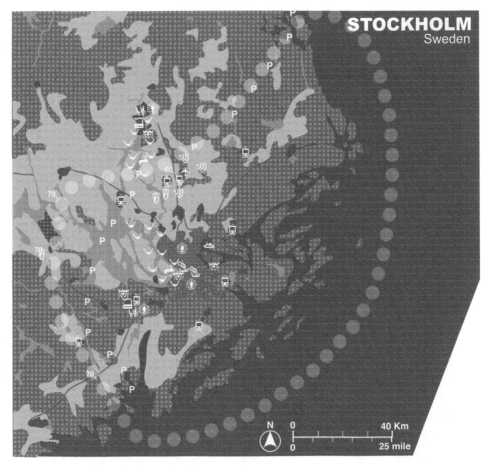

Color Figure 36 Stockholm, Sweden urban region. See Color Figure 1 for key to cover types and site labels.

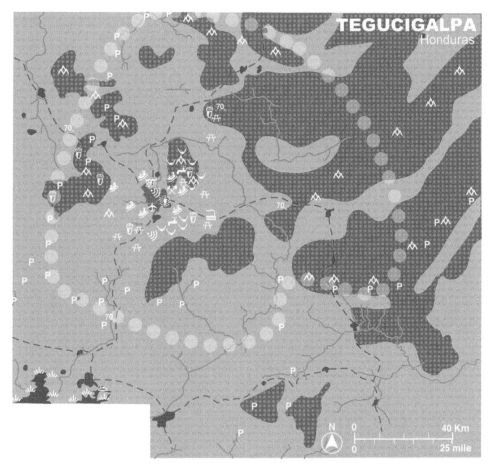

Color Figure 37 Tegucigalpa, Honduras urban region. See Color Figure 1 for key to cover types and site labels.

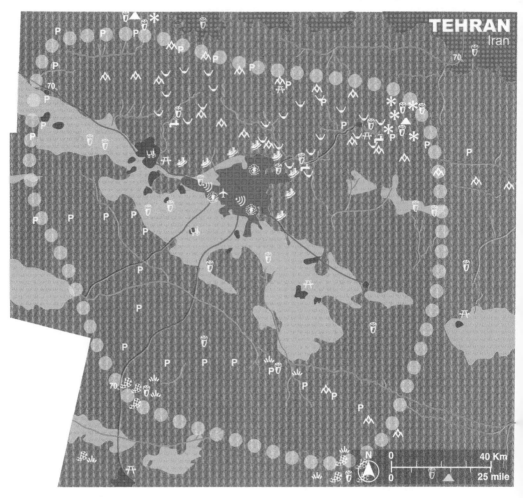

Color Figure 38 Tehran, Iran urban region. See Color Figure 1 for key to cover types and site labels.

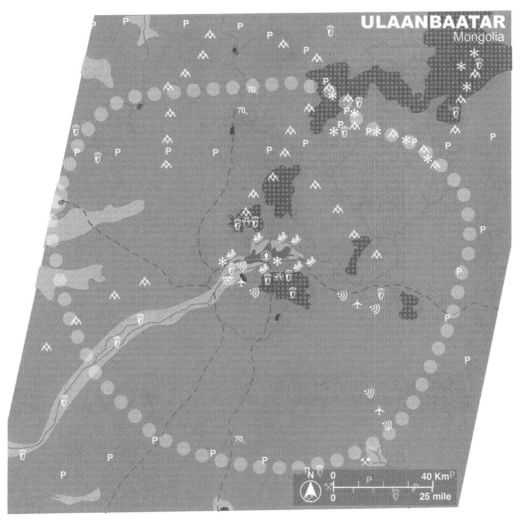

Color Figure 39 Ulaanbaatar, Mongolia urban region. See Color Figure 1 for key to cover types and site labels.

Color Figure 40 Mosaics of multi-colored broken ceramic pieces in Parque Guell, Barcelona, Spain; designed by Antoni Gaudi. R. Forman photo.

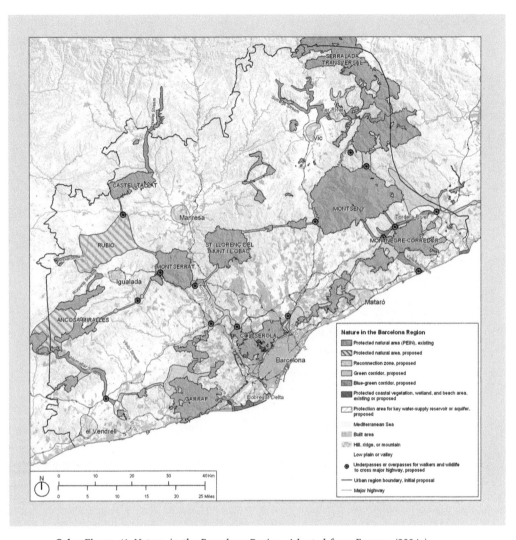

Color Figure 41 Nature in the Barcelona Region. Adapted from Forman (2004a).

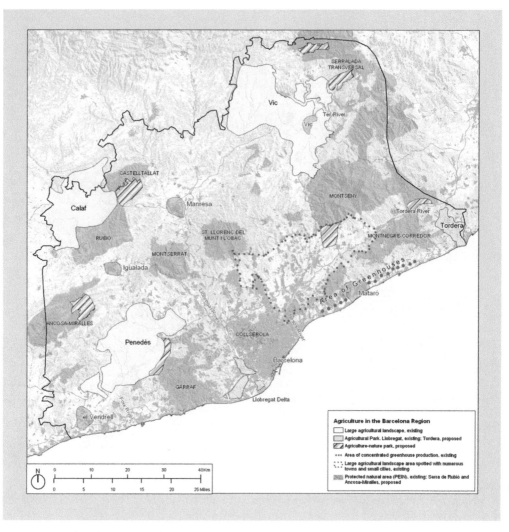

Color Figure 42 Food in the Barcelona Region. Adapted from Forman (2004a).

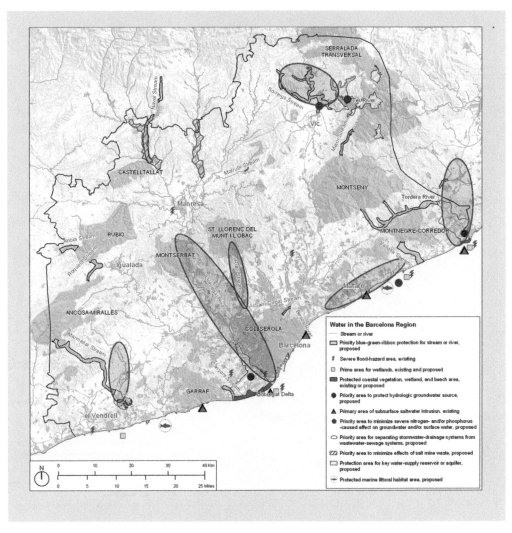

Color Figure 43 Water in the Barcelona Region. Adapted from Forman (2004a).

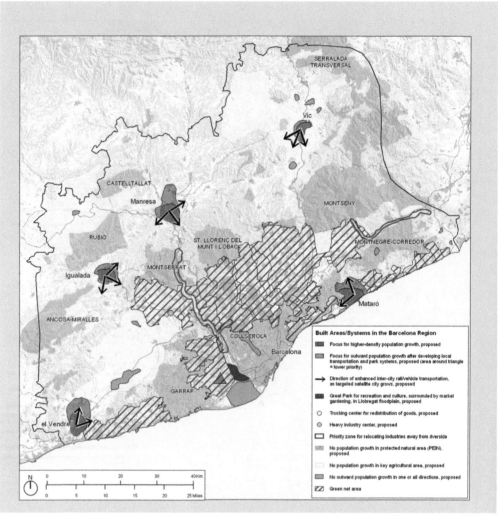

Color Figure 44 Built Areas and Systems in the Barcelona Region. Adapted from Forman (2004a).

types of information can and should be developed by local experts for urban region planning around every city.

Thirty-eight urban regions mapped

Thirty-eight brightly colored maps (Color Figures 1–39), a treasure-chest of information unraveled for the reader, have resulted from the sequence of detailed procedures and decisions above. Although measurements and spatial analyses were done directly with the large satellite images, these color maps are the flagship syntheses of the urban regions. They are designed for easy reference as we compare regions to discover patterns and principles in the next two chapters.

Place-name synopses of the regions

These cities and their urban regions are real places, with strikingly different natural systems and people living, working, and moving about. To help provide a "flavor" or "personality" to each place, an array of place names in and around a region is listed, often with key characteristics added. Years for population data and satellite images used in analyses are given, along with scattered ecology and planning references. Many of the place names offer insights into the land, history, people, culture, and feel of a place. Slowly reading through the names, even saying them aloud, should help avoid drowning in data-rich maps, tables, and figures. Better still, find and absorb some in-depth descriptions of these places, and visit them.

Abeche (Chad)

Sub-Sahara Sahel. Oum-Hadjer. Ati. Parc National de Zakouma. Am-Dam. Koulbo. Dopdopdop. Adre. Batha. Biltine. Osara Wall. Abougoudam. Arada. Haraz-Djombo. O. Enne. Gara. Am Humede. Aboy Goulem. Koulbo. Tiktike. Deressa. Am Sak. O. Rime. Am Zoer. O. Bitea. Saoue. Ougoune. Sudan and Darfur to east. Lake Chad, Nigeria, and Niger to west. Central Africa to south. Libya to north. Steeples and minarets. Moving dunes. Cutoff in all directions by rain-full wadis and *zones inondables*. Escarpment. Savanna to grassland to desert. (1993 population, adjusted to year 2000 using 5% annual growth rate; 2002 satellite image)

Atlanta (USA)

Stone Mountain. Decatur. Fulton. DeKalb. Reservoir Lake Lanier. Chattahoochee River. Marietta. Calhoun. Ludville. Pine Log. Red Top Mountain. Five Forks. New Hope. Chief Joseph Vann House. Agnes Scott. Cherokee. Starrs Mill.

Kennesaw Mt. National Battlefield. Sweetwater Creek. Rico. Lagrange. Griffin. Experiment. Indian Springs. McDonough. Oconee. Walnut Grove. Chestnut Mt. Coal Mt. Peachtree Street. Panthersville. The South. The longest commute. Cakes of red mud. (2000 pop.; 2000 image) (Odum and Turner 1990, Bullard *et al.* 2000, Frumkin *et al.* 2004, Burchell *et al.* 2005, Berger 2006)

Bamako (Mali)

Niger River. Chateau d'Eau. Sanankoroba. Koulikoro. Baguinedo. Foret de la Faya. Fana. Ouenia Lake. Nossombougou. Sirakorola. Fort de Koundou. Kati. Negala. Monte Manding. Sibi. Soussan. Reserve de Keniebaoule. Dlamba. Marche Rose. Kamablo. Kangaba. Pont des Martyrs. Torokorobougou. Submersible causeway. Ivory Coast to south. Guinea and Senegal to west. Mauritania to north. Burkino Faso to east. Timbuktu downriver to northeast. Watching boats. Mande music. (1999 pop.; 2002 image)

Bangkok (Thailand)

Chao Phraya delta. Ratchaburi. Nakhon Nayak. Suphan Buri. The Golden Mount. Ban Bung. Ayutthaya. Khao Yai National Park. Pattaya. U-Thong. Siam Square. Chachoengsao. Si Racha. Nam Tok Heo Narok. Saraburi. Nakhon Pathom. Sanut Saklon. Gulf of Thailand. Wat Amphawan. Sal Noi. Phutthamonthon. Tha Maka. Fulbright Office. Bang Pla Ma. Na Di. Bang Pa-In. Samphran. Lam Lukka. Burma to west. Cambodia to east. Linear villages. Aquacultured crustaceans. (2000 pop.; 2002 image) (Beesley and Cocklin 1982, Takaya 1987, Stubbs and Clarke 1996, Tonmanee and Kuneepong 2004, Hara *et al.* 2005, McGrath and Thaitakoo 2005)

Barcelona (Spain)

Parque Guell. Las Ramblas. Sagrada Familia. Llobregat River. Agricultural Park. La Tordera. Vic sausages. Ter reservoirs. Sabadell. El Vendrell. Les Cuatro Gats. Igualada. Manresa. Mataro. Garraf. Castelltallat. Montseny. El Corredor. Calaf grain. Collserola. Montserrat. Besos. Foix reservoir. Mercat St. Joseph. Marmellar gorge. Penedes wine. Valles. Mediterranean beaches. Montjuic. Catalunya. Sardana. Segnale. France to east. Wetland. Pig farms. Salt mines. Greenhouses. Dolmens. (2001 pop.; 2002 image) (Vallejo and Alloza 1998, Forman 2004a, Pauleit *et al.* 2005, Mayor Farguell 2005, Mata and Tarroja 2006)

Beijing (People's Republic of China)

Great Wall. Forbidden City. Temple of Heaven. Tsinghua University. Tiananmen Square. Confucius Temple. Great Hall of the People. Dadu/Khanbalik. Ming Tombs feng-shi. Summer Palace. Peking University. Two large reservoirs, one for drinking. Dust from Inner Mongolia. Ring roads. Summer Olympics. Air pollution. Grid city. Mountain range. Friendly beloved dragons. Bicycles. (1999 pop.; 2000 image) (Sit 1995, Gu and Kesteloot 1998, Chen *et al.* 2004, Yang 2004)

Berlin (Germany)

Brandenburg. Potsdam. Spreewald. Kostrzyn. Ruiner Canal. Naturpark. Schwedt. Buch. River Oder. Kyritz. Baumgarten. Lindenberg. Frankfurt. Plan. Petersdorf. Luckenwalde. Karlshof. Bad Saarow-Pieskow. Krausnicker Berge. Waldstadt. Rathenow. Britz. Hammer. Blankenfelde. Bretzsee. Zixdorf. Poland to east. France to west. The Wall, East and West. Industrial/agricultural collectives. Sudgelande railroad exotics. Biotopes. Tiergarten transformations. (2001 pop.; 2000 image) (Sukopp and Werner 1983, Sukopp and Hejny 1990, von Stulpnagel *et al.* 1990, Godde *et al.* 1995, Breuste *et al.* 1998, Kuhbler 2000, Mauerer *et al.* 2000, Von Krosigk 2001, Bahlburg 2003, Girot 2004, Rink 2005, Pauleit *et al.* 2005, Kowarik and Langer 2005)

Brasilia (Brazil)

The Cerrado. Lake Brasilia. Goiania. Anapolis. Serra Geral do Pasana. Formosa. Luziania. Gama. Sobriadiaho. Barragem do Descoberto. Preto. Sao Bartolomeu River. Planaltina. Corumba de Goias. Brazlandia. Padre Bernardo. Mimoso. Lake Brasilia. Salto de Itiquira. Bom Sucesso. Rodeador Peak. Cd. Eclectica. Pedra Fundamental. Bananal. Amazonas to north. Atlantic to east. Rio de Janeiro and Sao Paulo to south. Planned city. (2000 pop.; 2001 image) (Epstein 1973, Starling 2000, Hall 2002, Oliveira and Marquis 2002)

Bucharest (Romania)

Carpethians. Danube River. Giurgiu. Buftea. Titu. Snagov. Ploiesti. Oltenita. Slobozia. Ialomita River. Lake Dambovita. Ruse. Vedea River. Gaesti. Cucurugu. Sarbeni de Jos. Dragomeresti Vale. Baleni Romani. Balotesti. Strandul Titan. Fireta Mare. Dumbrava. Iszerul Mostistei. Baba Ana. Floreasca Park. Black Sea and Ukraine to east. Serbia to west. Bulgaria to south. Clejani Gypsy music. (2002 pop.; 2000 image)

Cairo (Egypt)

The Nile. Al-Qahira. Giza Pyramids. Suez Canal. Rosetta, the stone location. The Fifth Aggregation. Heliopolis. Birket Qarum. El Faiyum. El Gharaq el Sultani. Beni Suef. Crocadilopolis. Philadelphia. El Wasta. Sol. El Lisht. Garza. The Sphinx. Saqqara. Warraq el Arab. Sadat City. Gebel Qatrani. Wadi el Natrun. Kulet el Qrein. Great Bitter Lake. Isma'iliya Canal. Zagazig. Benha. Ramsis. Shibin el Kom. Aswan Dam to south. Sahara to west. Mediterranean to north. Queen. (2002 pop.; 2000 image) (Vigier 1997)

Canberra (Australia)

ACT. Cooma. Togganoggera. Yass. Bunyan. Doughboy. Jinglemoney. Cotter River. Snowy Mts. Lake George. Collector. Gundaroo. Goulburn. Yarrangobilly Caves. Lake Eucumbene. Kosciuszko. Tumut. Cockington Green. Boro. Goobarragandra Wilderness. ANU. Wee Jasper. Murrumbidgee River. Tidbinbilla. The Thunderer. Victoria to south. Tasman Sea to east. NSW to north. Planned Capital city. Brindabella Dreaming. Waves of invasives. Stockmen. Griffin geometry. Yellow box-red gum. 2003 wildfire. Kamberri country. (2002 pop.; 2001 image) (Troy 1995, Hall 2002, Houston 2005, Rigby 2006)

Chicago (USA)

Lake Michigan. O'Hare. Sears Tower. Des Plaines River. Gary. Stickney Wastewater Treatment. Johnsburg. Ivanhoe. Batavia. Saint Charles. Algonquin. Lake Zurich. Frankfort Square. Indiana Dunes. Calumet. Oswego. York Center. Hickory Hills. Romeoville. Joliet. Cook County. Black Oak. Minooka. Manhattan. National Tallgrass Prairie. The Loop. Deer park. Midwest shipping to the Atlantic. Agribusiness. Steel. Chicago Wilderness. Brownfields. Bluff beaches. 1871 fire. City Hall green roof. (2000 pop.; 2001 image) (Witham and Jones 1987, Cronon 1991, Nowak 1994, Heisler *et al.* 1994, Cityspace 1998, Benfield *et al.* 1999, Calthorpe and Fulton 2001, Greenberg 2002, Daley and City of Chicago 2002, Hall 2002, Perlman and Milder 2005, Dwyer and Chavez 2005, Berger 2006)

Cuttack (India)

Kataka. Orissa. Mahanadi delta. Kathajodi River. Puri. Paradip. Bhubaneswar the Capital. Khurda. Sunak-halla. Konark beach. Hindal. Atri hot sulfur spring. Bay of Bengal. Chilika Lake. Gop. Khandagiri Caves. Angul. Nandankanan Biological Park. Indipur. Altiri. Babaan Bazaar. Pipli. Fort Barabati. Subhash Chandra Bose. Calcutta and Bangladesh to northeast. Monsoonal sea of river water. Cane. Cricket. Rich filigree. Rice research. High railway. Tigers and elephants. (2001 pop.; 2000 image)

East London (South Africa)

Mgwali. Gonubie Mouth. King William's Town. Hamburg. Keiskammahoek. Wavecrest. Oos-Londen. Morgan's Bay. Fort Hare. Bolo Reserve. Qoboqobo. Potsdam. Kiva-Pita. Beacon Bay. Great Fish Point. Great Kei River Bridge. Mpetu. Roolkrans Dam. Committees. Zwelitsha. Amatola Mountains. Kidd's Beach Nature

Reserve. Braunschweig. Haga-Haga. Indian Ocean. Bonza Bay. Haga-Hago. Wild Coast. (1999 pop.; 2001 image)

Edmonton (Canada)

Native People's land. Stony Plain Moraine. Beaver Hills. Riverbend. Fort Saskatchewan. Leduc. Yellowhead Trail. Terrace Heights. Sturgeon. Petrochemical Alley. Big Lake. Saskatchewan to east. Canadian Rockies to west. River corridor. Long freight trains. Flat ranchland and cropland. Cold wind. Big river. Treasured poplar–aspen–spruce stands. Royal Alberta Museum. Shallow prairie lakes. The bison herd. (2001 pop.; 2002 image) (Saley *et al.* 2003, Wein 2006)

Erzurum (Turkey)

Cifte Minare. Palaneloken peaks. Hunis. Tekman. Cat. Bayburt. Aras River. Coruh Nehri. Askale. Ispir. Ilica Spa. Pasinler. Dogu Karadeniz Mountains. Bingoze wetlands. Koprukoy. Tercan Reservoir. Dumludagi. Tortum Selalesi. Attaturk University. Kavurmacukuru. Haho. Narman. Horasan. Abdullahkomu. Toptepe. Hyspiratus. Black Sea to north. Armenia and Georgia to northeast. Kurdish centers and Iran to southeast. Treeless hills and steppe. Riparian vegetation. Dadas area. Skiing. (2000 pop.; 2002 image)

Iquitos (Peru)

Nauta. 12 km wide powerful turbulent muddy Amazon. Rio Nanay. Mazan. Francisco de Orellana. Rio Napo. Tamshiyacu. Tigre. Lake Quistococha. Itaya River. Ex Petroleros. Trece de Febrero. El Dorado. Paujil. Rainforest-squeezed long narrow city. Boats coming, going. Native Peoples from la selva. Little-river transportation routes. Logging, deforestation, wood products up rivers. Road construction. Meanders. Erosion. Coca. Palm products. (1998 pop., adjusted to year 2000 using 5% annual growth rate; 2001 image) (Browder and Godfrey 1997, Laurance *et al.* 2000, Maki *et al.* 2001)

Kagoshima (Japan)

Sakurajima Volcano. Tarumizu. Sata Misaki. Osumi Hanto. Kushima. Shibushi Wan. Kagoshima Bay. Pacific Ocean. Kanoy. Onajime. Hi Zaki. Osaki. Aira. Fukuyama. Kokubu. Kirishima Yaku National Park. Togo. Yokogawa. Izumi. Kushikino. Havato. Iriki. Ouza. Makurazaki. Ikedo Ko. Bono Misaki. Satsuma Hanto. Kuta Shima. Kaimon Dake. Kiire. Kawanabe. Kyushu, Japan's subtropical South. Ryukyu Islands and East China Sea to south. Ferry boats and fishing. Volcanic puffings. (2002 pop.; 2002 image) (Yamamoto 1930)

Kuala Lumpur (Malaysia)

Straits of Malacca. Klang. Petaling Jaya. Secemban. Genting Highlands. Temerloh. Raub. Port Dickson. Karak. Bentong. Sabak Bernam. Slim. Melaka. Gn. Benom. Tembeling River. Kuala Selangor. Sri Menanti. Batu Caves. Taman Negara Park. Mah Meri Orang Asli. Pelabuhan Klang. Blue Lagoon. Kg. Jawi-jawi. Java and Indonesia to southwest. South China Sea to east. Tallest building. Active rainforest canopy. Oil palm snakes. (2000 pop.; 2001 image) (Rieley and Page 1995)

London (UK)

Tower Bridge. Westminster. Thames River. Slough. Windsor Castle. Luton. Heathrow. Weston-on-the-Sea. Cambridge. Bishop's Stortford. Ascot. Aldershot. King George V Reservoir. Canterbury Cathedral. Isle of Grain. Sevenoaks Weald. Letherhead. Kew Gardens. Brent. Wimbledon. Harrow. Oxford. Eel Brook Common. Paddington station. Letchworth. Bradwell defuelled. Central congestion-control zone. The London Ring Main. Doubledecker buses. Elizabeth. Greenbelt. 1666 Great Fire. (2001 pop.; 2000 image) (Davis 1976, Munton 1983, Gilbert 1991, Turner 1992, Bartuska 1994, Parsons and Schuyler 2000, Ishikawa 2001, Howe 2002, Hall 2002)

Mexico City (Mexico)

Toluca. Cuernavaca. Pueblo. Pachuca. Sierra Nevada. Ajusco. Texmelucan. Volcan Popocatepetl. Volcan Iztaccihuatl. Guadaloupe Zaragoza. Teotihuacan. Calpulalpan. Apan. Laguna Tecocomulco. Tula. Ecatepec. Coacalco. Villa de Carbon. Cuantlan. Tlanepantla Ocampo. Jilotepec de Molina. Xochimilco. Mina Vieja. Santa Cruz. Contreras. Milpa Alta. San Juan Ixtayopan. Cerro Cervantes. Santa Barbara. Barranca Grande. Llano Grande. La Finca. Parque A. de Humboldt. Cerro El Coyote. (2002 pop.; 2000 image) (Rapoport 1993, Losada *et al.* 1998, Pezzoli 1998, Hall 2002)

Moscow (Russia)

Obninsk. Moskva River. Sergiev Posad. Stupino. Klin. Pokrov. Elektostal. Losiny Ostrov National Park. Vodokhraqnilishche. Dmitrov. Izmaylovsky Lesopark. Troize-Sergieva Lavra. Kremlin. Klimkinskoe. Nudol. Krasnobogatyrskaya. Mow. Bittsa. Balashikha. Ruza. Biryulevsky Dendropark. Archangelskoe. Domodedova. Oka River. Mytisci. Cehov. Ukraine to south. Belarus to west. Finland and Barents Sea to north. Two-lane outer ring road. Weekend dachas. (2002 pop.; 2002 image) (Barker and Sutcliffe 1993)

Nairobi (Kenya)

Ngong Hills. Mt. Kinango. Masai Gorge. Kabira. Nyeri. Murang'a. Kikuyu. Athi River. Dandora. Saba Saba. Gitutu. Machakos. Thika. Lake Naivasha. Suswa. Gilgil. Hyrax Hill. Soda Lake. Magadi. Olorgasailie. Konza. Fourteen Falls. Santamor Halt. Nairobi National Park. Kikoro. Kithimani. Embu. Kipipiri. Rift Valley to west. Mt. Kenya to north. Mt. Kilimanjaro and Tanzania to south. Indian Ocean to east. (1999 pop.; 2000 image)

Nantes (France)

The Loire. Lac de Grand-Lieu. Canal de Nantes a Brest. Etang de Pin. La Chapelle-Basse-Mer. La Montagne. Paimboeuf. Etang du Grand Moulin. Charbonnieres. The Atlantic. Cote de Jade. Pont de St-Nazaire. La Pommeraye. St-Mars. Foret d'Ombres. Le Pin. Menhirs des Loueres. Missillac. Ste-Anne. Maine Riviere. Chateau des Ducs de Bretagne. La Grande Haie. Notre-Dame-des-Landes. Piqueniques. (1999 pop.; 2001 image) (Pauleit *et al.* 2005)

Ottawa (Canada)

Ontario. Quebec. Ottawa River. Rideau Locks. Hull. La Peche. Gloucester. Algonquin. Rockland. Parc de la Gatineau. Buckingham. Aylmer. Carleton Place. Brockville. St. Andrews. Lac Simon. Cornwall. Kilmamock. Papineau-Labelle. Moose Creek. Fitzroy Park. New Dublin. Tincas. Smith's Falls. Ogdensburg. Limoges. Stittsville. Brightside. Pike Lake. Fort Coulonge. Quebec to north. Montreal to east. New York and USA to south. Sewage-lagoon birds. Greenbelt and greenways. (2001 pop.; 2001 image) (Hough 2004, Billington and Tozer 1977)

Philadelphia (USA)

The Schuylkill. Delaware River. Valley Forge. Pine Barrens. New Jersey. Levittown. Camden. New Hope. Bristol. Chester. Chestnut Street. Main Line. Haverford College. Bryn Mawr. Swedesboro. Mullica River. George School. Fort Dix/McGuire AFB. Pennsylvania Turnpike. Liberty Bell. Wilmington to southwest. New York to northeast. Quaker William Penn atop City Hall. Benjamin Franklin's university. Paddling through cedar swamp magic. (2000 pop.; 2002 image) (McHarg 1969, Forman 1979a, Pinelands Commission 1980a)

Portland (USA)

Oregon City. Columbia River. Bonneville Dam. Willamette. Capital Salem. Trojan Nuclear. St. Helens. Timber. Scappoose. Vancouver. Washington State. Beaverton. Mt. Hood. Turner. Silverton. Scotts Mills. Elk Lake. Mt. Jefferson. Bull Run. Woodburn. Zigzag. Cedar Hills. Rhododendron. Cougar. Milwaukie.

Orenco Station. Happy Valley. Forest Park. The Pacific to west. Urban growth boundary. Lewis and Clark. (2000 pop.; 2000 image) (Platt *et al.* 1994, Diamond and Noonan 1996, Benfield *et al.* 1999, Houck and Cody 2000, Calthorpe and Fulton 2001, Hulse *et al.* 2002, Avin and Bayer 2003, Ozawa 2004, Irazabal 2005)

Rahimyar Khan (Pakistan)

Indus River floodplain. Sadiquiped. Ubauro. Punjab's south end. Maumubarik. Bahawalpur. Cholistan Desert. Rajanpur. Mithankot. Allahabad. Siri natural gas. Khanpur. Liaquatpur. Thar Desert. Ghani Goth. Sadiqabad. Kashamor. Derawar Fort. Rojhan. Kandhkat. Jhil Marav. Loti. Daharki. Khan Bela. Jamaluddinwali. Kot Samaba. India to southeast. Karachi-Lahore railway. Mandi towns. Town Hall's tower. Camels. Desert crossing. Fish-migration ladders. (1998 pop., adjusted to year 2000 using 5 % annual growth rate; 2000 image) (Masud-Ul-Hasan *c.* 1965)

Rome (Italy)

Mediterranean. Tirreno Sea. Tiber River. Colosseo. Ostia. Tivoli Gardens. Teatro Adriano. Albani Hills. Lago di Bracciano. Cisterna di Latina. Frosinone. Roccamassima. Grotta del Pianoros. Leonardo da Vinci Airport. Tarquinia. Civitavecchia. The Vatican. Ladipoli. Vigna di Valle. Macchia della Manziana. Castel S. Elia. Viterbo. Avezzano. Campo Felice. Cava di Pietre. Mola di Bassano. Villa S. Giavanni. Allumiere. Gran Sasso d'Italia. L'Aquila. Rieti. Campagna Romana. Beaches. Roman roads. (2001 pop.; 2001 image) (Barker and Sutcliffe 1993, Blondel and Aronson 1999)

Samarinda (Indonesia)

Borneo. East Kalimantan. Pulau Terentang. Balikpapan. Bontag port. Ayu. Pemarung. Mahakam River. Samboja. Tg. Bayur. Santan. Muarajawa. Teggarong tourism. Sepasu. Danau Melintang. Kutai National Park. Klampo. Sengata. Bangsalsepulun. Equator. Lohjanan. Pandang Luwai Reserve. Java and Java Sea to south. Sulawesi (Celebes) to east. Malaysia and South China Sea to northwest. Rainforest logging outlet. River bridge. Multi-channel delta. Mangrove swamps. Offshore oil field. (1990 pop., adjusted to year 2000 using 5 % annual growth rate; 2001 image) (Lawrence 2004)

San Diego/Tijuana (USA/Mexico)

Oceanside. Rio Tia Juana. Rosarito. The Pacific. Bronco Flats. Bear Canyon. El Mezquito. Escondido. Thing Valley. Ruidoso. Guadalupe. Mission Beach. Ecological Reserve. Temecula. Neji. Colorado River water. Palomar Mt. Ramona. Cayapaipe Indian Reservation. Lake Henshaw. El Descanso. El Condor. Barbara Terrace.

Camp Pendleton. Poptla. Oak Grove. Baja to south. Imperial Valley and Arizona to east. Los Angeles to northwest. Chaparral gnatcatcher. Multiple Habitat Conservation Plan. Wildfire. TOD. Water sources. (2000 pop.; 2000 image) (Soule 1991, Beatley 1994, Bloom and McCrary 1996, Swenson and Franklin 2000, DiGregoria *et al.* 2006)

Santiago (Chile)

The Andes. Vina del Mar. Los Condos. Maipu. Rancagua. San Antonio. Valparaiso. Colina Hot Springs. Conchali. Acudo Lake. Rio Maipo. Cerro San Cristobal. Penaflor. Paine. Laguna Negra. Naturaleza Yerba Loca. Melipilla. Nos. Park La Campana. Pomaire. 6070 m Cerro Juncal. Esmeralda. Valle Allegre. Tiltil. Rio Clarillo Park. Tierras Blancas. Santa Sara. Pueblo Hundido. Villa Seca. La Rana. El Melocoton. Pacific to west. Argentina to east. Skiing and beaches. (2002 pop.; 1999 image) (Atkinson *et al.* 1999, Simmonds and Hack 2000)

Sapporo (Japan)

Promenade O-dori. Tokei-dai Clocktower. Sea of Japan. Tomakomai. Muroran. Bibai. Yoichi Dake. Lake Toya. Lake Shikotsu. Otaru. Takikawa. Yubari. Rumoi. Ainu Village. Hiroshima. Shakitan Peninsula. Noboribetsu Spa. Chitose Gawa. Niseko skiing. Ski Jump Hill. Atsubetsuku. Nopporo Forest Park. Hassoan Teahouse. Shokanbeetsu. Ishikari. Asahikawa to north. Japan's big northern island. Ice sculptures extravaganza. (2002 pop.; 1999 image)

Seoul (South Korea)

Goyang. Bucheon. Anyang. Uijeongbu. Guangju. Incheon. Demilitarized Zone. North Korea. Seongnam. Gunpo. Chugyop-san. Yellow Sea. Asan Bay. Kanghula Island. Namhan River. Soyang Lake. Mt. Ch'iaksan National Park. Concentrated militaries. Greenbelt skylines. Dust from Inner Mongolia. Mountain and rice valleys. (2000 pop.; 2001 image) (Im 1992, Park and Lee 2000, Song and Jin 2002, Hong *et al.* 2003, Lee *et al.* 2005, Bengston and Youn 2006)

Stockholm (Sweden)

Saltsjon. Kronobergs-Parken. Vallhallavegen. Strandvagen. Tennisstadion. Norra Station. Ringvagen. Gamla Stan. Kungliga Slottet. Lake Malaren. The Archipelago. Gustavsberg. Boo. Katrinaholm. Riddersholm. Upplands-Vasby. Norrfjarden. The Baltic and Finland to east. Poland to south. Norway to west. Green wedges. Boats and fishermen. Norway spruce and Scots pine. Moose. (2002 pop.; 2000 image) (Bolund and Hunhammar 1999, Mortberg 2001, Hall 2002, Elmqvist *et al.* 2004, Clark 2006)

Tegucigalpa (Honduras)

Golfo de Fonseca. Danli. Comayagua. Comayaguela. Parque Nacional La Tigra. Parque El Obrero. Ojojona. Escuela Agricola Panamericana. El Sauce. Cerro Grande. Rio Hondo. La Cuesta No. 2. Talanga. Mata de Platano. Santa Ana. Sabanagrande. El Paraiso. Montana El Chile. Las Mesas. El Oro. Rio Choluteca. El Uyuca. San Antonio. Angalteca. Pacific, Nicaragua, and El Salvador to south. Caribbean to north. Romantic Parque La Leona. (2001 pop.; 1999 image)

Tehran (Iran)

Karaj. Namak Lake. Alborz Mountains. Kuh-e Damavand volcano at 5601 m. Qom. Garmsar. Varamin. Lar River Reservoir. Eslamshar. Saran. Baqer Abad. Shemshak. Rey. Shariyar. Bumehen. Dasht-e-Kavir Desert. Caspian Sea, Azerbaijan, and Armenia to north. Mountain ridges. Salt flats. Irrigating the blooming desert. Palm shade. (1999 pop.; 2000 image)

Ulaanbaatar (Mongolia)

Ulan Bator. Druumod. Chentejn. Nuruu. Altanbulag. Darhan. Tuul Gol. Jargalant. Mandal. Ihsuuj. Tov. Nalayh. Zuunmod. Assait Hairhan. Ikh Khorig. North Gobi Desert. Kerulen River to east. Chinese Inner Mongolia to south. Water that ends up in Lake Baikal of Siberian Russia to north. Tundra-covered mountain ridges. Rushing rivers. Felt walls (yerts) on the move. Zud. Dung for life. Wolves. Camels. Cold. Grassland. (2000 pop.; 2001 image)

Broad patterns of the urban-region set

The 38 cities are no random sample of the world's cities. Nor are they explicitly a representative sample. The selection process mainly eliminated cities near other major cities. It eliminated outliers (e.g., Rio de Janeiro, Miami, Johannesburg, New York, Venice) whose patterns might not readily extrapolate to other cities. Putting those two constraints aside, the process did choose a range of at least somewhat representative urban regions worldwide so that results should be widely extrapolable. As still more regions are analyzed, the central tendencies and variability will become clearer.

Before dissecting the regions individually and comparing them, a brief overview of the total set is useful. City populations ranged from *c.* 260 000 (Rahimyar Khan) to 11 million (Beijing), with a fairly even distribution in multiples-of-two size classes (0.25–0.5, 0.5–1, 1–2 million, etc.) (Table 5.1). The six broad geographic areas (somewhat similar to continents) each has five to eight urban regions represented. Thirty-two nations are represented. Latitudes range from 56 °N (Moscow) across the equator (Nairobi, Samarinda) to 35 °S

Table 5.1 Statistics for the thirty-eight urban regions analyzed. P/A = perimeter to area; L/W = length to width. See Figure 5.1 map.

City (Nation)	Population of city	Latitude (°)	Elevation (m)	Temperature (°C)	Precipitation (cm)	Metro perimeter (km)	Metro area (km²)	Metro P/A ratio	Metro L/W ratio
EUROPE									
London (UK)	7 127 000	52N	5	11.1	59	381	1 672	0.23	1.5
Berlin (Germany)	3 388 000	52N	50	9.4	56	221	905	0.24	1.9
Rome (Italy)	2 733 000	42N	115	16.1	65	155	388	0.4	1.6
Bucharest (Romania)	1 922 000	44N	82	10.9	58	105	286	0.37	1.2
Stockholm (Sweden)	1 850 000	59N	44	6.1	57	183	500	0.37	1.7
Barcelona (Spain)	1 527 000	41N	95	15.7	60	97	219	0.44	1.6
Nantes (France)	278 000	47N	30	12.3	79	87	171	0.51	1.3
NORTH AMERICA									
Chicago (USA)	2 896 000	42N	185	10.6	84	748	3 993	0.19	1.4
SanDiego/Tijuana (USA/Mexico)	2 373 000	33N	4	17.2	25	354	1 138	0.31	1.7
Philadelphia (USA)	1 518 000	40N	2	12.4	104	538	2 459	0.22	1.4
Ottawa (Canada)	774 000	45N	79	6.3	73	133	372	0.36	1.8
Edmonton (Canada)	616 000	54N	656	2.6	44	176	550	0.32	1.3
Portland (USA)	529 000	46N	9	12.8	101	325	1 325	0.25	1
Atlanta (USA)	416 000	34N	307	16.1	120	707	3 133	0.23	1.1
LATIN AMERICA									
Mexico City (Mexico)	8 590 000	19N	2231	15.6	58	287	1 606	0.18	1.5
Santiago (Chile)	4 668 000	33S	529	14.1	37	219	525	0.42	1.3
Brasilia (Brazil)	2 051 000	16S	1061	21.3	154	248	478	0.52	2
Tegucigalpa (Honduras)	820 000	14N	1004	20	94	51	81	0.63	1.3
Iquitos (Peru)	368 000	4S	116	25	274	68	85	0.8	5.8

(cont.)

Table 5.1 (cont.)

City (Nation)	Population of city	Latitude (°)	Elevation (m)	Temperature (°C)	Precipitation (cm)	Metro perimeter (km)	Metro area (km²)	Metro P/A ratio	Metro L/W ratio
AFRICA									
Cairo (Egypt)	6 789 000	30N	116	21.1	3	193	543	0.36	2.1
Nairobi (Kenya)	2 143 000	1S	1657	19.6	87	82	159	0.52	2.1
Bamako (Mali)	1 083 000	13N	527	28.1	112	87	201	0.43	1.8
East London (South Africa)	332 000	33S	125	18.6	81	59	85	0.69	1.8
Abeche (Chad)	263 000	15N	295	27.9	50?	21	24	0.88	1.7
WEST to EAST ASIA									
Beijing (P. R. China)	10 820 000	40N	52	11.8	62	230	1 243	0.19	1.4
Moscow (Russia)	10 360 000	56N	154	4.4	63	318	1 111	0.29	1.1
Seoul (Korea)	9 900 000	38N	34	12.2	134	252	971	0.26	2.1
Tehran (Iran)	6 935 000	36N	1217	16.5	25	196	575	0.34	1.4
Sapporo (Japan)	1 823 000	43N	17	8.2	113	165	370	0.45	1.5
Ulaanbaatar (Mongolia)	760 000	48N	1155	−2.2	19	62	80	0.78	3.2
Erzurum (Turkey)	578 000	40N	1757	6	44	31	33	0.94	2
Kagoshima (Japan)	554 000	32N	5	16.6	224	77	113	0.68	2.8
SOUTH ASIA to AUSTRALIA									
Bangkok (Thailand)	6 355 000	14N	2	27.6	140	195	751	0.26	1.7
Kuala Lumpur (Malaysia)	1 379 000	3N	22	26.6	237	331	871	0.38	1.6
Cuttack (India)	588 000	20N	15	27.6	152	55	72	0.76	2.2
Samarinda (Indonesia)	546 000	0S	230	25.9	214	34	41	0.83	2.3
Canberra (Australia)	309 000	35S	558	13.3	58	113	240	0.47	2.2
Rahimyar Khan (Pakistan)	258 000	29N	82	26.9	12	29	26	1.12	1.8

(Canberra), though, like land masses, they are skewed to the northern hemisphere. Elevations range from nearly sea level (Bangkok, Philadelphia) to 2200 m (Mexico City); the median is 116 m. Average annual temperature ranges from −2 °C (Ulaanbaatar) to 28 °C (Bamako), while nearly 60 % of the cities fall between 10 and 20 °C. Average annual precipitation ranges from 3 cm (Cairo) to 274 cm (Iquitos); only five places record <50 cm and only four places >155 cm.

Latitude and elevation are central determinants of average temperature and precipitation, which primarily determine the natural vegetation type for a region. Of the world's major vegetation types, no cities chosen are in tundra or boreal forest (taiga). Cities are reasonably well distributed over the six major vegetation types present: six in *boreal–temperate transition forest* (Berlin, Stockholm, Ottawa, Portland, Moscow, Sapporo); eight in *temperate deciduous-evergreen forest* (London, Nantes, Chicago, Philadelphia, Atlanta, Beijing, Seoul, Kagoshima); five in *tropical rainforest* (Iquitos, Bangkok, Kuala Lumpur, Cuttack, Samarinda); five in *Mediterranean-type woodland* (Rome, Barcelona, San Diego/Tijuana, Santiago, East London); six in *savanna-woodland* (Mexico City, Brasilia, Tegucigalpa, Nairobi, Bamako, Canberra); five in *grassland* (Bucharest, Edmonton, Abeche, Ulaanbaatar, Erzurum); and three in *desert-woodland* (Cairo, Tehran, Rahimyar Khan).

City location relative to water bodies is particularly important in urban regions. Fourteen cities are on a single major river, and another four are on the intersection of major rivers. Four cities are on the shore of a lake or reservoir, one associated with a river and one with a sea/saltwater bay. Four are on a seacoast and one on a saltwater bay. Ten cities have no adjacent major surface water body, though all have streams or seasonal water flows in gullies. Again a wide range of city and water-body locations is included.

Cities were chosen so their urban region would be minimally affected by another nearby city. Thus 13 cities have no nearby major city (>250 000 population) within 200 km, and another 19 have no city within 100 km. In one case (Kagoshima) the nearest major city is smaller in population and 90 km away. In the other five cases (Mexico City, Cairo, Seoul, Tehran, Stockholm) with closer nearby cities (20–80 km distant), the other major city is much smaller.

The perimeter and area of metropolitan areas vary from tiny Abeche to huge Chicago (Table 5.1). In general, US cities have the largest metro areas, both in perimeter and area, though the much-more-populous London and Moscow have long perimeters, and London, Moscow, Mexico City, and Beijing have large areas. The perimeter-to-area ratio of metropolitan areas tends to be inversely proportional to city population size. Seventy percent of the metro areas have a length-to-width ratio <2; only Iquitos at 5.8 is very long and narrow.

In effect a broad cross-section of geography, climate, land use and culture is represented in this set of 38 urban regions. The next two chapters will examine these regions in some depth to discover valuable patterns and principles.

6

Nature, food, and water

The color maps of 38 urban regions represent a treasure chest for the curious. Opening this chest to discover intriguing and important patterns is the delight of this chapter and the next. Here we look rather directly for patterns in the areas of nature, food, and water. In chapter 5 we explore built systems, built areas, and whole regions to find significant natural systems and human use patterns.

Although nature, food and water are separated for sequential presentation purposes, clearly broad overlaps exist among the three categories. Nature often thrives both in food-producing areas and in water-bodies (Figure 6.1). Food products are harvested from both natural vegetation and aquatic ecosystems. And water is often abundant in both natural areas and farmland. Still, from a spatial planning perspective, providing for viable natural-vegetation areas, agricultural landscapes, and streams/rivers/lakes/aquifers/marine areas is fundamental.

With the treasure chest open before us, first we must consider how to find the nuggets. How do we sort through the mass of material and decipher key patterns. We are only looking for major patterns or results. Minor results, as well as major ones that we miss, will await discovery by others. So we start by briefly considering the important spatial-analysis process used to reveal the nuggets.

Spatial analysis for patterns

Intelligence agencies know that if you know what you are looking for, the chance of finding it increases enormously. That does not work here because, as a scientist, I attempt to put my views and wishes aside, and let the patterns appear through objective analysis.

Figure 6.1 Nature, food, and water as overlapping concepts and interacting spaces on land. Forest, crop fields, and a 20 hectare (50 acre) dammed reservoir are linked by flows and movements of energy, water- and air-borne materials, animals, and people. Tennessee, USA. Photo courtesy of US Department of Agriculture.

Still, lots of possible approaches exist. Perhaps directly measuring the diverse parameters of natural systems and human uses in all urban regions using consistent, rigorous methods would be optimal. Then multivariate statistics and other detailed quantitative analytic methods could be used. Alas, I would get old before the measurements were done. A second choice might be to directly measure some parameters, and obtain dependable peer-reviewed data for the rest. Unfortunately few such data exist, and at the current rate I would have to sit around for centuries.

As suggested in the previous chapter, a different approach was used. The best data I could find were located from articles, books, and maps, and additional information was absorbed from consultations with knowledgeable persons. Data from computer searches were rarely included (due to the prevalence of data-source agendas, ephemerality of some data, and especially the scarcity of independent peer review needed for dependable scholarship).

Large detailed satellite images (30 m cell size) of urban regions were printed. Number, type, distance, form, and area of objects were directly measured with ruler and planimeter on the images by a single observer. In almost all cases each type of measurement was made on all 38 urban regions within a few days. Spatial measurements for patterns in this chapter required an average of 20 minutes per urban region, and those for Chapter 7 required 35 min. To evaluate observer consistency or variability, periodically five repeated measurements of the same attribute were made. Compared with the first measurement, the average of the five measurements varied on average 3 % (range 0–9 %), suggesting that this methodology produced reasonable consistency.

The degree of variability and resulting confidence in data and measurements can be further understood from the following examples. Population data for urban regions were for slightly different years (1999 to 2002) and probably differed in quality and area coverage (Chapter 5). Perimeter and surface area measurements of the metropolitan areas depend on determining cut off points for house-lot density and width of greenspace, plus their consistent application. Perhaps, on average, 5 % of a metro-area perimeter was near the cut-off points, with the percentage ranging from about 0–10 % over the set of regions. Attributes such as rivers, lakes, reservoirs, airports, shipping/ferry ports, and large mine sites were clear on the images, so confidence in the associated numbers was very high. Some biodiversity sites and recreation/tourism sites, for example, were clear on an image, but many were not, and thus had to be marked in approximate locations. In addition, considerable interpretation was necessary in estimating whether certain sites had the appropriate attributes (rare species or rare natural community, significant recreation or tourism from the city in a single day), in locating a site on the image, and in deciding whether the site was a major one to be included. In another case, urban region boundaries, which were invisible on the images, were mainly delineated using six variables (Chapter 5), with different variables being primary in different portions of a boundary. On the other hand, many measurements varied negligibly, such as number of marked objects and distance between two points. These examples attempt to provide insight into the degree of confidence appropriate for different portions of the extensive information base underlying the color maps and their spatial analyses.

In essence, most of the patterns and results highlighted in these two chapters were based on data and measurements with the high degree of confidence quite normal in science. In cases where more variability existed in the measurements or data, a conservative approach of only pinpointing the clearest major results was taken. With this approach, errors discovered or new data added are unlikely to have a significant effect on the results highlighted, though of course caution in interpretation is always warranted. Many minor results, and even major results not pinpointed, are evident in the figures and color maps, and are likely to be discovered and highlighted in the future.

A spatial-analysis technique using *landscape metrics* bears mention (Fortin 1999, Klopatek and Gardner 1999, Leitao *et al.* 2006), though it was not used here. Landscape metrics are measures (and equations) that quantify *spatial attributes* of a large area, such as connectivity, patch density, total interior habitat, boundary length or density, and association of types of objects. The beauty of these is that important ecological characteristics are known to correlate with spatial attributes. These *ecological characteristics* at the landscape scale include interior species, large-home-range species, aquifers, wildfire hazard, wildlife movement routes, animal dispersal, species-rich sites, stream-network flows, fish migration, and so forth. Thus a particular quantitative level for a spatial attribute is an indicator or surrogate for conditions of an ecological characteristic on the land.

The idea of spatial attributes and ecological characteristics will be used periodically in this book. Landscape metrics do not replace direct, detailed measurements of ecological characteristics at the landscape scale that may be time consuming, or experiments that may be impossible. Rather, because of the previously documented relationship between spatial attributes and ecological characteristics, the landscape metrics may represent a useful handle to ecological understanding and planning of landscapes.

In landscape ecology the *landscape* is a kilometers-wide area over which local land uses and ecosystems are repeated in similar form (Forman 1995). For operational convenience in analyzing urban regions, the *landscape* concept is narrowed slightly and refers to a compact (<2:1 length-to-width ratio) area of >100 km^2 with repeated internal heterogeneity, such as a cropland landscape or a wooded landscape.

The first section below, "Nature in urban regions," will emphasize natural vegetation areas and connectivity for species movement. The second section, "Food in urban regions" will focus on cropland areas, including regional diversity and stability. The third section, "Water in urban regions" will highlight water bodies and areas that affect them. These are explored outside the metropolitan area, that is, in the surrounding urban-region ring.

Thirty-seven major patterns and results emerged from spatially analyzing the 38 urban regions. These are presented in the nature-in-urban-regions section as

N1, N2, N3, etc., food section as F1, etc., and water section as W1, etc. Brief elucidations under each result may: (1) discuss or interpret it; (2) illustrate its implications ecologically or for society; and (3) identify useful approaches or solutions for land planning, protection, restoration, and other objectives. Some solutions are labeled as *priority* and some as *high priority*, relative to the total set of patterns presented in Chapters 6 and 7.

Each point in the following graphs represents an urban region. Unlike the usual scientific graphs that focus on and highlight close correlations between variables, axes here are chosen to maximize information presentation and to spread out the data in order to detect variability and contrasting patterns. Thus most insights and results emerge from examining the four sections or corners in a graph.

Nature in urban regions

Presence of natural patches and landscapes

[N1] At least three scales of natural area in urban regions can be readily and usefully recognized: (1) natural landscapes >100 km²; (2) large natural patches, i.e., the largest widely distributed ones, which average about 16 km²; and (3) small wooded patches averaging 0.4 km², but mostly <0.2 km² (<200 ha) (Figures 6.3 and 6.10).

These three size categories emerged qualitatively from extensive perusal of large (*c.* 70 × 100 cm) images of dozens of urban regions. Geographers, landscape ecologists, and others have quantitative techniques to determine the frequency of patch sizes as spatial scale changes (Milne 1991a, 1991b, Klopatek and Gardner 1999). Results often show a series of peak patch-size frequencies; the above three scales seem to correspond to three peaks. The natural-landscape scale probably mainly reflects the geomorphology of, e.g., hilly areas, mountain ridges, and major valleys. The large-natural-patch scale may reflect the combination of geomorphic features, soil types, and human activities on the land. The small-wooded-patch scale probably reflects farmers' crop-production practices, as these patches are overwhelmingly in cropland areas. Natural landscapes are of a size to protect aquifers, large-home-range species, and so forth (Forman 1995). The large patches are generally sufficient to protect populations of some interior species, and serve as sources of these species for the surrounding land. The small natural patches may support isolated rare plants, but are especially useful as stepping stones for species movement across a cropland landscape.

[N2] No natural landscapes are present in 8%, and no wooded landscapes are present in 16%, of the regions (Figures 6.2 and 6.3).

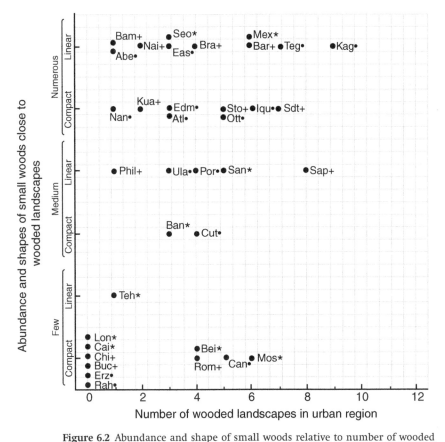

Figure 6.2 Abundance and shape of small woods relative to number of wooded landscapes in an urban region. Wooded landscapes (forest or woodland) are > 100 km^2 and compact in shape. Small woods within 5 km of a wooded landscape average 0.4 km^2, but are mostly <0.2 km^2 (<200 ha). Relative abundance (few, medium, numerous) is estimated for small woods <5 km from a wooded landscape. * = city population >4 million ("large"); + = 1 to 4 million ("medium"); • = 250 000 to 1 million ("small"). City abbreviations: Abe (Abeche, Chad); Atl (Atlanta, USA); Bam (Bamako, Mali); Ban (Bangkok, Thailand); Bar (Barcelona, Spain); Bei (Beijing, China); Ber (Berlin, Germany); Bra (Brasilia, Brazil); Buc (Bucharest, Romania); Cai (Cairo, Egypt); Can (Canberra, Australia); Chi (Chicago, USA); Cut (Cuttack, India); Eas (East London, South Africa); Edm (Edmonton, Canada); Erz (Erzurum, Turkey); Iqu (Iquitos, Peru); Kag (Kagoshima, Japan); Kua (Kuala Lumpur, Malaysia); Lon (London, United Kingdom); Mex (Mexico City, Mexico); Mos (Moscow, Russia); Nai (Nairobi, Kenya); Nan (Nantes, France); Ott (Ottawa, Canada); Phi (Philadelphia, USA); Por (Portland, USA); Rah (Rahimyar Khan, Pakistan); Rom (Rome, Italy); Sam (Samarinda, Indonesia); Sdt (San Diego/Tijuana, USA/Mexico); San (Santiago, Chile); Sap (Sapporo, Japan); Seo (Seoul, South Korea); Sto (Stockholm, Sweden); Teg (Tegucigalpa, Honduras); Teh (Tehran, Iran); Ula (Ulaanbaatar, Mongolia). See Table 5.1 and Chapter 5 for further urban region information.

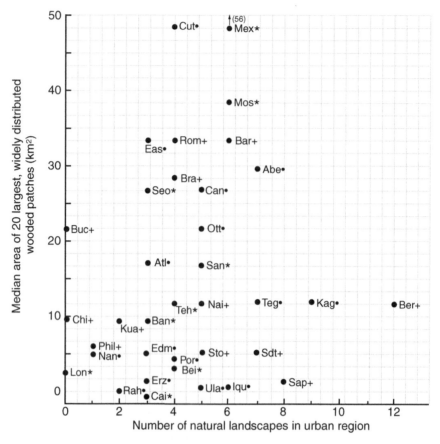

Figure 6.3 Size of large wooded patches relative to number of natural landscapes. Natural landscapes are >100 km^2 and compact in shape. All landscapes are wooded except for: grassland in Erzurum urban region; grassland and wooded in Nairobi, Ulaanbaatar, and Abeche; desert in Cairo and Rahimyar Khan; desert and wooded in Tehran. Patch size is the median estimated area of the 20 largest wooded patches (<100 km^2) widely distributed across agricultural and built areas (i.e., where clustered, only one patch from the cluster is included). 10 km^2 = 3.9 mi^2. See Chapter 5 for natural vegetation types, and Figure 6.2 caption for city information.

This suggests problems for aquifer protection, large-home-range species, and for the viability and movement of key forest/woodland species. Wooded landscapes also are especially important for diverse recreational uses. In these urban regions a *high priority* is to reestablish natural landscapes, especially wooded ones, by reconnecting pairs or clusters of large natural patches.

[N3] Eighty-two percent of the regions have natural landscapes, and 71% has wooded landscapes, near the city (<50 km from city center); 11% and 13%, respectively, only have more distant ones (Color Figures 2–39).

Natural and wooded landscapes close to a metropolitan area are subject to diverse and often intense human impacts, which are likely to increase as urbanization proceeds. Thus flood-control benefits, nature conservation, cooling sources reducing urban heat-island effects, and so forth in these important landscapes represent a particular challenge. Indeed wooded landscapes are especially valuable for diverse recreational activities in proximity to a city. In the outer portion of an urban-region ring, natural and wooded landscapes are normally less subject to degradation, and hence are a particularly good investment for maintaining aquifer/water-supply protection, nature conservation, clean surface water bodies, and many other environmental and socioeconomic benefits for society.

[N4] For half the regions, the largest natural patches present (outside of natural landscapes) are relatively small (median sizes 0–12 km²) (Figure 6.3).

This suggests lower biodiversity, lower connectivity, more degraded habitat, and more risk of patch shrinkage and disappearance, than if these ecologically critical patches were larger.

Stability related to number and types of natural landscapes

[N5] Most regions have 3–12 natural landscapes, 8% have only one or two, and another 8% have no natural landscapes remaining (Figure 6.3).

A low number indicates little stability plus high risk for the many ecological values of whole natural landscapes. Also the presence of few natural landscapes may include some diversity of types, but little stability exists due to the low redundancy of types. A higher number of natural landscapes should provide both a diversity of types and reasonable stability.

Regional connectivity for nature

[N6] A fully connected emerald network of natural landscapes, with no major gaps separating them, is present in a quarter of the urban regions (Color Figures 2–39).

The *emerald network* is one of the most important patterns in this book. Effectively it is a set of large natural interconnected patches, in this case natural landscapes of >100 km². These landscapes are large enough to provide the range of large-patch benefits, such as protection of aquifers, large-home-range species, and viable populations of interior species. Also they are numerous enough to provide some diversity and some redundancy of types, and therefore stability. The connections or corridors permit effective movement of species and walkers throughout the network. Multidirectional corridors for movement also provide stability, for example, in the face of global climate change or the spread of a

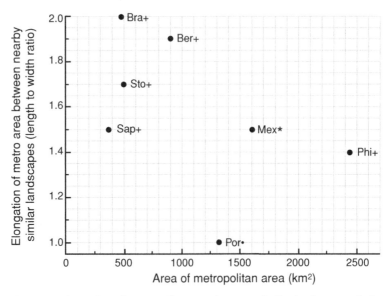

Figure 6.4 Elongation of metropolitan area between similar landscapes relative to size of metropolitan area. Metro-area size is measured with a planimeter on large 1:200 000 satellite images. Landscapes are >100 km² and compact in shape; all are wooded except Philadelphia which is agricultural. Long axis of metro-area separates the landscapes. $100 \, \text{km}^2 = 38.6 \, \text{mi}^2$. See Chapter 5 for natural vegetation types, and Figure 6.2 caption for city information.

disease or pest. Rather than a group of patches, the emerald network functions as a system of patches and corridors for both biodiversity and recreation. Protecting the connections, the most vulnerable places, is especially important. A viable emerald network is a key long-term *high priority* objective for many regions.

[N7] In a sixth of the regions, an elongated metropolitan area lies between two nearby similar, usually wooded, landscapes (Figure 6.4).

In these cases the metro area doubtless serves as a significant barrier to regional species movement between the nearby landscapes (Pauleit *et al.* 2005). In one case (Philadelphia), the metro area separates cropland landscapes containing scattered small woods, and across which many forest birds may readily move (Knaapen *et al.* 1992, Forman 1995). The only really long and narrow metro area is Iquitos, which, however, is nearly surrounded by close-by rainforest.

A dual approach to overcoming the barrier effect of elongated metro areas may be best. Strengthening or creating rows of parks or linear greenspaces across the metro area would help reconnect the landscapes on opposite sides. Establishing a large elongated greenspace at each end of the metro area, and perpendicular to it, would provide connectivity for the future as development spreads outward.

[N8] Small woods are common in open areas close to the forest/woodland landscapes in two-thirds of the regions; however, in a sixth of the regions few small woods are present around wooded landscapes (Figure 6.2).

The presence of small woods around a large natural patch not only enhances the viability and persistence of key interior species, but increases connectivity for regional species movement (Opdam and Schotman 1987, Opdam *et al.* 1992, Forman 1995). The basic concept probably also applies here to whole landscapes, though less strongly so. Many of the regions are satisfactory for this pattern, though in a small number, increasing the density of small natural patches near natural landscapes should improve ecological conditions (see Figure 2.2).

[N9] In about half the regions with abundant small woods around wooded landscapes, the woods are mainly linear (Figure 6.2).

This pattern provides greater connectivity value as well as, usually, stream-corridor protection benefits. Except along streams, in habitat restoration it is not worthwhile elongating woods, because too much of the wooded area is lower-quality forest-edge habitat. It would be better to increase the size or number of patches.

Connections and gaps as strategic points

[N10] Most regions have few connections and gaps between natural landscapes, mainly because few natural landscapes are present (Figure 6.5).

In these cases connections are key strategic points for land protection, and gaps are key points for land restoration, particularly in the outer urban-region ring (where patterns are more likely to persist in the face of urbanization). Where natural landscapes are widely separated, establishing stepping stones between them may be desirable. Creating whole new natural landscapes, e.g., by inter-connecting clusters of natural patches, would be quite significant.

[N11] Most regions have more major gaps than connections between natural landscapes (Figure 6.5).

This indicates poor regional connectivity for species movement. Restoring corridors that reconnect the natural landscapes, especially in the outer urban-region ring, is a *priority*.

[N12] In nearly 30% of the regions, at least one natural landscape is isolated by a single major gap (Figure 6.6).

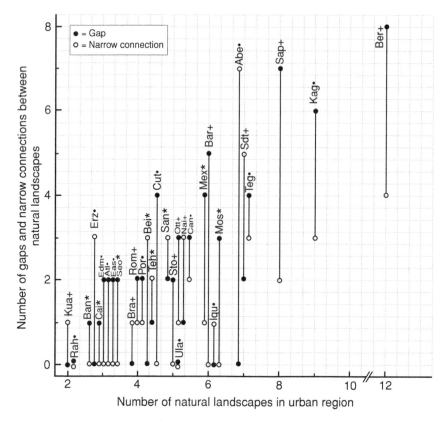

Figure 6.5 Number of gaps and connections between natural landscapes relative to number of natural landscapes. Natural landscapes are >100 km^2 and compact in shape. Vegetation gaps or breaks separate landscapes and are caused by human clearing, not an intervening river, ridge, or valley. A connection refers to continuous vegetation in a narrows or "isthmus" or corridor between landscapes. London, Bucharest, Chicago, Philadelphia, Nantes, and Samarinda have 0 or 1 natural landscape present. See Chapter 5 for natural vegetation types, Figure 6.3 caption for landscape types, and Figure 6.2 caption for city information.

Where several natural landscapes are present, regional connectivity may be only slightly reduced. However, where few natural landscapes are present, this is an acute problem for regional connectivity. Here reestablishing the connection is a *high priority*.

[N13] At least half of the major gaps between natural landscapes are located near the metropolitan area (Figure 6.6).

These gaps are particular problems in disrupting regional connectivity for wildlife, both because of present human usage and potential future urbanization.

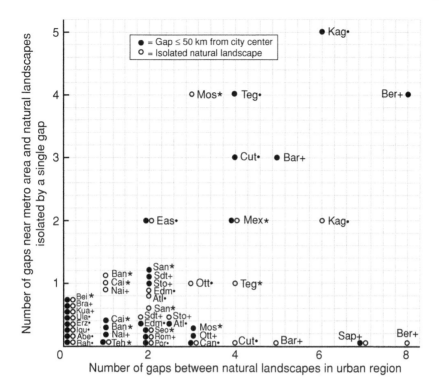

Figure 6.6 Number of gaps near cities and number of isolated natural landscapes relative to total number of gaps between natural landscapes. Vegetation gaps or breaks separate natural landscapes (>100 km^2 and compact in shape) and are caused by human clearing, not an intervening ridge, river, or valley. Isolated landscapes are those whose only connection to another natural landscape has been cut, leaving a vegetation gap; this also considers possible connections outside the urban region. See Chapter 5 for natural vegetation types, Figure 6.3 caption for landscape types, and Figure 6.2 caption for city information.

Wetlands

[N14] Major wetlands are absent in urban regions with cities of >8 million population, generally scarce in regions with cities of 2 to 8 million, and sometimes frequent around cities of <0.5 million (Figure 6.7).

Wetlands in urban regions were often drained or filled long ago for farming purposes. As the human population rose, wetlands disappeared under roads and buildings, and often to reduce mosquito and other insect populations as pests and disease vectors. However, wetlands provide many benefits to society, such as reducing floodwaters, absorbing and breaking down chemicals, supporting many special wetland species, and providing recreational sites.

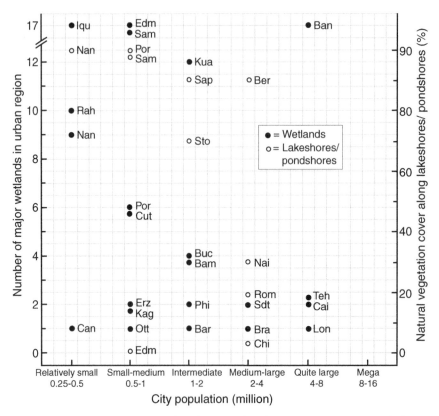

Figure 6.7 Major wetlands and natural lakeshores relative to city population size. Major wetland is > approx. 4 km^2; an unusually long or large wetland is counted twice. Wetlands include salt marsh, mangrove swamp, freshwater marsh, and freshwater swamp, and are present along coasts, along rivers, and in depressions. Lakeshores include major pondshores but exclude shorelines of reservoirs. Natural lakeshore vegetation refers to the predominant land cover within approx. 1 km of a shoreline. City population is in year 2000 ± 2 yr (Chapter 5 and Table 5.1). See Figure 6.2 caption for city information.

Re-creation of wetlands, large and small, and their benefits is a *high priority* in urban regions. Especially suitable locations are at the base of certain hills and mountains, on floodplains of rivers and streams, along coastal areas, and in the outer urban-region ring. Tiny wetlands, some seasonal, may be produced at the ends of stormwater drainage pipes.

Food in urban regions

Diversity and stability relative to agriculture

[F1] Cropland is the predominant land cover of the urban-region ring in half of the regions (Color Figures 2–39).

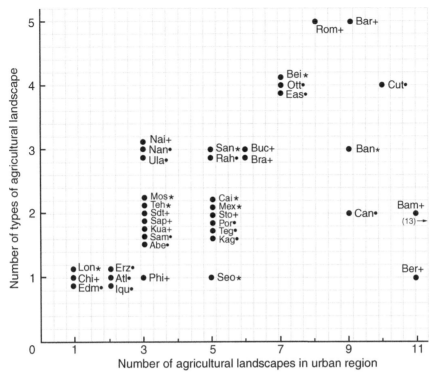

Figure 6.8 Number of types of agricultural landscape relative to number of agricultural landscapes. Landscapes are >100 km^2 and compact in shape. Agricultural includes cropland and pastureland. See Figure 6.2 caption for city information.

Communities that became cities often started where good soil and a water body come together. Spreading farmland sends sediment and agricultural chemicals into water bodies. The frequency of cropland around cities offers a significant opportunity to recognize the values of urban agriculture, from farm communities and economies to food products for the city at low transport cost (Chapter 3).

Today prime agricultural soil near cities often sprouts buildings. The spread of buildings in future urbanization should generally be concentrated along or near the boundary between landscape types, such as agriculture and built land or agriculture and natural land. As illustrated in the aggregate-with-outliers model (Forman 1995, Forman and Collinge 1996), these locations minimize degradation to either agricultural or natural landscapes.

[F2] Three or more types of cropland landscape are present in a third of the regions, in contrast to a single type of cropland landscape in a quarter of the cases (Figure 6.8).

More landscape types suggest a diversity of food products available, plus flexibility for the future. It also indicates a high diversity of farmland species, plus diverse farming communities and economies. A *priority* is to encourage the production of different crops within a landscape, and eventually different types of cropland landscapes.

[F3] Most regions have several separate agricultural landscapes of a type (Figure 6.8).

This provides stability for that agricultural type in the event of a spreading pest or disease. The farm support system is likely to be strong. Also stability is provided for the particular group of farmland species present.

[F4] Both kinds of stability, several cropland types and several landscapes of a type, are present in over a quarter of the regions (Figure 6.8).

This is the optimum arrangement providing flexibility and stability for long-term agriculture. It also is optimum for sustaining a high diversity of farmland species.

Proximity of cropland landscapes to city center

[F5] A cropland landscape <20 km from city center is present in over half of the regions, while the nearest cropland landscape is >45 km distant in an eighth of the cases (Figure 6.9).

A nearby cropland landscape, especially for market-gardening, can provide convenient food products for the city with low transport cost. It also provides open vistas and clean air, and if valued by society and urban planners, can help prevent outward urbanization. However, it is particularly susceptible to existing human impacts as well as future outward development. Downwind of the city and industrial areas, the cropland landscape is subject to air pollution, and if near the city's main airport, it is subject to aircraft noise.

The somewhat more distant cropland landscapes largely escape these negative effects. More distant croplands can serve for market-gardening with only slightly higher transport cost. And they are easier to maintain and more likely to persist.

[F6] If the nearest cropland landscape were degraded or destroyed, another such landscape is <45 km from city center in 80 % of the regions; however, local food products would have to be transported >65 km in a fifth of the regions (Figure 6.9).

The proximity of at least two cropland landscapes provides flexibility and stability for market-gardening. Also farmland species should remain relatively near the metro area. The more distant landscape adds transportation costs, and increases the likelihood of a strip-development corridor across the region, which reduces regional connectivity for wildlife movement and walkers.

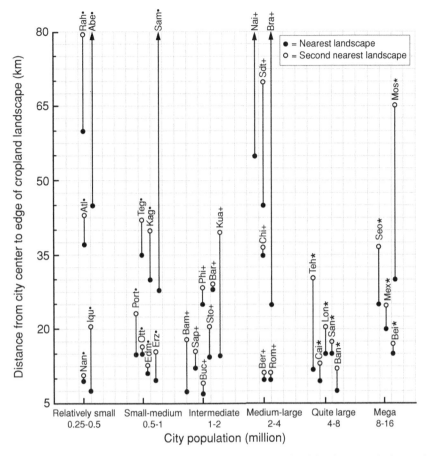

Figure 6.9 Distance from city to two nearest cropland landscapes relative to city population size. Landscapes are >100 km^2 and compact in shape. 10 km = 6.2 mi. City population is in year 2000 ±2 yr (Chapter 5 and Table 5.1). See Figure 6.2 for city information.

[F7] Although data on market-gardening was difficult to find and thus only occasionally mapped on the urban-region images, almost all consultants agreed that at least one market gardening area was present outside a city (Color Figures 2–39).

These areas provide food products in proximity and low transport cost to the metropolitan area. Little trucks (lorries) can load up early morning, snake their way through narrow city streets, and provide fresh vegetables and fruits to markets and restaurants. The agricultural park adjacent to Barcelona is a nice example (Chapter 10; Acebillo and Folch 2000, Forman 2004a). However, market-gardening usually intensively uses water, fertilizer, and pesticides. A *high priority*

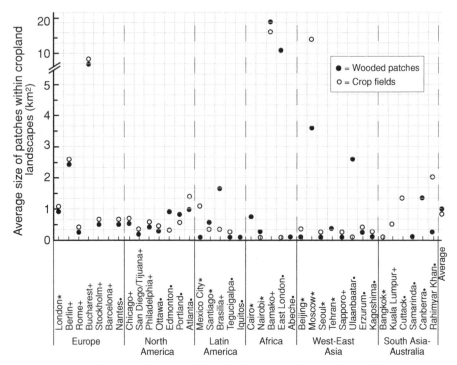

Geographic area, with cities from large to small population

Figure 6.10 Average size of woods and fields within agricultural landscapes relative to geography and city size. Average sizes of woods and fields are estimated in all agricultural landscapes (1 km² = 247 acres). Within each geographic area, cities are in decreasing order of population (Table 5.1). See Figure 6.2 caption for city information.

is to maintain at least two major market-gardening areas in proximity to the metropolitan area.

Woods and field sizes in cropland landscapes

[F8] In most regions, average field size and woods size are about the same within a cropland landscape no matter the type of cropland (Figure 6.10).

Some woods grew up from former fields, and some fields are likely to become woods over time. The farmer has flexibility, and is also likely to change the crops over time, as markets and owners and interests evolve. Dispersing small woods evenly across the cropland landscape may spread predators over the area sufficient to reduce crop pest populations, thus increasing crop production.

A regular grid on the land is normally considered ecologically undesirable, in part because it favors only a subset of the natural species complement for

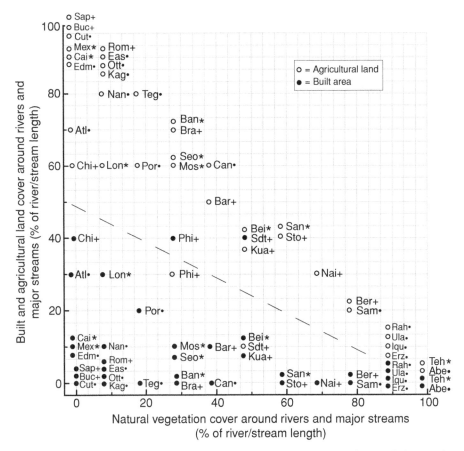

Figure 6.11 Agricultural and built land relative to natural land around rivers and major streams. Predominant cover type is estimated for the area between approximately 100 and 3000 m from the water along all major streams and rivers (marked on Color Figures 2–39) in urban regions; stream/river corridor vegetation and unmapped small or seasonal streams are excluded. See Figure 6.2 caption for city information.

the area. Thus aggregating the small woods to mimic larger woods, or making lines of woods as stepping stones for movement in appropriate directions, is ecologically useful.

Water in urban regions

Land cover around rivers and streams

   [W1] Most urban regions have <33 % natural vegetation cover near rivers and major streams (Figure 6.11).

Thus water degradation due to sedimentation, agricultural chemicals, and solar heating is probably widespread across urban regions. Where fields are irrigated in dry climates, water tables are often lowered and streamflow much reduced. Ecologically effective vegetated stream corridors are likely to be few and fragmented.

Reestablishing the stream corridors with attached patches of natural vegetation, especially upstream of the city or far downstream, is particularly valuable. Reducing agricultural inputs to streams and rivers, both by fine-scale land-use changes and farming practices, is also valuable.

[W2] Only a fifth of the regions has >80% natural vegetation cover around rivers and major streams, while another fifth has 40–70% natural cover (Figure 6.11).

In the former case, relatively good overall water quality, aquatic ecosystems, and fish populations are likely to be present over major portions of the region. In the latter case, stream and river degradation is probably widespread, but some high-water-quality streams and rivers exist in the region, which can serve as valuable species sources for rapid restoration. Reestablishing ecologically effective vegetated stream corridors to help reconnect high-water-quality streams and rivers is a *priority*.

[W3] Cropland is the predominant human land use around rivers and major streams in almost all urban regions (Figure 6.11).

Food products for the nearby city and viable farming communities and economies are real benefits that often need emphasis. Farmland species are well established. The widespread cropland also means extensive warm water, soil erosion, muddy water, sediment-covered bottoms, degraded fish habitat, and agricultural chemicals, including varied pesticides and nitrogen/phosphorus from fertilizer. A variety of improvements can be made. Better fine-scale farming practices and land uses would help. Future outward development should avoid degrading streams and rivers, as well as disrupting large prime-agricultural-soil areas.

[W4] Considerable built area (10–40% land cover) surrounds rivers and major streams in 40% of the regions (Figure 6.11).

Consequently, urban runoff and human impacts are likely to be widespread. Various solutions include relocating (often old) industries away from the stream/riverside to industrial parks with efficient water, power, and waste-disposal availability. Increase the length and width of vegetated stream/river corridors. Limit future outward development to sites adjacent to existing built areas.

Lakeshores

[W5] Half of the regions with a lake(s) present have natural vegetation around 90–100 % of lake shorelines, while nearly half have <30 % of the shoreline length with natural vegetation (Figure 6.7).

Generally lakes, unlike dammed reservoirs, are scarce in urban regions, and therefore of special importance for recreation, visual quality, and biodiversity. Continuous natural vegetation around lakeshores is particularly important for maintaining high water quality. In those regions with low protection around shorelines, avoiding development, even dispersed development, is a key to maintaining clear water, natural aquatic ecosystems, fish populations, and associated recreation.

Slopes around cities

[W6] More than half of the cities with nearby hillslopes or mountain slopes facing the city have 90–100 % natural vegetation cover on the slopes (Figure 6.12).

Maintaining natural vegetation on these slopes provides many benefits, including high visual quality, good recreation opportunities, and rich biodiversity, all close to the metropolitan area. The vegetation also minimizes soil erosion, mudslides, sediment accumulation, and flood hazard. Forest on slopes provides cool air that on still nights drains downward and helps ventilate a city by pushing out hot air and pollutants. In dry climates, vegetation on nearby slopes increases the fire hazard. "Skyline conservation" on slopes surrounding the city has deep roots and cultural significance in Korea and Japan (Im 1992, Bengston and Youn 2006). Where few nearby city-facing slopes are present, protecting them with natural vegetation is a *priority*.

[W7] Nearly 30 % of the cities with nearby city-facing slopes have them only 25–50 % covered with natural vegetation (Figure 6.12).

As implied in the preceding case, the paucity of protective natural vegetation on these slopes reduces benefits and poses increased hazards. Avoiding additional development is a priority. Fine-scale improvements on the developed slopes involving water, soil, and vegetation should have a significant effect. Also, gradually removing buildings on the most inappropriate sites should help.

[W8] Cities with more surrounding city-facing slopes generally have a greater percentage cover of natural vegetation on them, whereas few nearby slopes near a city tend to be much built up (Figure 6.12).

In the case of few nearby city-facing slopes, their scarce natural resources are much degraded. Avoiding future development, rigorously implementing

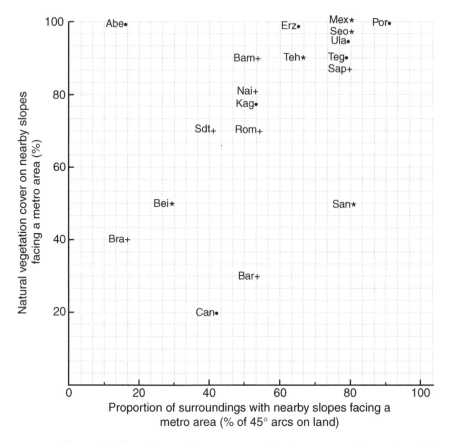

Figure 6.12 Natural vegetation cover on nearby slopes surrounding a city relative to the proportion of the surroundings with city-facing slopes. Proportion of surroundings with nearby slopes is the percentage of the 45° arcs on land (i.e., excluding major water bodies) (Color figures 2–39). Natural vegetation cover is the average for hillslopes/mountain-slopes within approx. 15 km of a metro area in each 45° arc. See Chapter 5 for natural vegetation types, and Figure 6.2 caption for city information.

fine-scale improvements, gradually removing buildings, and perhaps establishing large parks, is a *priority* for these scarce valuable sites.

Nearby major water bodies and coastlines

   [W9] A major surfacewater body (river, sea, bay, lake) is adjacent to half the cities (Figure 6.13).

This is convenient for shipping/ferry transport, local recreation, and removal of urban pollutants. But, if the difference in elevation between city and water

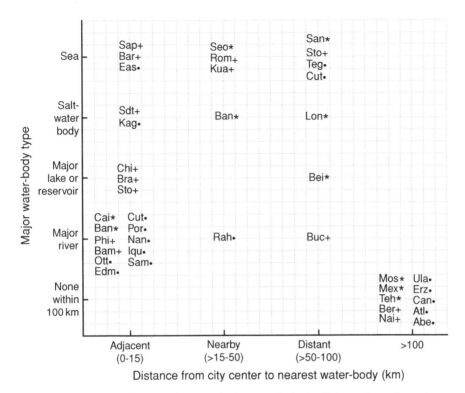

Figure 6.13 Nearest major water-body type relative to distance from city center. Major is in a global rather than regional context. Three cities (Bangkok, Stockholm, Cuttack) are near two major water-body types. 10 km = 6.2 mi. See Figure 6.2 for city information.

body is small, a major flood hazard for the metro area is likely. Also nearby coastal or riverside recreation sites are likely to be degraded and polluted, especially close to and downriver of the city. Fish migration on the river is apt to be blocked by urban water pollution.

[W10] A quarter of the regions has no major water body present (i.e., within 100 km of city center); another quarter has the nearest major water body 15–100 km distant (Figure 6.13).

Aquifers are likely to be especially important in these regions, and require nearly continuous natural vegetation cover to protect water quality. These regions are likely to have little flood hazard, less-polluted recreation sites serving fewer people, and a less-convenient shipping/ferry port. A strip of development between city and port, which forms a barrier to regional wildlife movement, is apt to be present. Major breaks in the strip development should be established and maintained for regional wildlife movement.

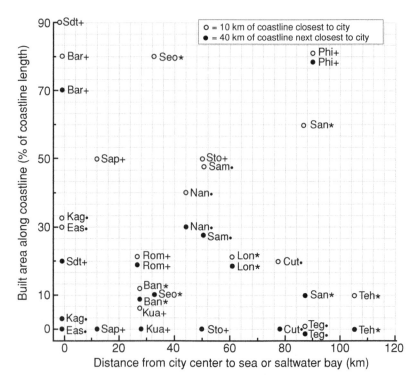

Figure 6.14 Built area along coastline relative to distance from city center. Predominant land cover within approx. 1 km of a coastline is measured for the 10 km of coastline closest to a major city, and for the next closest 40 km (20 km in each direction) of coastline. 10 km = 6.2 mi. See Figure 6.2 caption for city information.

[W11] The coastline closest to a city is much more built up than that somewhat further away, with the difference being most pronounced where city and coast are close (Figure 6.14).

This suggests that clean water, natural vegetation, attractive beach areas, and good recreation are most likely along coastlines far from a city, and in coastline locations not directly opposite the city. Strip development between city and nearest coastline should have major breaks to maintain regional wildlife connectivity. Protecting coastline stretches beyond the portion closest to a city, and avoiding dispersed coastline development, should maintain the rich natural and societal benefits of relatively natural coastlines.

Water supply and drainage area

[W12] In general, large cities use reservoirs for water supply, medium cities mainly use lakes and reservoirs, and small cities use rivers, and in a few cases, streams or groundwater (Figure 6.15).

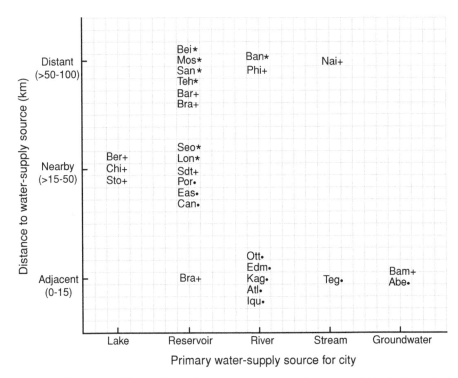

Figure 6.15 Distance from city center to water-supply source relative to type of water supply. Brasilia is listed twice because water pollution in an adjacent large reservoir has effectively caused the construction of distant small reservoirs. 10 km = 6.2 mi. See Figure 6.2 for city information.

Since cities typically grow in population and spread outward, alternatives for water supply are important for stability. Two suitable sources are much better than one, and three are somewhat better than two. Large reservoirs for large cities are hard to locate and create, so their protection is especially important. Medium cities typically have more options, but probably should plan for when they may be large cities. Small cities normally have still more options, but may have limited resources to locate and create water supplies for a future medium-sized city. Providing adequate alternatives to supply a future larger city population with clean water is a *priority*. Some cities doubtless benefit from a nearby large aquifer which must be protected with vegetation cover.

[W13] On average, reservoirs have the best drainage-area protection by natural vegetation, while drainage-area protection for lakes and rivers varies widely (Figure 6.16).

Reservoirs usually have an upstream drainage basin and headwater stream network that should be protected. Without adequate land protection, rapid stream erosion, reservoir sedimentation, and loss of reservoir water capacity is

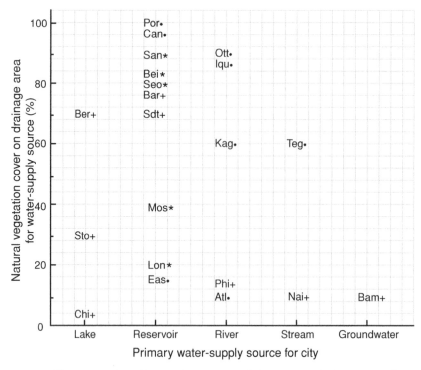

Figure 6.16 Natural vegetation cover on drainage area relative to type of water-supply source. Drainage area refers to the general area around or upslope of a water source (see Color Figures 2–39) and may not correspond with a specific drainage basin (watershed/catchment). See Chapter 5 for natural vegetation types, and Figure 6.2 caption for city information.

a familiar situation around the world. Also chemical pollutants can essentially eliminate the usefulness of a reservoir (e.g., see large reservoir to the northwest in Color Figure 7).

A water-supply source on a river is essentially always upriver of a city, since the downriver stretch is subject to urban pollutants. As urbanization spreads upriver, the water-supply source may move further upriver (or elsewhere). Therefore protection of extensive continuous natural-vegetation cover in areas surrounding upriver and tributary stretches is important. Water treatment costs, e.g., for lake and river sources, vary widely depending on how well or poorly the upslope land is protected with natural vegetation. Stream-corridor vegetation helps, especially for erosion/sedimentation and phosphorus inputs, but covering the surrounding land with vegetation helps much more.

The loss of vegetation protecting a water-supply drainage basin also increases the cost of water. Indeed, a sharp rise in cost may occur when vegetation cover drops to some 70 % (Peter Kareiva, personal communication).

[W14] The few cities with water supply from stream or groundwater have considerable built area in the general drainage area of the water supply (Color Figures 2–39).

Tegucigalpa apparently largely uses streamwater (supplemented by some lakewater), while apparently Bamako and Abeche mainly use large numbers of groundwater wells to provide water supply for the city population. Spreading development around the somewhat-fragile streams poses a severe threat for erosion/sedimentation, chemical pollution, and water loss. Avoiding additional development in the general drainage area is important. In the case of extracting groundwater close to the city, urban pollutants of numerous types can readily percolate into the groundwater. Neither case is stable. Achieving viable water-supply alternatives for a future larger city is a *high priority*.

[W15] Forty percent of the cities has an adjacent water supply, and a sixth of the cities has a distant water supply (Figure 6.15).

In the former case, adjacent relatively high-density development plus air pollution degrades the water supply. In addition, urbanization, both densification and outward spread, occur around the water supply, worsening the problem. Establishing alternative water supplies at a distance is a *high priority* (Color Figure 9).

Cities with a distant water supply have little urbanization or air pollution threat. However, they have a higher maintenance and water-transport cost. Perhaps more importantly, the city must usually depend on other political/administrative units to adequately protect the drainage basin. Establishing strong long-term land protection measures with dependable policing is important.

Thus 37 "major" patterns and results focused on nature, food, and water have emerged from analyzing the urban regions (Color Figures 2–39). Useful guidelines and priorities doubtless apply widely for other cities worldwide. The next chapter continues this analysis by focusing on built systems and built areas.

7

Built systems, built areas, and
whole regions

Following the preceding chapter on nature, food and water, perhaps surprisingly
we now turn to built areas to learn about natural systems and their human uses
in an urban region. Consider two diplomats attempting to negotiate a treaty.
If wise, each spends considerable time in advance learning about the other's
perspectives and concerns. With those dual insights, agreement is more likely
and the product should be more solidly constructed. In the present case, we
must understand nature in built areas, and how built systems and built areas
affect their surroundings.

Built systems are basically for the transport of people and goods. Radial highways, ring highways, commuter-rail lines, airports, shipping/ferry ports, and
transportation corridors to ports and airports in urban regions are the emphasis
here. In contrast, built areas are mainly for locating people and their activities
on land. Here we focus on the metropolitan area, satellite cities, towns, certain
small sites, strip development, and adjacent land covers. Evidence of regional
planning is also highlighted.

Several attributes relate to whole urban regions. Those presented include the
context of the metropolitan area, border lengths of various built-area types,
other major political/administrative units in the region, and the effect of other
major cities surrounding an urban region.

Interactions of an urban region with other regions, especially the surrounding regions, are usually quite significant. Effects may cover essentially the whole
urban region, the metro area or city, a particular land-cover type, or specific sites
or types of sites. This subject is explored in more depth in Chapter 11.

Because the intermixing and arrangement of built areas and greenspaces is so
important for natural systems and their uses, we begin this chapter by exploring
the subject. Built areas have big effects on natural areas, and vice versa. Yet,

the negative effects differ markedly from the positive effects in amount and direction.

At the core of this chapter are patterns and results from the spatial analyses and comparisons of the 38 urban regions, as described at the beginning of the preceding chapter. Forty-one major patterns or results highlighted here are presented under "built systems" as S1, S2, S3, etc., "built areas" as A1, etc., and "whole regions" as R1, etc. Brief elaborations under each result may: (1) interpret or further discuss it; (2) identify its implications ecologically or for society; and (3) point out useful approaches or solutions for land protection, planning, and other objectives. Some solutions are labeled *priority* and some *high priority*, relative to the total set of patterns presented in Chapters 6 and 7. Graphs highlight variability and contrast (compare the four quadrants) more than central trend.

Natural systems within and next to built areas

Greenspaces within built areas are introduced at the outset. This leads to a key subject, the effects of proximity of built and natural areas.

Greenspaces within built areas

Greenspaces in built areas are particularly significant for recreation, inspiration and the general well-being of residents. For instance, biophilic design specialists point out the value of nature, or even a tree, to improving recovery rates of hospital patients, reducing illness, and so forth (Kellert and Wilson 1993, Kellert 2005). Because people are so diverse in a city, designing diverse recreation opportunities is a challenge. A particular park could have numerous types of recreation resources, or each park could provide a different resource (Chapter 1). Also, distributing the resources appropriately in parks across the city is difficult. Some compromise is typical, which often results both from responding to influential pressure groups and from an overall logical plan.

But greenspaces in a city provide many other benefits to society, especially for natural systems. Large or medium-large spaces, such as Berlin's Tiergarten, New York's Central Park, and Tokyo's palace grounds, may contain a good facsimile of natural ecosystems of considerable educational value for residents hemmed in by buildings. Reasonable biodiversity may be present, such as rare meadow plants in Tiergarten (Caroline Chen, personal communication). These relatively large greenspaces therefore serve as important sources of species that spread to small parks, gardens, and other greenspace areas across the city, keeping them somewhat species-rich.

The question of how to design the city's parks for diverse recreation applies also to nature. Should each greenspace be a fair representative of a different natural community, or should each green area have a similar mix of many natural communities? How should the different ecosystems be distributed across the city, and how can the answer be effectively meshed with recreational opportunities? The questions are tractable and important, but answers are scarce.

For instance, ponds and lakes, much-valued by the public for visual quality values, often exist in some city greenspaces. Large and small, shallow and deep, ringed by walkways or almost inaccessible, clean or polluted, these water bodies support very different aquatic ecosystems and species. They also provide many societal functions from water supply and sewage overflow to boating/fishing recreation, solid-waste disposal, and flood control. Mosquitoes and other insects are often pests or disease vectors. Meshing these characteristics over a heterogeneously distributed array of large, small, squarish, and elongated city parks is a worthy goal.

Large and small natural patches provide somewhat different benefits in rural than urban areas. In the city large greenspaces typically serve as major hydrologic sponges against flooding, maintain the best facsimile of a natural community, and cool the summer city temperature for hundreds of meters outward (Forman and Hersperger 1997). On the other hand, large patches in the urban-region ring may help protect an aquifer or lake, connect headwater streams, and sustain large-home-range vertebrates.

The other major role of urban greenspaces is functional, facilitating natural flows and movements across the city (Figure 7.1). Linear spaces usually reflect the presence, or previous presence, of a stream or river. Often streams have become transformed into rushing water in large straight underground pipes. A green strip on the ground over a former stream normally facilitates walking and wildlife movement. Riversides and floodplains, except during flood stages, may also be effective for the movement of people and animals. Urban floodplain areas, however, also commonly serve as major infrastructure conduits for cities. Gas, oil, electricity, water supply, stormwater, sewage wastewater, commuter trains, intercity trains, trucks, and cars flow along different floodplains. Furthermore, little roads commonly cross and go along floodplains for the maintenance and repair of the infrastructure conduits.

Effects of proximity of built and natural areas

Diverse flows and movements between a built area and an adjacent natural area may be both positive and negative for each receiving side (Forman 1995, Harris et al. 1996, Hersperger and Forman 2003, Hersperger 2005). The strongest interactions are the negative effects of built areas on natural areas. Very few effects of built areas on natural areas are positive for natural systems and the human activities that depend on them. In the opposite direction,

Figure 7.1 Linear greenspaces serving as separating barriers or filters and as channels for flows and movements. The river corridor separates neighborhoods and land uses, while facilitating floodwater flows and wildlife movement. Vegetation strips, and to a lesser extent, street tree lines channel birds and mammals along them. Note that: street trees shade and cool impermeable surfaces; wildlife can readily cross under the bridge on both sides; the dam outlet fixes the channel in place so that normal river migration across a floodplain cannot take place; and stormwater runoff from pipes enters downriver. Milwaukee, USA. Photo courtesy of US Federal Highway Administration.

i.e., effects of natural areas on built areas, both positive and negative, appear to be intermediate in impact.

The *negative effects of built areas on natural areas* are highlighted in more detail because these have the greatest importance for planning and the urban region. Heat, air pollutants, chemical pollutants in stormwater, human wastewater, vehicular traffic, non-native and invasive species, domestic animals, and especially people dispersing outward from a built area are all familiar. The last of these, movement of people, produces varied effects, such as tree cutting, overfishing, livestock grazing, road construction, soil erosion, damaging sensitive habitats and disturbance of wildlife (Luck *et al.* 2004). Each of the diverse effects often has a major degradation effect, generally extensively analyzed in the literature, on an adjoining natural ecosystem, whether terrestrial or aquatic.

More modest interactions go in the opposite direction, from natural to built area. Here positive effects include cooling during hot periods and a source of species and biodiversity. A natural area adjacent to a built area can absorb and break down pollutants, and provide ready access to nature and nature-related recreation for residents. Negative interactions of natural areas on built areas include being a source of mosquitoes and other insects as pests and disease vectors, floodwater, wildfire in dry climates, pest animal populations, and dangerous large predators.

If everyone lived in an enormous skyscraper and the rest of an urban region were nature, people would have ready access to a small area around their building, and natural systems would be the best ever. Conversely, if the homes of everyone were evenly distributed across the region, natural systems would be extensively degraded, probably the worst possible design. The planning trick then is to create a land with people aggregated enough to sustain widespread natural systems, in order that their uses for society remain vibrant for the long term.

With all urban regions little-planned and urbanization ongoing, one could focus on mitigation or restoration. Yet those approaches have usually been piece-by-piece site-by-site activities. While valuable, especially for the most sensitive or strategic places, a different *two-step planning approach* seems more promising. First plan the region for big areas, then mold small places to fit the big vision.

Built systems

Key patterns and results identified in the figures are grouped into the following categories: (1) ring highways; (2) radial highways, ring highways, and commuter-rail lines; (3) airports and aircraft noise; (4) shipping/ferry ports, airports, and development corridors; and (5) wildlife underpasses and overpasses.

Ring Highways

[S1] Half of the urban regions have no external ring highway, only radial highways (Figures 7.2 and 7.3).

In these cases, strip (ribbon) development tends to spread along and near the radial highways producing car-dependent communities, interrupting stream/river corridors, and forming a barrier to regional wildlife movement. However,

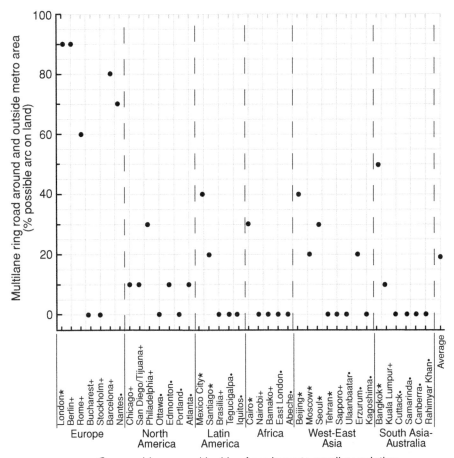

Figure 7.2 Length of multilane ring highway around a metropolitan area relative to geography and city size. Coastal cities have a surrounding arc <360°. Ring road length is measured outside the metro area; the absence of a ring road portion indicates that the road is within the metro area or does not exist. Cities are in decreasing order of population (Table 5.1) within each geographic area. * = city population >4 million ("large"); + = 1 to 4 million ("medium"); • = 250 000 to 1 million ("small"). City abbreviations: Abe (Abeche, Chad); Atl (Atlanta, USA); Bam (Bamako, Mali); Ban (Bangkok, Thailand); Bar (Barcelona, Spain); Bei (Beijing, China); Ber (Berlin, Germany); Bra (Brasilia, Brazil); Buc (Bucharest, Romania); Cai (Cairo, Egypt); Can (Canberra, Australia); Chi (Chicago, USA); Cut (Cuttack, India); Eas (East London, South Africa); Edm (Edmonton, Canada); Erz (Erzurum, Turkey); Iqu (Iquitos, Peru); Kag (Kagoshima, Japan); Kua (Kuala Lumpur, Malaysia); Lon (London, United Kingdom); Mex (Mexico City, Mexico); Mos (Moscow, Russia); Nai (Nairobi, Kenya); Nan (Nantes, France); Ott (Ottawa, Canada); Phi (Philadelphia, USA); Por (Portland, USA); Rah (Rahimyar Khan, Pakistan); Rom (Rome, Italy); Sam (Samarinda, Indonesia); Sdt (San Diego/Tijuana, USA/Mexico); San (Santiago, Chile); Sap (Sapporo, Japan); Seo (Seoul, South Korea); Sto (Stockholm, Sweden); Teg (Tegucigalpa, Honduras); Teh (Tehran, Iran); Ula (Ulaanbaatar, Mongolia). See Table 5.1 and Chapter 5 for further urban region information.

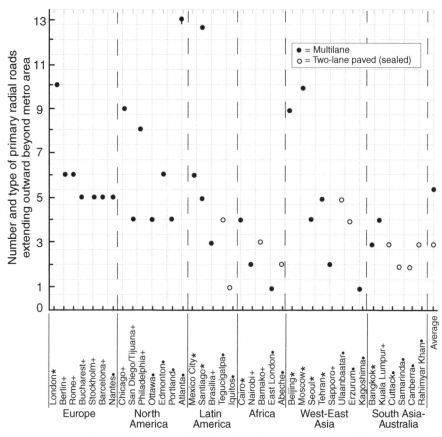

Figure 7.3 Radial roads extending outward from a metropolitan area relative to geography and city size. Radial roads extend ≥20 km beyond the metro area border. See Figure 7.2 caption.

green wedges between the radials and extending into metropolitan areas are likely to be present. Connected greenspace in wedges facilitates the movement of species inward and nature recreationists outward from the city. Breaks in the strip development for stream/river corridors and wildlife movement are important.

[S2] In Europe, ring highways are widespread and extend an average 56 % of a complete ring outside the metro area, whereas elsewhere all ring roads are <50 %, and the average is 25 % (Figure 7.2).

An effect of geography is evident for this pattern. Europe, which combines a dense population with high vehicle use, has mainly chosen the outer-ring-highway design. Europe's cities tend to spread concentrically, or dispersed towns and villages become nuclei for urbanization which later threatens to coalesce.

Some metropolitan areas have engulfed ring roads, but most regions have avoided, or not yet followed, the European approach. Instead, the varied benefits and relatively few shortcomings of a star or spoke design of radial roads are maintained.

[S3] Excluding Europe and North America, all large cities have ring highways and almost no small or medium city does (Figure 7.2).

Both city size and geography effects are present for this design, one of the very few such cases in the 78 major patterns highlighted in Chapters 6 and 7.

Ring roads are a response to large metropolitan areas and help stimulate surrounding urbanization. Or perhaps cities in other geographic areas are copying Europe and North America where vehicles and traffic first became dense. Irrespective, large cities (>4 million population) have strong development pressure inward and outward from ring roads, which make rural areas and villages seem easily accessible to the metro area. In contrast, small and medium cities are mainly surrounded by nearby connected greenspace. Thus the difference between large and small/medium cities for average distance of city residents to connected greenspace is great and growing.

Radial highways, ring highways, and commuter-rail lines

[S4] Large cities tend to have more primary radial roads than do small cities, and those of large cities are multilane, whereas radial roads of medium and small cities are about half multilane and half two-lane (Figure 7.3).

A city size effect is evident. Large cities are directly connected to more satellite cities and distant towns and cities, whereas small cities are more indirectly connected. More radial roads create more strip-development barriers to regional walking, wildlife movement, and stream/river corridors, and subdivide the region into more and smaller sections. Strip development tends to spread over a wider area. In contrast, with few radials, large segments of greenspace escape development pressure. Two-lane highways tend to roadkill more wildlife, but multilane highways cause a greater barrier effect, wider wildlife-avoidance zones, and more habitat fragmentation, which normally is much more ecologically serious than roadkills (Forman *et al.* 2003).

[S5] All European and North American cities have at least four primary radial roads, while cities in Africa and South Asia-Australia average 2–3 radials, many being paved two-lane (Figure 7.3).

A geography effect is present. Fewer radial roads mean that larger areas escape development pressure. Traffic and development are channeled in only two or three directions from the city.

[S6] Cities without ring roads have few primary radial roads, and paved two-lane radial roads are essentially limited to cities without ring roads (Figures 7.2 and 7.3).

With few widely separated radial highways, perhaps less pressure builds for a ring road. Two-lane radials may suggest mainly local traffic rather than long-distance intercity transport travel. Housing and jobs are local rather than dispersed, so commuter traffic and associated commuter residential areas are limited. The adage, "If you want to work in a city, move and live there," seems to apply. Erzurum, the one exception to the pattern, has two-lane radials with a partial multilane ring road (Color Figure 17).

[S7] Cities with considerable ring-road length normally have relatively few primary radial roads, and cities with many radials have little ring-road length (Figures 7.2 and 7.3).

In general, cities seem to either invest in many primary radial roads that may protect more nearby greenspace or invest in a ring road. More radials mean more strip development, more interrupted stream/river corridors, more barriers to wildlife movement, more connected greenspace, and more in-and-out greenspace access between city and countryside. More radial roads also imply more human dependence on the metro area, rather than dispersed movements in surrounding landscape areas.

Beijing is an interesting exception, with both a relatively complete ring road and many radials (Color Figure 7). Beijing has heavily invested in ring roads, with the fourth ring relatively complete and the seventh ring beginning in places (Yang Rui and Laurie Olin, personal communications). No large greenspace is apt to remain for future residents or for nature near the metropolitan area.

[S8] Cities with more radial roads may also have more radial commuter-rail lines extending beyond the metropolitan area (Figure 7.4).

Although limited data on commuter-rail systems were collected, apparently commuter-rail service is mostly within metropolitan areas. Radial commuter-rail lines, however, serve separate communities outside certain metro areas. Cities with many radial highways plus outside commuter-rail lines (London, Philadelphia) may have a large outside commuter population that demands an alternative to radial vehicular-traffic flows. Large connected nearby greenspaces could be maintained with such an emphasis on radial transportation routes.

Airports and aircraft noise

[S9] Most urban regions have one or two major airports, while 10% have five or more airports (Figure 7.5).

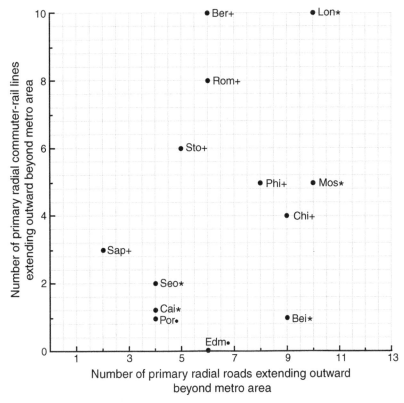

Figure 7.4 Number of radial commuter-rail lines relative to radial highways extending from a metropolitan area. Number of commuter-rail lines is determined based on 16 possible compass directions. All radial roads are multilane and extend ≥20 km beyond the metro-area border. See Figure 7.2 caption.

All regions have a major airport near the city for passenger air travel, and commonly also for cargo freight. A second major airport provides flexibility if the nearby one becomes inadequate, or if passengers and cargo are concentrated in different airports, or if international and local/regional travel is separated. Most regions probably have a major military airport. Some regions have many airports (Moscow, London), presumably with different major uses. Some airports may be decommissioned (Edmonton), but remain conspicuous on the aerial images and are therefore mapped.

[S10] Most cities have the nearest major airport <20 km from the city center, and most cities with two or more major airports have the second closest one >20 km distant (Figure 7.6).

The primary airport is close and convenient to the city for passengers and business cargo, and may be connected to the city by rail. Where the second

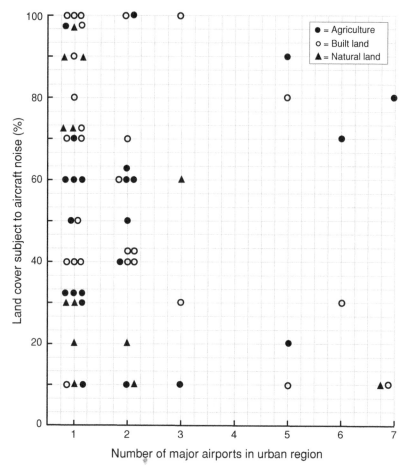

Figure 7.5 Land-cover types subject to aircraft noise relative to number of major airports. Percentage cover is the average for land extending approximately 5 km (3 mi) out from both ends of major runways at major airports in the urban region. The 5 km distance is based on Miller (2005).

closest one is rather distant, it is less convenient for travel to and from the city and would tend to catalyze a connecting strip of development, serving as a barrier to stream/river corridors and regional wildlife connectivity. However, from air traffic and air pollution perspectives being further from the city is good for a local/regional airport. Being distant is also good for a military airport assuming that the many temporary personnel live nearby.

[S11] Most major airports are in cropland; a third of the regions have major airports mainly in built areas, and 15 % mainly in natural land (Color Figures 2–39).

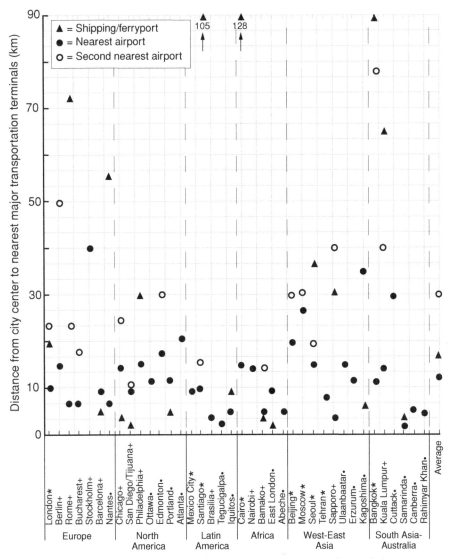

Figure 7.6 Distance from city center to nearest coastal port and two nearest airports relative to geography and city size. 10 km = 6.2 mi. See Figure 7.2 caption.

Cropland is the primary land use in most urban regions so land lost to airport construction, transportation, and associated facilities is normally insignificant regionally. Typically built areas around an airport grew up around it. Runway and infrastructure expansion of a major city or military airport is a strong force, so the airport should be located distant from key natural and cultural resources.

If important resources are nearby, such as a key aquifer and an agricultural park (Barcelona; Chapter 10), strategic planning for a second location, e.g., to emphasize local/regional travel or cargo transport, is important.

[S12] Aircraft noise in the urban regions occurs in built land most, cropland next, and natural land least (Figure 7.5).

Aircraft noise in and by built land is well known to create problems for people, from stopping a conversation or class to physiological stress or disruption of park recreation. Avoidance and mitigation are commonly attempted. Animals, like people, hear and are stressed, but additionally they respond to predator-prey interactions. Wildlife may not hear a predator and get eaten, or may go hungry if the prey are gone, or may see a sibling get eaten and then avoid the area. Avoidance and mitigation are also important for nature. Cropland is probably least affected by aircraft noise. Wildlife species there are mostly tolerant generalists and may be common where cropland is extensive and connected.

Shipping/ferry ports, airports, and development corridors

[S13] About half of the shipping/ferry ports are close to the city center, while the other ports are evenly distributed from about 20 to 140 km distant (Figure 7.6).

Ports at a distance require transport of goods and people to and from the city. This stimulates strip (ribbon) development, with associated disruptions of stream/river corridors and wildlife crossing. The aquatic zone around a port is polluted and, in addition, adjoining shoreline stretches tend to be degraded. Port areas are also likely to be sources of non-native species, some of which may become invasive in the region.

[S14] About one-sixth of the cities has two of the three major transportation facilities (port and two closest airports to center city) >20 km out (Figure 7.6).

This suggests a considerable length of strip-development corridors, with negative consequences for natural systems as described above. Busy flows of truck and car traffic continue throughout the day, unlike the typical pulses of commuter traffic.

Wildlife underpasses and overpasses

[S15] Most urban regions have busy multilane highways passing between or cutting through major natural areas, where wildlife underpasses or overpasses are especially valuable to provide connectivity for walking and wildlife movement (Color Figures 2–39).

Wide underpasses and overpasses designed with vegetation for wildlife crossing across major highways are widely and successfully used in Europe, and are present in North America and elsewhere (Trocme *et al.* 2003, Iuell *et al.* 2003, Forman *et al.* 2003, van Bohemen 2004). The best designs facilitate crossing by targeted species as well as a large portion of the fauna. Local residents and walkers on regional trail systems often also use the crossings. Wildlife-crossing structures mitigate the barrier and habitat-fragmentation effects of highways, and enhance regional connectivity for movement among natural areas. Thus major wildlife-crossing structures are a *priority* in the outer urban-region ring to facilitate regional wildlife movement, walking on regional trail systems, and connectivity for local residents.

Built areas

For built areas the following groups of patterns and results are highlighted: (1) green patches and corridors within metropolitan areas; (2) metropolitan-area form; (3) evidence of regional planning; (4) satellite cities; (5) towns in the urban-region ring (or urb-region ring); and (6) "natural" disasters.

Green patches and corridors within metropolitan areas

[A1] All combinations of high, medium, and low densities of green patches and green corridors are present within the metropolitan areas. Most common is a high density of patches and low density of corridors; a sixth of the metro areas has a low density of both greenspace types (Figure 7.7).

Small and medium greenspaces dispersed over a metro area provide nearby access to nature and recreation for residents. A high density provides stepping stones for many species to move across built areas, and even tiny patches can be effective for such movements. Linear green patches are corridors that further enhance species movement, especially in the direction a corridor is oriented. Green corridors also line and protect streams, though in metro areas most streamwater is in underground pipes so the green corridors are over former streams. In the urban regions greenspace patches are considerably more abundant than corridors.

Increasing the number and area of parks in metropolitan areas with little greenspace per person is a valuable goal. Mapping the average distance between housing units and parks (e.g., as in Chicago and London) pinpoints priority areas for new parks. These internal greenspaces facilitate accessibility of people to nearby parks, but do not provide connected greenspace accessibility for walking and bicycling to greenspace outside the metro area, a much-valued weekend activity (e.g., around Dutch cities).

[A2] About 30 % of the metro areas lack a greenspace $\geq 1\,km^2$ (250 acres), whereas about a fifth of the metro areas has a $\geq 10\,km^2$ (4 mi^2) greenspace present (Figure 7.8).

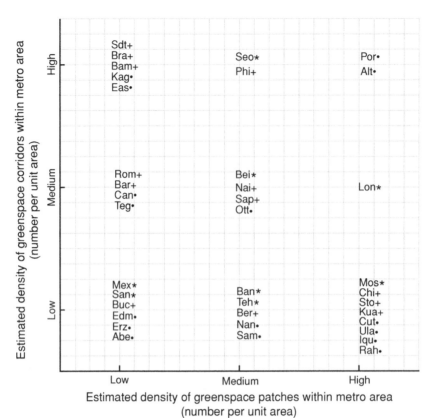

Estimated density of greenspace corridors within metro area (number per unit area)

Estimated density of greenspace patches within metro area (number per unit area)

Figure 7.7 Estimated density of greenspace corridors relative to greenspace patches within a metropolitan area. See Figure 7.2 caption.

The presence of a medium-to-*large greenspace in a city*, such as Berlin's Tiergarten, New York's Central Park, and Tokyo's palace area, provides several ecological and societal benefits unavailable with only small greenspaces. Areas surrounding the large green area, especially downwind, are cooler in summer (Schmid 1975, von Stulpnagel *et al.* 1990). Wetlands and a higher, more natural watertable may be maintained. Flood hazard warrants evaluation, but might be lower. A facsimile of natural ecosystems can be supported that provides educational and inspirational values. Rare native species as in the meadows at Tiergarten may be present from time to time (Caroline Chen, personal communication), though probably not sustained. The large greenspace is a key source of species that populate the city's small parks, gardens, and other spaces. As in the three examples mentioned, large greenspaces are of major interest to a large population and may remain stable, may evolve, or may undergo convulsions along with society.

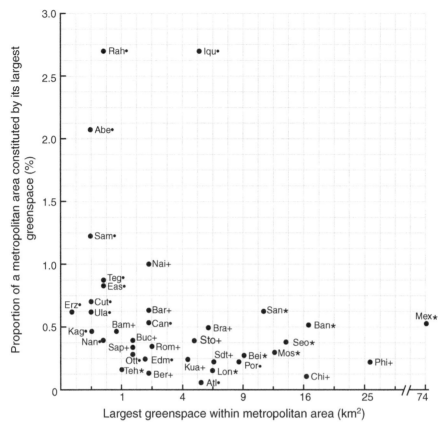

Figure 7.8 Proportion of metropolitan area constituted by its largest greenspace relative to area of the largest greenspace. Note that the horizontal axis changes with the square of a number. $1\,km^2 = 0.39\,mi^2$. See Table 5.1 for area of metro area. See Figure 7.2 caption.

Creating a large greenspace, such as converting an aging industrial area to a large park, requires commitment to a long-term plan. Alternatively, a cluster of smaller greenspaces might be partially linked and enhanced to provide some of the benefits mentioned.

[A3] Small cities, especially in developing nations, generally have only small greenspaces, which, however, are the largest greenspaces relative to the surface area of metropolitan areas (Figure 7.8).

Both city size and geographic area effects are evident here. Small urban greenspaces are mainly of value for nearby residents' recreation and as stepping stones for species movement across built areas. Still, in small cities these small parks are close to the extensive greenspaces around the metropolitan area, which may provide accessible recreation opportunities and are major multi-directional sources of species moving into the city.

The problem to be overcome is that the city almost certainly will grow, e.g., from small to medium. About a century ago Frederick Law Olmsted designed Boston's much-heralded emerald necklace near the edge of the city, and Antoni Gaudi designed Parque Guell (Color Figure 40) on the outskirts of Barcelona. Analogously, the establishment of greenspaces, including large ones, on the edge of small cities is a *priority*. Such parks will well serve the larger city of the future.

Metropolitan-area form

[A4] Half of the metro areas adjoin major physiographic features, often on two sides, that constrain outward urbanization spread (Figure 7.9).

Adjacent seacoasts (San Diego, Barcelona) and lakeshores (Chicago) tend to produce elongated metropolitan areas, and consequently long stretches of polluted seawater, of damaged or destroyed coastal ecosystems, and of degraded recreational resources. Adjacent mountain ranges (Sapporo, San Diego/Tijuana) tend to have considerable development on nearby slopes, with associated flood hazards, erosion/sedimentation problems, reduced visual quality, and so forth. The regional geometry also makes major transportation networks particularly difficult. Protecting nearby slopes with natural vegetation and establishing coastal natural areas and parks long enough to maintain clean water are *priorities*. Focusing development around satellite cities or other locations away from slopes and coastlines is also a key part of the solution.

[A5] About half of the metropolitan areas are compact with 0–2 major built lobes, while metropolitan areas with several (4–9) lobes have a highly convoluted form (Figure 7.10).

A *compact metropolitan area* has a minimal perimeter and hence degrades outside greenspaces the least. Without strip-development corridors it also poses the least disruption of regional connectivity for wildlife and major stream/river corridors. With people of a compact metro area being closer, on average, to the city center, public transport can be efficient. In contrast, several built lobes projecting outward from a metro area suggest prominent strip development along major radial transportation routes, with associated shortcomings as described above. The long metro-area border means that much surrounding greenspace is degraded. Maintaining a relatively compact form provides important benefits.

[A6] Metropolitan areas, on average, have 3.5 major built lobes and 2.3 major greenspace wedges on their perimeter, with lobe and wedge number positively correlated (Figure 7.10).

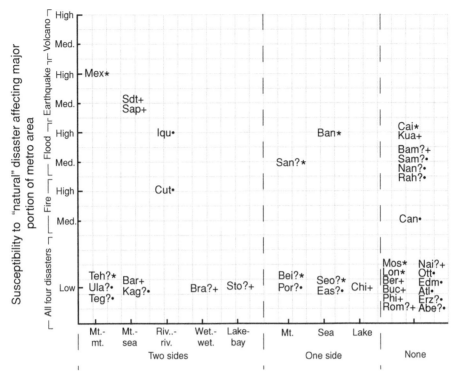

Figure 7.9 Estimated susceptibility of a metropolitan area to "natural" disasters relative to the presence of physiographic features constraining the spread of a metro area. Rough susceptibility estimates are based on a known 200 yr history for the city or being located in a known longer-term disaster zone. "Natural" disasters or catastrophes have significant property damage over a major portion of the metro area, a pattern often due in part to inappropriate preceding human activities. Hurricane/cyclone, tsunami, and avalanche/mudslide may occur, but apparently are not frequent major problems in the 38 urban regions selected. Hills are not considered to be a major constraint on urban expansion. Mt. = mountain range; riv. = large river; wet. = major wetland; bay = coastal saltwater bay. See Figure 7.2 caption.

As just indicated, major built lobes around a metro area pose diverse important problems for natural systems in the region. Major greenspace wedges (e.g., Stockholm, Copenhagen) projecting into a metro area, on the other hand, offer more benefits than shortcomings. The wedges are subject to degradation, sometimes intense, by being squeezed between built areas usually with high population densities. However, green wedges, like a high density of small parks, provide nearby access to greenspace for people. Unlike small parks though,

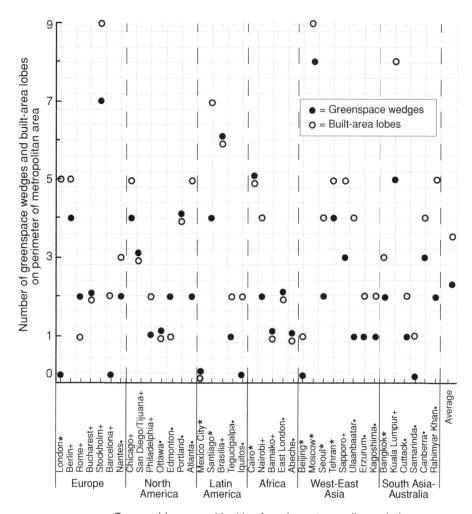

Figure 7.10 Greenspace wedges and built lobes of metropolitan area relative to geography and city size. Major greenspace wedges or coves projecting into a metro area are ≥40 % of the average radius of a metro area (wedge length measured from a line connecting the tips of adjacent metro-area built lobes). Some wedges included mainly result from a large river bisecting the metro-area (See Color Figures 2–39). Major built lobes are ≥40 % of the average radius of a metro area (lobe length measured from the ends of adjacent wedges or coves). See Figure 7.2 caption.

wedges provide connectivity for walking and bicycling, as well as accessibility to greenspaces outside the metro area. Furthermore, green wedges serve as major routes for species to enter and populate greenspaces in the metro area. Therefore, in addition to the remarkable recreation benefits provided, green wedges may be the best strategy to provide a richness of nature throughout the city.

As a long-term *priority* planning strategy, creating green wedges in existing built areas would normally be slow and very difficult. However, establishing major connected greenspaces near the edge of metro areas can create future wedges or a ring of major parks for the city as it expands outward.

[A7] A few metropolitan areas have mainly "scalloped" borders of medium-length lobes and coves (Color Figures 2–39).

The *scallop border* design (e.g., Bucharest; Color Figure 10) degrades slightly more regional greenspace than does a compact metro area, as described above, but it degrades noticeably less greenspace than does a lobed metro area. Furthermore, little strip development along major transportation routes is present to interrupt major stream/river corridors or regional wildlife movement. The border scalloping provides ready access to surrounding regional greenspaces for people living in the outer portion of the metro area. Indeed, examining the pros and cons of scalloped-edge natural patches is informative (Forman 1995). Overall though, combined with a high density of small green patches in the central portion, a metro area with scalloped borders seems to be a reasonably good design.

Evidence of regional planning

[A8] Ten extremely different metro-area-form and urban-region-ring attributes identified suggest that regional planning, emphasizing at least one of the attributes, has been used in 60 % of the urban regions (Figure 7.11).

Spatial planning is common for small spaces and rather uncommon for broad areas. The issue comes to a head in urban regions where so many people live and resources are so finite, but political/administrative units, often overlapping, typically are so many. Based on the regional attributes identified, apparently regional planning has been significant in a majority of the urban regions. Further work, or the reader, will have to decide whether the attribute or attributes chosen for emphasis has led to a wise plan and a suitable urban region. In fact, the degree to which regional attributes reflect planning or simply unplanned human and natural changes warrants evaluation. Several significant attributes may be embedded with unequal weight in a plan, so in this analysis only the heavyweight one(s) clearly emerges. Regional planning of urban regions and metropolitan areas is a *high priority*.

Beijing is noteworthy here because it has a single (strong) government for virtually its entire urban region, about 200 km in diameter. In this case, regional planning might be relatively easy, avoiding the paralysis often produced by many overlapping political/administrative units characteristically present in urban

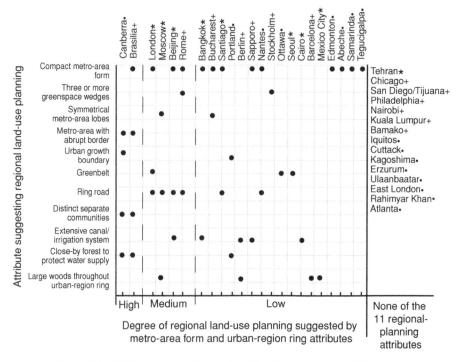

Figure 7.11 Attributes suggesting regional land-use planning relative to the degree of regional planning suggested. The conspicuous spatial attributes highlighted relate either to the form of the metropolitan area or to patterns in its surrounding urban-region ring (see Color Figures 2–39). Also see captions for Figures 7.2 and 7.10.

regions. However, without the checks-and-balances such units provide, regional plans could be brilliant, a disaster, or somewhere in between.

[A9] A quarter of the urban regions manifests two or three attributes that suggest regional planning (Figure 7.11).

Compared with the prominence of a single attribute suggesting regional planning, 2–3 attributes may indicate a greater role of regional planning in the evolution of a metropolitan area and urban region. They may also represent a better balance among competing strategies.

[A10] A compact metropolitan-area form in a third of the regions is the most frequent attribute suggesting regional planning, and a ring highway in a sixth of the regions is the next most-frequent attribute (Figure 7.11).

A compact metro-area form means the near absence of major built lobes, major greenspace wedges, and an elongated spread. As described above, major lobes are normally along transportation routes and elongated forms along coastlines and/or mountain valleys. Major greenspace wedges doubtless involve significant regional planning. A compact metro-area form typically results from concentric-zone-like spread from an initial nucleus (Chapter 8). Planning decisions were made to not elongate, to develop along spokes, or to protect green wedges, but rather to keep the focus on the city center.

A ring highway outside the metropolitan area reflects regional planning, not only for transportation flows, but also to develop areas between radial transportation strips. As described above, this has important negative consequences for natural systems and their human uses. Perusal of the other attributes suggesting regional planning emphasizes that some have mainly positive and some negative implications for these societal objectives. Thus regional planning per se is not the solution, but rather wise and strategic planning of large areas, which accords high priority to natural systems and their human uses, is the objective.

Satellite cities

[A11] Satellite cities around large cities are mostly within 50 km of the metropolitan area, whereas most satellites around small cities are 50–100+ km distant (Figure 7.12).

City size has an important effect here. The relative proximity of satellite cities to large metropolitan areas means that targeting their development instead of metro-area expansion is more feasible, since new residents are not too far from the big city. Also, while the length of highway strip development to the nearer satellite cities disrupts major stream/river corridors and wildlife movement in the urban region, it is not as serious as developed strips to more-distant satellites. However, more-distant satellite cities are more likely to be near natural areas, where outward urbanization may cause significant degradation to natural systems.

[A12] Large-population cities tend to have more satellite cities in the urban region than do small cities, with 5–16 around most large cities and 0–2 around most small cities (Figure 7.12).

The number of satellite cities depends on the size of the core city. Few satellites in small-city urban regions mean that most urban people live in the region's core city where urban resources are concentrated. Thus natural systems of the region tend to be little affected by the few dispersed concentrations of people. People in and around the regional greenspace mainly live in villages and towns, and, on average, are more attuned to nature, though may or may not protect it well. The abundance of satellite cities around large core cities means that almost all regional greenspace is near and subject to the diverse impacts of many cities. Since dispersed satellite cities also tend to expand outward, natural systems and

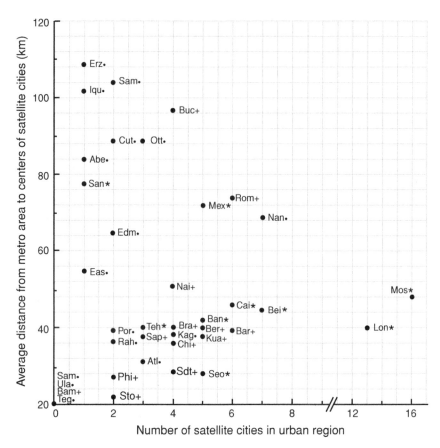

Figure 7.12 Distance from metropolitan area to satellite cities relative to the number of satellite cities in an urban region. Only satellite cities ≥20 km from the metro-area border are included. The area of a satellite city is almost always noticeably larger than that for any of the more numerous towns present (see Color Figures 2–39). 10 km = 6.2 mi. See Figure 7.2 caption.

their human uses all remain under a degree of degradation threat. Strategically protecting large greenspaces in key areas, especially in the outer urban-region ring, in the face of many expanding urban circles, is a *high priority*.

Towns in the urban-region ring

[A13] *In urban regions half the towns (excluding those by water bodies) are in agricultural areas, and a fifth near the boundary between agricultural and natural areas* (Figure 7.13).

Since the urban regions are overwhelmingly dominated by cropland, not surprisingly towns are most common in mainly cropland areas. These towns may

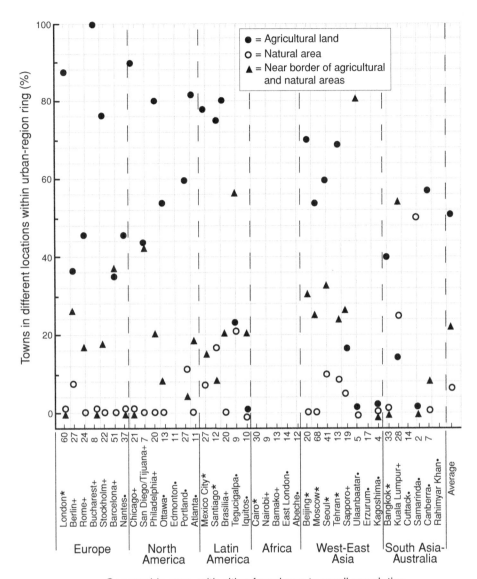

Figure 7.13 Locations of towns in agricultural and natural land relative to geography and city size. Towns are almost always noticeably smaller in area than satellite cities (see Color Figures 2–39). Percentage is relative to all towns (number listed near bottom) in an urban-region ring. Location means mainly surrounded by agricultural land or natural land (usually landscape-sized, i.e., ≥100 km² and compact in shape), or near the border between the land types. Towns not included are along rivers, major streams, and other major water bodies. See Figure 7.2 caption.

retain much of their rural appearance and culture, but almost all are near the metropolitan area or a satellite city. Other than the normal agricultural activities, effects of these towns on natural systems and human uses are usually limited. Towns near borders between cropland and natural land are discussed below.

[A14] *Most urban regions have no towns in natural areas, and hardly any region has >25% there* (Figure 7.13).

Within natural landscapes and large natural patches, towns commonly have significant negative impacts on natural systems, including water, soil, and wildlife. Analogous negative impacts also occur on human land uses, such as recreation and water supply, that depend on intact natural systems. Fortunately urban regions have few towns in natural areas. Limiting growth of the existing ones is a valuable investment.

[A15] *Towns near borders of agricultural and natural areas are widespread, i.e., present in >75% of the regions* (Figure 7.13).

Towns near the boundary between agricultural and natural areas are well located, because they displace little valuable land and their outward impacts are mainly on the edges of the natural and cropland areas. This pattern is consistent with the aggregate-with-outliers model for optimally meshing different land uses in a landscape (Forman 1995, Forman and Collinge 1996). If villages are to grow into towns in a region, limiting the growth of villages in natural areas, plus encouraging it in those near the border of agricultural and natural land, is a priority.

"Natural" disasters

[A16] *Many regions are susceptible to a "natural" disaster that affects a major portion of the metropolitan area, with flood hazard being the problem in about half of the cases* (Figure 7.9).

Although appropriate data on this important subject were difficult to find and interpret, it appears that metropolitan areas could be called hazardous places due to their location and their concentration of people and structures. Flooding from an old dam that fails or bombing by warplanes would be human-caused disasters. Natural disasters are earthquakes (e.g., Kobe, Japan), debris-flows from volcanic eruptions (Perera, Colombia), and hurricanes/cyclones (Darwin, Australia). Yet natural disturbances may become "disasters" due in major part to human activities. Thus high levees holding back a huge lake broke in a 2005 hurricane to cause widespread inundation of the adjacent lower-elevation

New Orleans. Disaster-preparedness planning is important, but tends to become high-profile following rather than before a disaster.

Flooding, following heavy rainfall and/or snowmelt, is the most widespread disaster hazard in urban regions. The process is accentuated by hard surface cover of buildings and associated roads that have spread on mountain- and hill-slopes. Linear roads and their stormwater pipes accelerate stormwater runoff to streams and rivers. But streams are commonly straightened, squeezed by legal and illegal structures in floodplains, and channeled to underground pipes with hydraulic rather than hydrologic water flows, further accelerating down-water flows. Pipes and streams lead to rivers, which commonly have normal low water flows and may seem inconspicuous or inconsequential. Yet rivers in urban regions are commonly straightened, channelized with rock or concrete barriers, squeezed from the sides, and pockmarked with bridges and other pilings. Wetland sponges and floodplain riparian woodland are largely long-gone. The combined result of these mountain/hillslope, stream, and river activities is periodic big floods – enormous water volumes zooming down a river channel, aimed directly at the metropolitan area. Flood disasters result. A package of solutions is the answer (Chapter 10; Forman 2004a).

Whole regions

Important results and patterns relative to whole regions are grouped as follows: (1) urban-region rings relative to metropolitan areas; (2) land cover near and far from metro areas; (3) unique features near metro-area borders; (4) border length of built area in urban regions; and (5) nearby major cities and political/administrative units.

Urban-region rings relative to metropolitan areas

[R1] The metropolitan area is almost always a small, centrally located portion of the urban region, averaging 8 % of the area, and being <1 % in a third of the regions (Figure 7.14).

The rather large area beyond the metropolitan area, the urban-region ring, means that typically a reasonable amount of space exists for multiple resources and human activities in an urban region. Also resources are generally convenient for the city, as the central nucleus of a region. However for two exceptions, Chicago and Philadelphia, the metro area exceeds 25 % of the urban region, suggesting that the urban-region ring and its resources are somewhat limited (Color Figures 13 and 27). Philadelphia is especially problematic because its region is hemmed in by urban regions of surrounding cities. Indeed, a competition for

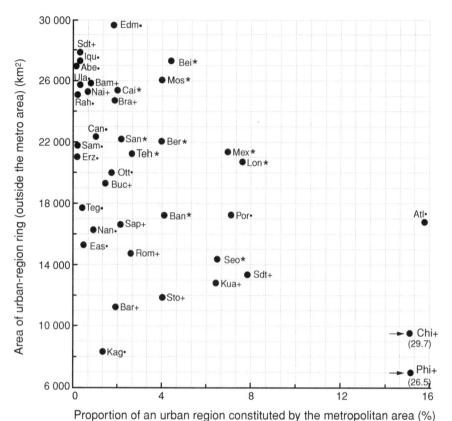

Figure 7.14 Area of the urban-region ring relative to the proportion of an urban region constituted by the metropolitan area. $1000\,km^2 = 386\,mi^2$. See Table 5.1 and Figure 7.2 caption.

space and resources, such as housing development and recreation opportunities, occurs in the outer portions of Philadelphia's urban region. Rapid expansion of protected natural resources around cities with relatively small urban rings is a *high priority*.

[R2] Most urban-region rings are rather large (12 000 to 30 000 km²), though a handful are quite limited in area (Figure 7.14).

Small urban-region rings have a relative shortage of natural and agricultural land. Chicago and Philadelphia have small urban-region rings in part because their metropolitan areas are very large, and as just mentioned, Philadelphia because of surrounding cities. But cities on coastlines, such as Kagoshima and Barcelona, may also have relatively small urban regions, due mainly to geometry and a water body (Color Figures 6 and 19). Small urban-region rings may have

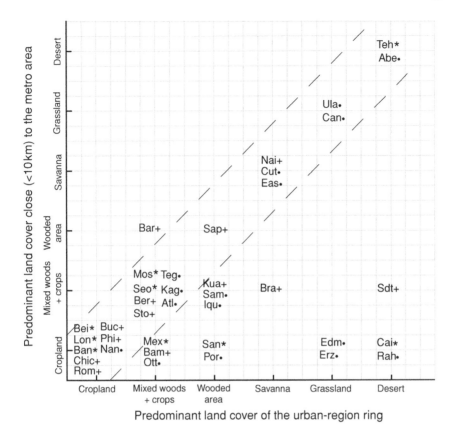

Figure 7.15 Predominant land cover close to the metro area relative to land cover of the urban-region ring. Built area was excluded from the estimates; probably in no case would its inclusion change a result. Grassland includes pastureland; desert includes desertified area. Cities in the diagonal band have the same land cover close to the metro area as across the urban-region ring as a whole. See Figure 7.2 caption.

a relative shortage of natural and/or agricultural land. For such cities, natural-resource protection is a *high priority*.

Land cover near and far from metro areas

[R3] Most urban regions have the same predominant land cover close to the metropolitan area as across the whole urban-region ring, and generally the larger the city the more likely cropland predominates close to the metro area (Figure 7.15).

A city size effect is evident. The prevalence of cropland close to large cities means that urbanization outward is mainly on cropland. But since cropland dominates most urban regions and is both near and far from the city, the loss of cropland to urbanization would normally be of minimal importance. However, protecting soils and locations close to the metropolitan area that are especially valuable for market-gardening (Chapter 3) is an important exception.

[R4] In nearly 40% of the urban regions, agriculture (cropland or mixed crops/woods) is more prominent close to the metropolitan area than across the urban-region ring as a whole, where natural land (desert, grassland, or forest/woodland) usually predominates (Figure 7.15).

In these regions with somewhat limited agricultural land, urbanization spread of the metropolitan area mainly covers this valuable land. Experience from some urban regions indicates that, since the original community began by prime agricultural soil, urbanization over time may cause a significant loss of the region's best soils.

Urbanization spread from satellite cities and towns in the urban-region ring, on the other hand, is more likely to degrade natural land. As described above, this may be quite significant depending on the amount and location of natural land.

Unique features near metro-area borders

[R5] Almost all metropolitan area border lengths are about 35 to 350 km (22–220 mi) long, though three exceptions have much longer convoluted borders (Figure 7.16).

The *border length* of a metropolitan area is a rough overall index of how much surrounding greenspace is degraded by the metro area. Cities with populations from 260 000 to over 10 million almost all have metro-area borders within a single order of magnitude. Three outliers, Chicago, Atlanta, and Philadelphia, have much longer boundaries, in part because of their convolutions of major built lobes and green wedges, and in part because of sprawl. The lobes generally, but not entirely, follow major transportation routes. The greenspace wedges generally, but not entirely, follow major stream or river corridors. Extensive low-density "unsatisfactory" outward urbanization spread, or sprawl, has pushed the overall metro-area border on flat or gently rolling terrain far outward from the city center. A greenbelt (e.g., London) or urban growth boundary (Portland) has been used to arrest further sprawl and its reverberating impacts.

[R6] Almost all metropolitan areas have at least one unique natural-system-related feature close to its border, while a quarter of the metro-areas have three or more such features nearby (Figure 7.16).

These diverse unique natural-system-related features (e.g., a scenic viewpoint, historical site, geologic feature) are threatened by existing human activities, as well as by potential urbanization spread near almost all metro areas. This is particularly evident by large cities. Identifying and protecting areas around these

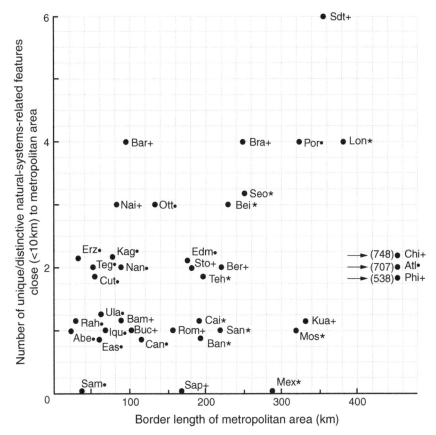

Figure 7.16 Unique/distinctive natural-systems-related features close to a metropolitan area border relative to the length of the border. The highly diverse features identified around the 38 metro areas are illustrated by: the main water-supply reservoir; a major archaeological and tourism site; rare coastal vegetation site; one of only two wetlands; and only major market-gardening area. 100 km = 62 mi. See Figure 7.2 caption.

locations should reduce degradation and maintain the features. Some features have combined cultural heritage, natural systems, and recreational values.

Border length of built areas in urban regions

[R7] Three-quarters of the urban regions have an average built-area border "density" of <5 km length per 100 km² (8 mi/100 mi²) (Figure 7.17).

High-border-density regions have widespread negative effects of built land on natural land, and both negative and positive effects of natural land on built land, as described at the beginning of this chapter. Most urban regions currently have rather low average border lengths, equivalent to a line less than

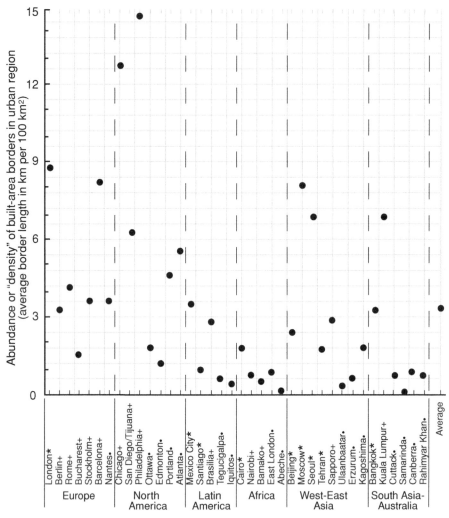

Figure 7.17 Abundance or "density" of built-area borders in urban region relative to geography and city size. Abundance or density equals total border length [metro area + inner satellite cities + outer satellite cities + towns] × 100, divided by area of urban region. 1 km/km² = 1.6 mi/mi². See Figure 7.2 caption.

5 m (16 ft) long in a football field. However the Chicago and Philadelphia regions, with >12 km/100 km² border density, are outliers (Color Figures 13 and 27). Much surrounding greenspace in these regions is doubtless degraded.

[R8] Overall, the metropolitan area contributes >40 % and towns <40 % to total border length in a region; satellite cities contribute the least and have the least-variable total-border-length from region to region (Figure 7.18).

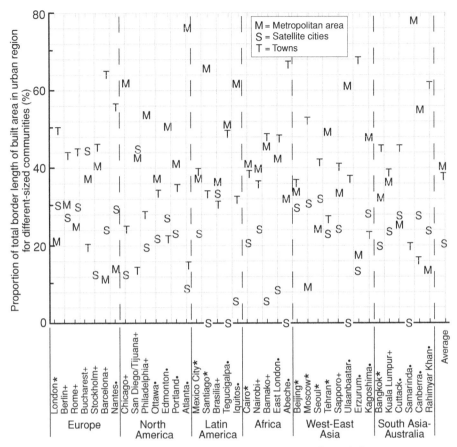

Figure 7.18 Proportion of built-area border length due to metropolitan area, satellite cities, and towns relative to geography and city size. Border length of the metro area and all inner and outer satellite cities was directly measured, while length for towns was estimated from a representative sample. Cities and towns are differentiated by area (see Color Figures 2–39). See Figure 7.2 caption.

This emphasizes that the border length of metropolitan areas matters, and is associated with degrading the greatest area of regional greenspace. The metro-area border plus that of nearby cities (in the inner urban-region ring) provides about half of the total border length for the region. Greenspace in this area tends to be close to the huge metro-area population, which also is likely to expand outward.

In the outer portion of the urban region, satellite cities and towns provide the other half of the total border length. In this area many town and small-city governments with different perspectives, mostly local, make decisions that affect

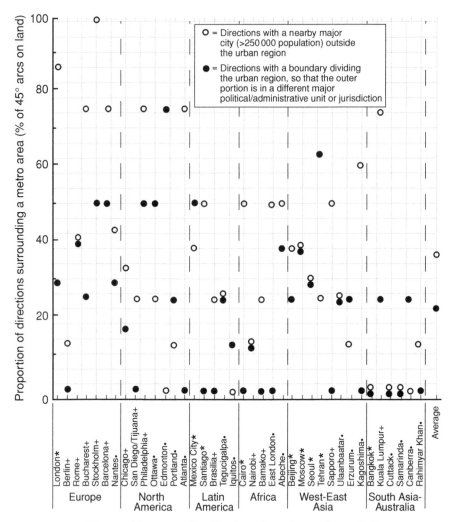

Figure 7.19 Proportion of surrounding directions with a nearby major city or other major political/administrative unit present relative to geography and city size. Proportion of surroundings refers to land, thus excluding sea, coastal bay, and major lake. Nearby cities outside the urban region are >250 000 population and are generally located <1.5 times the distance from the center of the focal city to its urban region boundary (more precisely they are the 60 % and 40 % distance cases described in chapter 5). A major political/administrative boundary indicates that another nation, state, province, department, county, or equivalent unit is present in the urban region. See Table 5.1 and Figure 7.2 caption.

urbanization and natural resources. Both present and future impacts around different towns are likely to vary and fluctuate. Targeting windows of opportunity for local natural-resource protection, and using planning and economic resources of the region as a whole, should provide for the future.

Nearby major cities and political/administrative units

[R9] Most metropolitan areas have nearby major cities outside the urban region in 25–50 % of the surrounding directions; a fifth of the metro areas are nearly surrounded by other cities, and a few have no nearby cities (Figure 7.19).

Nearby major cities outside an urban region are competitors, often for space and resources in the outer portions of a region, as illustrated above for Philadelphia. Metro areas surrounded by outside cities would do well to quickly focus on land protection in the outer portions of their urban regions.

[R10] Two-thirds of the urban regions contain land of a different major political/administrative unit (one-third does not), and for 20 % of the regions half or more of the directions surrounding a city include such land in the urban region (Figure 7.19).

This suggests considerable competition for space and resources in outer portions of urban regions, where political control and decisions by, e.g., another nation, province/state, or county, diminish the influence of the core city. For example, critical water-supply sources and drainage-basin protection for the metropolitan area are often in the outer portion of a region. The regions with different major political/administrative jurisdictions in several directions probably need some kind of an ongoing regional authority, in order to plan and sustain regional resources for all political/administrative units over time.

In summary, 41 "major" patterns and results emerge here from comparing the 38 urban regions worldwide. These patterns apply to built systems, built areas, and whole regions. Adding the results for nature, food, and water (Chapter 6) makes a total of 78 patterns identified from the global urban-region analysis. Of course, other patterns, including many minor ones, exist.

Surprisingly, very few of the patterns correlate with either geography or city population size (Chapters 6 and 7). Patterns for natural systems and their human uses seem to cut across culture, geography, and city size, and be inherent characteristics of urban regions themselves. As a consequence, the patterns suggest general principles and useful guidelines of wide applicability. Indeed that is a salutary conclusion for planning, natural systems, and society.

8

Urbanization models and the regions

Suppose you placed a small oval rug-sized aerial photo of your favorite city in the center of a huge room, and invited a group of children to paint a "mural" on the floor. Lots of trees and buildings and roads and playgrounds and people and farmland and water might appear. Some children may work together to portray a neighborhood, while others play follow-the-leader and create repeated patterns. Most would probably paint their own idiosyncratic visions. The resulting mural could be artistically delightful. But would it be a promising model for your city, as it spreads outward progressively meshing with surrounding land?

The preceding three chapters have highlighted existing patterns of urban regions worldwide. A rich array of sites and resources appears in the wide ring of land surrounding our metropolitan areas. In this chapter on urbanization we turn to change, especially changing spatial patterns and their consequences, as metropolitan areas spread outward (Godron and Forman 1983, Forman and Godron 1986, Forman 1995, Lindenmayer and Fischer 2006). This dynamic view of an urban region is central to ecological understanding and to wise planning. A static or constant world is impossible. Maintaining valuable human and natural resources over time requires flexibility, even adaptability, to get through gradually changing conditions. Gradual urbanization is one of the most conspicuous changes around cities (Turner *et al.* 1990, Meyer and Turner 1994, Germaine *et al.* 1998, Schneider *et al.* 2003).

In essence, *urbanization* refers to densification and outward spread of the built environment. People and buildings become denser within a metropolitan area in several familiar ways, such as infilling on vacant lots and converting single-housing units to multiple-unit housing, low- to high-rise residential, and residential to mixed commercial–residential areas (Vigier 1997, Kuan and Rowe 2004, Chen *et al.* 2004, Ozawa 2004). Shanghai, Hong Kong and Singapore are striking examples of *densification* (Kuan and Rowe 2004, Wu 2006). Other internal changes in urban land use, such as park establishment, gentrification, infrastructure construction, and industrial development, could be included in the urbanization concept even if they involve no change in population density (Schwartz 2004). These internal structural changes in a city or metropolitan area tend to be quite important in how it works (Hall 2002). The concept of urbanization used here is more specific than some definitions, such as the spatial diffusion of people that creates new landscape patterns, or the transformation of landscapes formed by rural life styles into urban ones (Antrop 2000).

With this book's focus on urban regions, particularly beyond the city, the *outward spread* or expansion component of urbanization is primary. Numerous expansion patterns of the built environment outside a metropolitan area have been described, and others are possible. Therefore, rather than attempting to briefly address numerous types of development, we focus on the few central widespread urbanization patterns for which others seem to be variations.

Straight-forward analyses and simple spatial models to understand urbanization are an important goal of this chapter. These analyses and models lead to quite interesting results and principles, the other primary goal. The process and results presented are based on a landscape ecology perspective, 38 large-to-small cities on all continents, a dual focus on natural systems and their human uses, and 18 useful informative attributes in urban regions. Additional perspectives, cities, major dimensions, and detailed variables should be explored in future work to see which, if any, patterns and principles should be refined or replaced.

Land-change patterns and models

Imagine kangaroos hopping around our kitchen, while we carefully pour a glass of wine, measure a bit of salt for the soup, and examine some small grape-sized objects on the floor. The place would change and not for the better. Surely we would notice and do something. Analogously, the urban region is changing. Unnoticed? Noticed too late? An inevitable result of the tyranny of small decisions (Odum 1982)? Growth and market economics will take care of any problems (Chapter 3)? Urban regions are too big and complex to do anything? Or? Looking at the process of urbanization is a useful start.

Common urbanization patterns on the land lead off this section. Spatial models are then introduced to help understand the urbanization patterns. Finally, a section on attributes for evaluating patterns and models highlights the importance of number, type, and breadth of measures for evaluating quite-different urbanization alternatives.

Common urbanization patterns

Two major factors, physiography and planning, provide constraints on urbanization and lead to the characteristic spatial patterns of urban expansion (Fainstein and Campbell 1996, Simmonds and Hack 2000, Antrop 2000, Berger 2006). The prime physiographic constraints are mountains and water-bodies, especially sea, estuary, major river, and large lake or wetland. Planning, using legal and/or enforced constraints, creates diverse spatial patterns such as green-belts, green wedges, and transportation corridors.

With no major constraints, concentric growth is the archetypal form of outward urbanization. From the rounded perimeter of a city or metropolitan area a relatively equal amount of expansion over time occurs in all directions. Three other common growth patterns should be mentioned in the context of no constraints. First *infill* builds on unbuilt spaces embedded in the metropolitan area. Second, for metro areas with green wedges present (Chapter 4), urbanization creates a compacting or rounding pattern that fills in the wedges (e.g., Seoul, Melbourne). Third, a sequence of expansions in different directions tends to maintain a rounded form with bulges (e.g., growth of London up to 1830 [Turner 1992]). The *bulges pattern* or model of urbanization seems more characteristic than the rounded pattern, because investment for development is more likely targeted to a specific area on a metro-area perimeter than to a narrow strip along the perimeter.

Still more common patterns result from urbanization in the presence of planning and physiographic constraints (Barker and Sutcliffe 1993, Warren 1998, Pandell *et al.* 2002, Ozawa 2004, Ishikawa 2001, Clark 2006). Green wedges, green-belt, urban growth boundary, or ring of parks may be present near a metropolitan area. *Green wedges* are unbuilt greenspaces projecting into the metro area (e.g., Stockholm, Copenhagen, Melbourne). A *greenbelt* is a protected band of greenspace separating major land uses, such as a city and its surrounding agricultural or forest land (London, Seoul). An *urban growth boundary* is a "line in the sand" beyond which urbanization is prohibited or can only proceed at a much reduced rate (Portland). Unlike an urban growth boundary, urbanization may occur outside a greenbelt zone. A *ring of parks* is analogous to a greenbelt except that transportation corridors and nearby development radiate outward from the city leaving, for example, four to eight large parks in a ring and separated by highway corridors (Budapest). The Seoul greenbelt might become a ring of parks (Color Figure 35) (Bengston and Youn 2006).

More-distant patterns of urbanization are characteristically near transportation corridors (Figure 8.1) or satellite cities (Browder and Godfrey 1997, Simmonds and Hack 2000, Antrop 2000, Schneider *et al.* 2003). Development

Figure 8.1 An early stage of strip development along a radial highway. Note that the existing clusters of buildings in lower left and upper center are now connected by a line of houses on very large lots that further bisect and degrade the central farm-field area. North of Gainesville, Florida. R. Forman photo courtesy of L. D. Harris.

along major transportation corridors radiating from a city produces an over-all pattern reminiscent of a star with long points or a wheel-hub with spokes (e.g., Chicago 1850 to 1967 [Schmid 1975]). Strip development along a major radial transportation corridor progressively subdivides the landscape or region. Focusing urbanization concurrently around satellite cities in an urban region increases somewhat the size and importance of these small cities, while helping to protect the land near a large metropolitan area (Barcelona), which may be of considerable ecological importance (van der Ree and McCarthy 2005). If development is targeted to several or many satellite cities, rather than a few, the population will be considerably dispersed and a much more complex transportation network will probably develop to criss-cross the land.

The dispersed pattern of urbanization is more complex (Jenks *et al.* 1996, Gordon and Richardson 1997, Theobold *et al.* 1997, Bullard *et al.* 2000, Hobbs and Theobold 2001, Jenerette and Wu 2001, Hobbs and Miller 2002, Lopez 2003, Berger 2006, Kahn 2006). Most characteristic is where dispersed development, rather than compact concentric-zone expansion, occurs outside a metropolitan area. This produces a wide zone of relatively low-density development, or sprawl, around the city (e.g., many North American cities). But dispersed development can also be combined with the bulges, greenbelt, transportation corridors, and satellite cities patterns. Irrespective, the dispersed development pattern is associated with a massive fine-scale road net, which in turn is connected to the major radial (and ring road, if present) transportation network.

Various factors affect whether development is dispersed or compact (Yaro *et al.* 1990, Troy 1995, Theobold and Hobbs 1998). For instance, the use of septic systems for human wastewater facilitates dispersed residential development. In contrast, in Australia, Germany and certain other nations, homes in a development near a city cannot be built until after the sewer and other utilities have been installed, an effective way to create compact growth and neighborhoods. With a percieved financial gain, typically any town or municipality can expand, whereas in Britain, to protect especially valuable land, many communities receive a government subsidy not to expand.

The much-discussed *causes of urbanization* bear mention since some strongly affect the location and degree of development (Theobold and Hobbs 1998, Hansen 2002, Hall 2002, Burgi et al. 2004). Consider a large flood or cyclone (hurricane) or earthquake that causes extensive building destruction (e.g., San Francisco 1906, New Orleans 2005). The area destroyed is usually rebuilt, though perhaps in a form better able to deal with a repeat disturbance. Some of the residents affected, however, may relocate to a more secure location outside the metropolitan area. Or consider rising sea level associated with climate change. Low-lying urban areas near coasts may become largely inundated, causing some residents and businesses to relocate to the outskirts of the city (e.g., Bangkok, San Diego/Tijuana). Immigration may stimulate urbanization around certain communities where arrivals from particular nations or cultures wish to aggregate. Immigrants arriving without financial resources often live in squatter settlements within a city or near its perimeter (Perlman 1976, Main and Williams 1994), and may later move outward as land prices rise. Transportation corridors, both radial and ring-road, are especially associated with urbanization. For instance, adding a radial highway, or even new traffic lanes to an existing one, is apt to produce an urbanized bulge in that direction. Constructing a ring highway around a metro area facilitates concentric-zone or dispersed development over extensive areas.

Models for urbanization

Since the causes of urbanization vary widely and numerous recognizable repeated patterns are produced by the process, modeling has been frequently used to analyze such a complex system. A sampling of several types of models to understand urban expansion is outlined below. Much of the literature focuses on the causes, mechanisms, and processes of urbanization. An example is the set of *spatial processes*, i.e., perforation, dissection, fragmentation, shrinkage, and attrition/disappearance, that act on the existing land mosaic to produce changing landscape patterns (Forman 1995, McIntyre and Hobbs 1999, Lindenmayer and Fischer 2006). In contrast, a prime interest here is to provide a foundation and framework to understand the implications of different urbanization patterns. This will facilitate analyses of urban regions worldwide later in the chapter. Furthermore, we will attempt to identify the optimum urbanization pattern for people and nature.

Concentric zones around a population center are the usual starting point for modeling land-use pattern and change (Christaller 1933, Losch 1954, Covich 1976, Haggett *et al.* 1977, Antrop 2000), an approach based on central-place theory and its different-width zones of influence. Population centers or nodes are interconnected by transportation routes and a hierarchy of nodes (village to city) and routes (local roads to major highways) develops. Competition for space among population nodes produces a regular pattern sometimes reminiscent of, and modeled as, a multi-scale hierarchy of hexagons.

Elaborations and competing models have inevitably evolved (De Blij 1977, Forman and Godron 1986), including: a sector model of land-uses organized like wedge-shaped pieces of a pie; a multi-nodal model with separated growth nodes superimposed on sectors or concentric zones; and multiple star-shaped nodes in a hierarchical pattern produced by heterogeneity of the natural environment, regional history, and communication networks (Antrop 2000). A "rotating-sector model" where land uses rotate or alternate in pie-shaped slices around a point mimics the open-field landscape in Europe's Iron Age (Orwin and Orwin 1967, Rackham 1980, Forman and Godron 1986), as well as a sequence for forest cutting (Harris 1984, Peterken et al. 1992). Many other spatial models of change have been used in landscape ecology (Baker 1989, Sklar and Costanza 1990, Zonneveld and Forman 1990, Mladenoff 2005, Verboom and Wamelink 2005).

Several early types of urban-growth models, some mathematical, link transportation and land use (Forman *et al.* 2003, Berling-Wolff and Wu 2004): (1) gravity models for interaction between cities; (2) location-of-work models; (3) journey-to-work models; and (4) 1960s transportation models for Detroit and Chicago, still used in the USA to evaluate potential regional air-pollution effects. Newer approaches offer a richness of ways to think about changing land (Berling-Wolff and Wu 2004, Wiens and Moss 2005), though few have been directly used to model urbanization: (a) dynamic modeling techniques; (b) cellular automata; (c) spatial-statistics models; (d) GIS and visualization techniques; (e) ecological process models; (f) fractals (Milne 1991a, 1991b); (g) ecological energetics; (h) fuzzy-logic theory; and (i) neural-network theory. Another approach used in a Colorado (USA) study found that the likelihood of development correlates better with local patterns of existing development (described by spatial transition models) than with the traditional factors of proximity to highways, towns, and urban areas (Theobold and Hobbs 1998).

Simulation models that mimic changing landscape patterns offer particular promise for understanding urbanization (Franklin and Forman 1987, Li et al. 1993, Swanson et al. 1994, Collinge and Forman 1998, Forman and Mellinger 2000). Modeling *mosaic sequences*, i.e., changing spatial patterns, permits one to directly compare alternatives, and hence identify optimum or best options. This approach, which will be applied to urbanization sequences later in the chapter, is illustrated by the process for identifying the ecologically optimum spatial sequence for changing a large landscape from a more suitable land-use type (e.g., forest) to a less suitable type (e.g., desertified area).

First a literature review of forest cutting, suburbanization, desertification, agricultural spread, and other broad-scale phenomena pinpointed 30 actual mosaic sequences or changing spatial patterns in land transformation (Forman 1995, in collaboration with George F. Peterken). Five simple spatial models simulated the bulk of these land transformations: (1) edge model; (2) corridor model; (3) nucleus model; (4) few-nuclei model; and (5) dispersed-patches model. These mosaic-sequence models were directly compared by recording levels of several spatial attributes of ecological importance related to patch size, connectivity, and boundary length in a hypothetical landscape.

The *edge model*, whereby parallel strips are progressively degraded from one side to the opposite side of the landscape, was found to be the best ecologically (it retained attributes of the more-suitable initial land type furthest through the land transformation process). In contrast, the *dispersed-patches model*, where degradation occurs in small patches evenly and progressively dispersed across the land, was worst ecologically. These results apply to most of the actual land-use transformations observed in landscapes worldwide.

However, the edge model is not ideal. Two changes effectively create an ecologically better "jaws model" (Forman 1995, Forman and Mellinger 2000). First, instead of degrading parallel strips progressively from one side of the landscape, L-shaped strips beginning from two adjacent sides are progressively degraded toward the opposite corner (analogous to wide-open jaws moving across the land). Second, instead of creating a progressively larger continuous degraded area, scattered small patches and corridors to be protected across the land are established at the outset, and remain until the final phase.

Finally, one improvement in the jaws model makes what currently seems to be the ecologically optimum model for shaping landscape change, the so-called *jaws-and-chunks model* (Forman and Collinge 1996, Forman 2002a). In the middle phase of the mosaic sequence (e.g., from 70 % to 20 % of the initial more-suitable land type), instead of progressively shrinking a single huge natural patch, a few large natural patches are created (to spread risk), which are then sequentially rather than synchronously shrunk and eliminated. Thus the overall jaws-and-chunks mosaic sequence resembles wide-open jaws progressively moving from a corner and consuming more-suitable land, then creating and sequentially removing chunks of suitable land in the middle phase, and finally consuming the last chunk and the scattered bits of suitable land in the last phase.

Attributes for evaluating patterns and models

Change is the norm, a process illustrated by sunrise, spring, the first snowfall, and a warthog consumed by a crocodile. Urbanization, like the warthog–crocodile transformation, has advantages and disadvantages. A town in an urban region becomes jammed with traffic and loses its rural character, yet concurrently road access, cultural diversity, and economic opportunity grow. A wetland that cleans streams of pollutants and maintains rich biodiversity is drained, yet that loss eliminates malaria-carrying mosquitoes and reduces problems related to soggy soils in adjoining neighborhoods.

Thus evaluating an urbanization pattern, or model of it, requires both identifying the most important variables affected, and balancing the pros and cons. The other key decision to make in choosing variables is how general or specific they should be. Specific detailed variables are more numerous and individually may provide more precise response patterns. But determining the right variables to measure and interpreting large numbers of dissimilar specific responses normally are extremely difficult. Using broader indices that combine or integrate specific ones reduces the complexity of detail and thus may lead to discovering broad patterns. However, teasing apart the detailed causes or mechanisms may be difficult. Nevertheless, a lucid big picture, as well as major variables and handles for planning, will probably never appear with a detailed-variables approach. Broader variables or indices are required for that.

The process used in identifying ecologically optimum landscape change (see previous section) appears useful for analyzing urbanization. Rather than comparing the five mosaic sequences of landscape change using numerous specific ecological characteristics, such as species richness, stream structure, groundwater quality, interior species populations, and so forth, broader spatial attributes were measured for evaluating landscape change (Forman 1995). The ten spatial attributes chosen included total patch-interior area, grain size of the landscape, ability to cross the landscape, total boundary length, and patch density. Conveniently the spatial attributes fell into three broad useful categories: (1) patch size or landscape grain size; (2) connectivity; and (3) boundary length.

Based on considerable landscape ecology and other literature, numerous specific ecological characteristics were known to correlate with these broad spatial attributes. Therefore the limited number of broad spatial attributes served as convenient informative indices or surrogates for groups of detailed ecological characteristics.

For comparing and evaluating alternative urbanization patterns, this two-tiered approach, using broad spatial attributes as indicators of detailed ecological characteristics, appears promising. However, in the case of urbanization the attributes chosen must effectively measure effects on two key dimensions, natural systems and their human uses.

Outward expansion of the built environment produces obvious effects in the urban-region ring, but also some effects within the metropolitan area that warrant evaluation. For example, a semi-natural area in the city is enriched by periodic arrivals of species from the countryside. If urbanization expands outward in concentric zones, the semi-natural area becomes progressively more isolated, less enriched by species arrivals, and more impoverished (Soule 1991). Also, concentric-zone urbanization outside of green wedges is likely to simply convert them into embedded elongated greenspace patches. That severs connection to surrounding countryside for species, and truncates recreational opportunities for people living nearby in the city. The degradation of both biodiversity and recreational use in patches progressively more embedded in urban land is widely illustrated: Torrey Pines in San Diego/Tijuana; Batu Caves in Kuala Lumpur; Stone Mountain in Atlanta; Mill Creek Park in Edmonton (Saley *et al.* 2003, Wein 2006); desert tortoise habitat in Las Vegas (Beatley 1994); and Montjuic in Barcelona (Boada and Capdevila 2000, Forman 2004a).

Finally, the area of the urban-region ring itself is relevant in evaluating urbanization options. Outward urbanization in small urban regions (e.g., Philadelphia) may quickly reach the boundary of the urban region. In some cases an urban region boundary can expand outward, but in other cases it is constrained in place by an adjacent urban region or a high mountain range. Indeed, urbanization from a city may affect an adjacent urban region, and vice versa.

Four urbanization models

From the range of urbanization patterns and models introduced in the previous section, and based on accumulated literature, map-and-image analyses and personal observations, four models emerged for evaluation. These four – (1) concentric zones; (2) satellite cities; (3) transportation corridors; and (4) dispersed sites – are quite different, and may represent nearly all major global trends in outward urbanization pattern (Figure 8.2). The four core models have

Concentric-Rings Model   **Satellite-Cities Model**

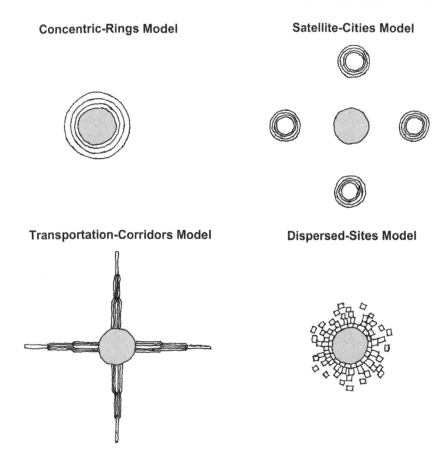

Transportation-Corridors Model   **Dispersed-Sites Model**

Figure 8.2 Four major spatial models of urbanization spread. Metropolitan area is the central circle; small open circles in upper right are satellite cities including their adjacent built area. As illustrated in a model, the total area urbanized after each of three time stages equals approximately 1.5 times, 2 times, and 3 times the initial metropolitan-area size. For illustration convenience in lower right, each new developed area equals about 2.5 % of the initial metro area. See text for quantitative rules (algorithms) of urban expansion.

parallels, both in mosaic sequence and analytic approach, to the edge, nuclei, corridor, and dispersed models used to understand landscape change (Forman 1995). In the succeeding sections, these urbanization models will be applied in some detail to two case studies (Mexico City, Nantes), and then more broadly to all of the 38 cities.

A few key assumptions or parameters apply to all four models. The mosaic sequences or land-cover changes for each model are illustrated at three stages or time periods, i.e., when the total urbanized area reaches: (1) 1.5 times, (2) 2 times, and (3) 3 times the initial size of the metropolitan area (an increase in built area by 50 %, 100 %, and 200 %, respectively). Urbanization of course produces quite different patterns around different cities, and tends to occur in pulses rather than at a constant rate. The three stages or levels of urbanization chosen are based, in part, on several US and Canadian cities that showed a 1.5 to 2 times increase in built area in a recent two-decade period (Forman *et al.* 2003). Another consideration was the growing interest in raising the time horizon for urban plans from the characteristic 3, 5, 10, or 20 yr to the decades-to-generations timescale of sustainability.

Additional model assumptions are important. All land, including mountains, parks, military bases, and large wetlands, can be urbanized; only major water bodies cannot be urbanized. Only two types of land cover exist in the models, built space and greenspace. No infill building on greenspace patches or corridors enclosed in a metropolitan area occurs. A final assumption applies to cities with green wedges, here defined as greenspaces that project into a metropolitan area at least one-third of the radial distance to the city center. Since a strong reason normally exists for the persistence of a wedge, in the models existing greenspace wedge areas remain and cannot be urbanized. However, wedges are not extended outward as urbanization expands, so in certain models development beyond green wedges leaves them as elongated greenspace patches within an expanded metro area.

The extreme complexity of existing greenspace boundaries and patchily dis-tributed development around the 38 worldwide urban regions precluded the feasibility of precise computer mapping of mosaic sequences on the region maps (Color Figures 2–39). Consequently, precise areas were calculated for each model, and approximate areas and boundaries were hand-drawn on maps derived from the Color Figures with sites (pictograms) and most land covers removed. This provided an efficient and effective way to directly visualize and compare the four broad urbanization models.

The mosaic sequences of the concentric, satellite, transportation, and dis-persed urbanization models (Figure 8.2) to be used below are outlined as follows.

Concentric-zones model mosaic sequence

All urbanization, except as noted above for green wedges, is adjacent to and evenly distributed around the metropolitan-area border. In the first stage of the concentric-zones model (reaching 1.5 times the initial size of metro area), urbanization "parallels" the broad outline of major built lobes and intervening greenspaces (which are wide, irrespective of how far outward or inward they project) of the metropolitan area. Finer-scale lobes and coves of the metro-area border are ignored. In the second and third stages (2 times and 3 times, respectively), outward urbanization progressively smoothes the metro-area border creating a compact rounded or oval form.

Satellite-cities model mosaic sequence

All urbanization occurs around a few (four in this case) satellite cities at equal rates and in concentric-zones form. Consequently, each satellite receives a quarter of the total urbanization at each stage, and the metropolitan area and its border remains unchanged.

The selection of satellite cities for urbanization is governed by the following priorities in order. Satellites are: (1) in the four compass quadrants and relatively equidistant from one another; (2) in the outer half of an urban-region ring, i.e., nearer the urban-region boundary than the metro-area border; (3) connected to the metropolitan area by a main road; (4) chosen from existing small cities, or in their absence, towns.

Some explanations for the priorities are helpful. The distribution of all other features, sites, and characteristics in the region was ignored in selecting satellite cities and main roads. For coastal cities, satellites are relatively equidistant though often not in four compass quadrants. Selecting satellites far from the metropolitan area means that the satellite and metro area remain separate as urbanization proceeds, and that the satellite-city model is clearly differentiated from the concentric-zones and dispersed-sites models. The presence of a main road between satellite city and metropolitan area helps make the satellite a logical place for regional urban growth, and enhances direct comparison of the satellite-cities model with the transportation-corridors model which uses the same main roads. Finally, small cities may be better prepared for urban expansion than are towns, due to the presence of urban infrastructure, park system, and administrative structures.

Transportation-corridors model mosaic sequence

All urbanization occurs adjacent to a few (four in this case) main radial highways connecting the metropolitan area with satellite cities, and progressively extends at equal rates outward from the metro area. From a highway, development extends outward in parallel bands, half on each side.

Transportation routes connect the metro area with the relatively equidistant satellite cities selected in the satellite model. If no main road exists, a straight-line route is established connecting a satellite city either to an existing main road at a logical point or directly to the metro area. This is consistent with the expectation that if a satellite city grows, either existing secondary roads are likely to be upgraded or a new road will be built to the metropolitan area.

For the transportation-corridors model, the distance from metro-area border to the satellite city is divided into three equal segments. In the first time stage (reaching 1.5 times metro area), the road segment nearest the metro area receives twice as much development as the middle segment, which in turn receives twice as much as does the farthest road segment near the satellite. In the final third stage (3 times metro area) all three road segments receive the same amount of development. The second intermediate stage (reaching 2 times) is half way between the first and third stages in distributing development along the trans-portation corridor.

Thus in the model, 41 % of the urbanization occurs along the road segment nearest the metro area, 32 % along the middle segment, and 27 % along the distant segment. This mosaic sequence of progressively extending outward, both radially from the metro area and in parallel bands from the highway, reflects the initial prominence (or source effect) of the metro area followed by the growing prominence of strip development along the road and influence of the satellite city.

Dispersed-sites model mosaic sequence

All urbanization occurs in equal-sized small patches dispersed outside the metropolitan area, and number of development patches per time period decreases with distance from the metro-area border. This dispersed-patch, low-density development model mimics sprawl.

For the model, five "concentric" bands around the metropolitan area are out-lined, somewhat analogous to zones in the concentric-zones model. Band sizes are scaled to the initial metro-area size, so the area of the first band equals half the metro area, and area of each of the four outer bands equals the metro area. The size of development patches is also scaled to the metro-area size, reflecting an assumed rough correlation between city size and average development size. In the model, each patch of development equals 2.5 % of the initial metro area. Doubtless this is much larger than reality, but it significantly facilitates calculation and mapping, and effectively illustrates the dispersed-sites model in contrast with the other urbanization models.

The first time stage (which reaches 1.5 times the metro area) distributes the small dispersed patches as follows: 50 % in band one, 25 % in band two, and 25 % in band three. The second stage (2 times the initial metro area) distributes 25 % of the development patches in each of the bands one, two, three, and four. The final third stage (reaching 3 times) distributes 25 % of the patches in each of bands one, three, four, and five. Development patches are dispersed within a band and between bands so that they are relatively similar distances apart.

Thus at the end of the third final stage the innermost band one is 100 % developed, band two 50 %, band three 37.5 %, band four 25 %, and the outermost band five is 12.5 % developed. The innermost band in this dispersed model is completely urbanized after the third stage, just as the equal-sized inner zone of the concentric model is fully developed after the first stage.

Models applied to case studies

What can we discover by applying the four urbanization models just described to the patterns of real cities? Which is the best way to spread outward – as concentric zones, satellite cities, transportation corridors, or dispersed sites? Which is worst? The answers here must be based on sustaining natural systems and their human uses. To answer these questions, this section examines two urban regions in some detail, and the following section compares all 38 urban regions more broadly.

Mosaic sequences generated by the four models are first mapped on the urban regions of Mexico City and Nantes (France). Mexico City has a large population (8 590 000) and a metropolitan area of 1606 km^2. Nantes is relatively small in both population (278 000) and metro area (171 km^2). Then the four major urbanization options are directly compared by recording the amount of key sites or areas (Color Figures 22 and 25) affected at each stage for each model. Two of the many nature-and-people variables on the urban region maps are selected for evaluating the urbanization models for these cities: (1) biodiversity sites, and (2) rivers/major streams. Biodiversity sites (described in Chapter 5) are simply counted. The second variable refers to the total length of rivers and major streams in the region that are surrounded (to c. 3 kilometers on each side) by natural vegetation. In short, the urbanization options are evaluated using four alternative models times three stages of urbanization times two urban regions times two variables of importance for natural systems and human uses.

Maps of the four mosaic sequences in the Mexico City region are strikingly different (Figure 8.3). Development in the concentric model completely covers the area close to the city. The satellite-model development covers four small outer areas. The transportation-model development covers four long radial spokes or finger-like areas. And the dispersed-model development extends at low density across a large area around the city. The same results are evident for Nantes, a relatively small city (Figure 8.4).

Chicago was originally included as a case study because, worldwide, it is a medium-size city in population (2 896 000), has the largest metropolitan area of all 38 regions analyzed (3993 km^2) (see Table 5.1), and contains considerable sprawl. However, the huge metro area posed problems for applying the

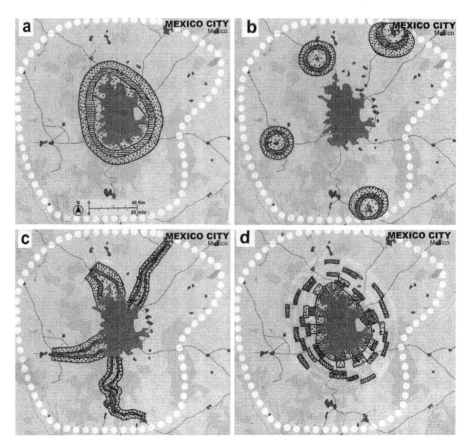

Figure 8.3 Four alternative urbanization models applied to the Mexico City Region: (a) Concentric rings; (b) satellite cities; (c) transportation corridors; and (d) dispersed sites. Scattered dots = urbanization in first time stage; shading = second stage; dense dots = third stage. See Figure 8.2, Color Figure 22, and text.

urbanization models, so Chicago is only briefly included in this section. To the north and northwest, the Chicago Region (Color Figure 13) is constrained by the region of nearby Milwaukee and by the State of Wisconsin. Only for the concentric-zones model will the three time stages logically fit within Chicago's urban region. For the satellite-city model and transportation-corridors model the first stage fits fine, but the second stage (2 times the initial metro area) only fits with "gerrymandering" (drawing a bizarre-shaped area so something fits in), and the third stage (3 times) with extreme gerrymandering. For the dispersed-sites model, the first stage fits fine, but the second and third stages cannot fit at all (stage two requires an area 4 times and stage three 5 times the initial metro area). Relative to the metropolitan area extent, the urban region is simply too small.

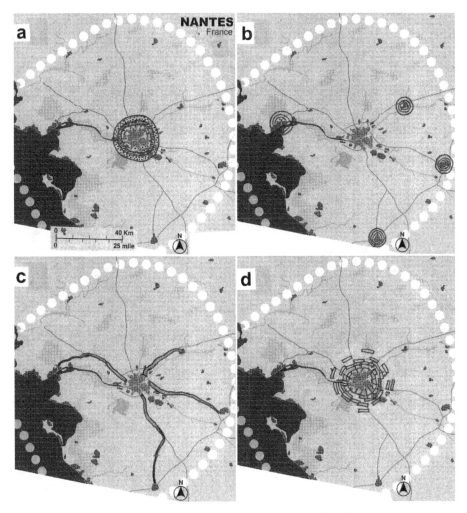

Figure 8.4 Four alternative urbanization models applied to the Nantes (France) Region: (a) concentric rings; (b) satellite cities; (c) transportation corridors; and (d) dispersed sites. Scattered dots = urbanization in first time stage; shading = second stage; dense dots = third stage (for clarity, only included in upper left). See Figure 8.2, Color Figure 25, and text.

Of the 38 worldwide regions, two other cities, both in the USA, have an unusually large metropolitan area relative to the urban region so the dispersed-sites model will not fit. Philadelphia, a medium-population-size city, has a particularly small urban region, and Atlanta, a small-population city, has an extensive metro area only exceeded in area by Chicago (see Table 5.1; Color Figures 3 and 27).

Table 8.1 Case study evaluation of urbanization models using two attributes in two urban regions. Mosaic sequence maps for Mexico City and Nantes (Figures 8.3 and 8.4) are superimposed on Color Figures 22 and 25, and the number of biodiversity sites and length of rivers/major streams covered by development is recorded for each of the three time stages portrayed by a model. For each attribute in each region the models are ranked from best to worst (**1** to **4**) based on the amount (low to high) of an attribute affected by urbanization, i.e., covered by development.

Attribute measured	Concentric-zones model				Satellite-cities model				Transportation-corridors model				Dispersed-sites model			
	Stages				Stages				Stages				Stages			
Urban region	1st	2nd	3rd	Total	1st	2nd	3rd	Total	1st	2nd	3rd	Total	1st	2nd	3rd	Total
Biodiversity sites (number affected by urbanization)																
Mexico City	1	2	5	8	1	0	1	2	1	2	5	8	3	4	3	10
Model ranking				**2.5**				**1**				**2.5**				**4**
Nantes, France	0	1	2	3	0	0	0	0	0	0	2	2	2	1	3	6
Model ranking				**3**				**1**				**2**				**4**
Rivers/major streams (km length affected by urbanization)																
Mexico City	0	0	0	0	23	11	14	48	1	1	11	13	0	0	2	2
Model ranking				**1**				**4**				**3**				**2**
Nantes, France	5	10	9	24	6	4	4	14	0	.5	.5	1	8	5	15	28
Model ranking				**3**				**2**				**1**				**4**

These apparent mismatches between metro-area and urban region may be harbingers of challenges ahead for the three cities.

For both of the case studies, Mexico City and Nantes, biodiversity sites in the region are most affected or degraded in the dispersed-sites model (Table 8.1). They are least covered by development in the satellite-cities model. Thus the satellite-cities model is deemed best, and dispersed-sites model worst, for biodiversity sites. Results for rivers/streams are different than those for biodiversity sites, and indeed effects on rivers/streams differ between regions. For Nantes the effects on biodiversity sites and rivers/streams are quite similar (Table 8.1). Based on these two variables, in this urban region the satellite and transportation models are better, and the dispersed and concentric ones worse strategies for urbanization.

The reasons for some of these results are revealed in characteristics on the land. Differences for biodiversity-site effects seem related to a concentration of sites by water close to the Nantes metro area, and a scarcity of sites in the mid and outer agricultural portions of the Mexico City Region (Color Figures 22 and 25). Differences in river/stream effects appear related to the location of Nantes at the confluence of rivers, in contrast to the lowered water-tables and dried-out river beds near Mexico City. Indeed cities began where dispersed natural systems were accessible in a day's walk, many now being threatened by urbanization.

In brief, the case studies have detected intriguing patterns in response to the alternative spatial models. Changing geometry is a central theme for understanding urbanization, but also specific physiographic and land-use patterns may help explain variations in response pattern. Yet how representative are the results based on two extremely different cities and two dissimilar variables? Broader response patterns at a global scale may transcend these detailed case studies.

Urbanization options evaluated with 18 attributes and 38 regions

Several land-cover types and >40 site types represented by pictograms are mapped on the 38 urban regions (Chapter 5 and Color Figures 2–39). From these, 18 diverse, particularly informative attributes are selected for evaluating the four urbanization models applied to all the regions. Most of the other promising attributes are excluded because of limited sample sizes. Many of the attributes chosen are direct measures of suitable conditions for natural systems or their human uses (e.g., biodiversity sites, recreation sites, flood hazard). Others are included as indirect measures known to correlate with direct factors (e.g., drainage area protection, development on nearby slopes, distance from major highway).

The big picture emerges by examining a summary of the best-to-worst urbanization models for the 18 attributes of 38 urban regions (Table 8.2). Certainly

Table 8.2 Summary effects of urbanization models on 18 attributes measuring natural systems and their human uses. Four alternative spatial models (Figure 8.2) were superimposed on the 38 urban region maps (Color Figures), and attributes covered by urbanization were counted or estimated at each of the three time stages in a model. Amount recorded is an index of effect or degradation due to urbanization, and the four models were ranked accordingly from best to worst (1 to 4) for an attribute in a region (see Appendix II with rankings for all attributes and regions). Number of urban regions measured excludes those where an attribute is absent or scarce on a region map. Attribute average ranking is for all urban regions measured. Best to worst model (Table columns) is based on comparing attribute average rankings. For asterisks, see further information in Appendix II.

Attribute measured	Number of urban regions	Concentric-zones model		Satellite-cities model		Transportation-corridors model		Dispersed-sites model	
		Attribute average ranking	Best to worst model	Attribute average ranking	Best to worst model	Attribute average ranking	Best to worst model	Attribute average ranking	Best to worst model
Biodiversity sites (%)	37	2.51	2	1.78	1	2.93	4	2.91	3
Recreation/tourism sites (%)	37	2.38	2	1.69	1	3.16	4	2.77	3
Forest/woodland*	35	1.81	1	2.53	2	2.94	4	2.71	3
Grassland/pastureland*	10	1.95	1	3.45	4	2.6	3	2	2
Desert/desertified area*	4	2	1	2.63	3	2.88	4	2.5	2
Nearby slopes facing city (% cover)	19	3.39	4	1.08	1	2.16	2	3.37	3
Rivers/major streams (km length)*	38	2.32	2	1.89	1	2.82	3	2.97	4

Attribute									
Major wetlands (% cover)	17	2.38	2	2.14	1	2.76	4	2.71	3
Flood hazard sites (%)	12	2.63	2.5	1.54	1	3.21	4	2.63	2.5
Marine coast (km length)*	14	2.04	1	2.71	4	2.61	2	2.64	3
Reservoirs/lakes (% & shoreline km)*	30	2.33	2	2.27	1	2.68	3	2.72	4
Drainage area for water (% cover)*	19	3.08	3	1.45	1	2.29	2	3.18	4
Market-gardening area (% cover)	10	2.75	3	1.95	1	2.5	2	2.8	4
Average distance to major highway (km)*	15	2.2	2	3.57	4	1.13	1	3.1	3
Degree of subdividing region (high to low)*	38	2	2	1	1	4	4	3	3
Edge density (km length per km^2)*	38	1	1	2	2	3	3	4	4
Center city to metro-area border (km)*	38	4	4	1	1	2	2	3	3
Other attributes combined*	10	3.2	3	1.4	1	1.8	2	3.6	4
Overall average		2.44		2		2.64		2.92	
Total number of 1s			5		12		1		0
Total number of 4s			2		3		7		6

results vary from attribute to attribute (Appendix II also shows variability from region to region) as expected in a relatively comprehensive analysis. Nevertheless, the total number of 1s (best models) suggests that the satellite-cities and concentric-zones models are better than the other two (12 and 5 versus 1 and 0, respectively) (Table 8.2). The overall averages furthermore suggest that the satellite cities approach is the optimum (2.00 vs. 2.44, 2.64, and 2.92).

Analogously the dispersed sites approach is apparently the worst for urbanization (no best-model 1s, six worst-model 4s, and the highest overall average, 2.92) (Table 8.2). Of the two intermediate cases, the concentric-zones model (five best-model 1s, two worst-model 4s, and overall average 2.44) is noticeably better than the transportation-corridors model (one best-model 1, seven worst-model 4s, and overall average 2.64). In essence, this broad analysis points to a rather clear ordering of the urbanization models from best to worst: (1) satellite cities; (2) concentric zones; (3) transportation corridors; and (4) dispersed sites.

These results suggest two guiding principles:

(1)   Regional urbanization in dispersed sites surrounding a metropolitan area, and to a lesser extent along transportation corridors, appears to cause extensive nature-and-human resource degradation, and thus should be avoided or minimized.

(2)   Urbanization focused around satellite cities, which causes the least overall resource degradation, appears to be the best regional development pattern, though factors specific to a region may indicate a preference for combining satellite-city development with concentric-zone development adjacent to a metropolitan area.

While the worldwide patterns provide a framework or foundation for understanding and planning, differences from the central pattern can be detected for certain cities. Based on the distribution of best and worst urbanization models (sums of 1s and 4s for each region in Table 8.2), no urban region is strongly at variance with the worldwide pattern. Half of the regions closely fit the broad pattern (Barcelona, Cairo, Canberra, East London, Edmonton, Erzurum, Kuala Lumpur, Moscow, Nairobi, Portland, Rahimyar Khan, Samarinda, Santiago, Sapporo, Seoul, Tegucigalpa, Tehran, and Ulaanbaatar).

Two regions (Chicago, Bucharest) only vary from the core pattern by having slightly more 1s for the concentric-zones than the satellite-cities model, though this is not inconsistent with the second guiding principle above. Four regions (Mexico City, Nantes, London, Brasilia) diverge from the central pattern by having slightly more 4s in the concentric and satellite category than in the transportation and dispersed category (though 1s are concentrated in the concentric and satellite category, as expected). Philadelphia perhaps diverges most from the worldwide pattern by having about the same number of **1**s and **4**s in both the concentric and satellite category and the transportation and dispersed category. Overall these divergences seem minor. The worldwide pattern for urbanization summarized in the guiding principles seems to apply well in virtually all the urban regions.

Table 8.2 also shows no correlation between urbanization-model response pattern (distribution of best and worst models) and either broad geographic area (Europe, Africa, Latin America, etc.) or city population size (see Table 5.1). These results parallel those found for numerous attributes associated with nature, food, water, built systems, and built areas in the urban regions (Chapters 6 and 7).

A closer look at the *specific attributes*, along with their response patterns, in this overall evaluation of urbanization options is useful (Table 8.2). The first two attributes, biodiversity and recreation/tourism sites, in general are widely distributed small areas of nature-and-people importance. Forest, grassland, and desert attributes primarily measure the broad urbanization effect on natural land in urban regions. Nearby slopes facing a city have diverse implications from aesthetics to landslides, erosion, sedimentation, city air ventilation, and proximity of species-source habitats. Six water-related attributes (rivers/major streams, major wetlands, flood hazard, marine coast, reservoirs/lakes, and drainage area for water supply) cover an extensive set of natural-systems-and-people issues central to the present and future of urban regions. Market-gardening areas in proximity to a city also reflect a range of benefits from food to economics and clean air. Distance from major highway reflects overall transportation costs and infrastructure effects on the land. Three attributes measuring diverse implications of urbanization (subdividing a region, edge density, and average distance from city center to metro-area border) produce identical patterns for all regions, which basically result from geometry (algorithms) in the four urbanization models. Subdividing a region with strip development disrupts regional connectivity for species movement. Edge density is an index of numerous effects of built areas on natural systems (Chapter 7). Average distance from city center to metro-area border measures how isolated city residents and city parks are from the surrounding countryside. Finally, "other attributes considered" is simply a mix of dissimilar characteristics measured – greenbelt, urban growth boundary, Native Peoples' land, aquaculture, and fire hazard – each with a fairly limited sample size. In short, this array of attributes evaluated (Table 8.2), like those highlighted in Chapters 6 and 7, nicely illustrates the regional focus on natural systems and their uses for society.

Especially high average rankings (1.0 to 1.9) or low rankings (3.0 to 4.0) for an attribute (Table 8.2) indicate considerable consistency, or low variability, in results from region to region. Thus the satellite-cities model is consistently the best among regions for 10 of the 18 attributes (biodiversity sites, recreation/tourism sites, nearby slopes facing cities, etc.). The dispersed-sites model, on the other hand, is consistently the worst among urban regions for seven of the attributes.

Is there a single attribute, or two or three, that produces results similar to that for the whole set of 18 variables? Rivers/major streams and reservoirs/lakes are the only two attributes with the same best-to-worst (**1**, **2**, **3**, and **4**) model results as for the whole set (Table 8.2). The actual averages, as well as the consistency in results across regions, for the *rivers/streams attribute* are closest to those of the total set. Measuring the length of rivers and major streams affected or degraded by development may be a particularly useful measure to differentiate mosaic-sequence models and to evaluate the overall effects of urbanization.

Assaying a global set of cities with 18 informative variables may seem ambitious, even sufficient. Yet other variables or considerations may be important for the models as the following cases illustrate.

Concentric-zones model: additional dimensions

Expanding outward from a metropolitan area would likely increase the heat-island effect for cities where that is a problem. The concentric-zones trajectory would also be environmentally detrimental for nearby slopes facing a city, and would leave city center residents increasingly isolated from greenspace of the urban-region ring (Table 8.2). Models that eliminate greenspace wedges in a metro area, or alternatively, continue green wedges outward as urbanization proceeds, would be interesting.

A bulges model instead of the concentric-zones model might produce little difference in response. However, if the bulges were on especially valuable greenspace, the concentric model would be better. Conversely, targeting development onto greenspace of low value should make the bulges approach even better than the concentric-zones model.

Satellite-cities model: additional dimensions

Concentrating growth around satellite cities is a dispersal process that may create new major cities in the outer urban-region ring where natural systems are most valuable. Or several rather than a few satellites could be nuclei for urbanization. This might create a pattern somewhat akin to dispersed sites, with a relatively extensive associated road network. The satellite-cities approach would normally be the best of the four models for maintaining greenspace wedges, a greenbelt, or an urban growth boundary.

A long-range plan for the Barcelona Region recommended concentrating expected future development around five satellite cities (Color Figure 44)

2004a). Selecting the satellite-cities approach and determining which cities should become nodes for urbanization were both done *after* the urban region had been analyzed, and areas valuable for natural systems and human uses had been mapped. The satellite-cities approach could avoid especially valuable areas and seemed least likely to cause significant environmental degradation. Similarly, the particular satellites were chosen from the small cities present in the Barcelona Region in part to minimize environmental problems.

In the alternative-models approach of this chapter where land of any sort is simply covered with development, the satellite-cities model emerged as best overall. No analysis of the relative value of resources and land uses across the region was involved. The analytical approach and results of the Barcelona plan reminds us that specific satellite cities can be chosen to minimize environmental degradation. As in the Barcelona case, satellites can also be chosen to include distinctiveness in character and enhance economic diversity, which provide flexibility and stability for the region as a whole. Finally, extensive growth around outer satellite cities is likely to affect nearby areas outside the urban region.

Transportation-corridors model: additional dimensions

The radial transportation-corridors approach (especially without a ring highway) may facilitate the maintenance of green wedges or a ring of parks by a metropolitan area. Several transportation corridors, instead of a few, would create lines of development that markedly subdivide a region. They would also tend to catalyze widespread fine-scale road networks supporting development, somewhat analogous to the dispersed-sites case.

Wide strips of development along highway corridors usually degrade streams and rivers crossing the region, and block movement patterns of certain key wildlife species, leaving them semi-isolated in smaller sections of the region. Overcoming these subdividing problems involves the establishment and protection of greenspaces that interrupt the strip development at appropriate intervals, as well as wide underpasses and/or overpasses along the highway for the undegraded passage of water courses and wildlife (Forman *et al.* 2003, Iuell *et al.* 2003, Trocme *et al.* 2003). Indeed, the transportation-corridors approach poses special problems for maintaining major greenspace corridors that interconnect an urban region as a whole for wildlife and walkers, and that connect effectively with adjoining regions in, e.g., the four cardinal directions.

Dispersed-sites model: additional dimensions

Results of the urbanization analyses are consistent with the awkward or unsatisfactory component of the basic sprawl concept (Chapter 1). Still, many variants of the dispersed-sites model may occur, including: a much higher density of tiny development patches; heterogeneity in patch size; patches dispersed further out from the metro area; and patches aggregated in various ways. Since this model mimics sprawl, these variants should be grist for modeling and discovering solutions for sprawl.

Finally, the four contrasting urbanization models were developed after considering a wide range of actual urban-growth patterns plus diverse types of possible change models (see preceding sections). The relatively clear-cut results from the four models seem to capture the primary patterns and alternatives for urbanization. Still, modifications of a pattern and different patterns warrant evaluation. In attempting to mimic reality with a model, selecting the most informative variables and determining the fewest that are sufficient are challenges. Avoiding oversimplification and avoiding misleading results are goals. Also excessive complexity is to be avoided so that the model enhances understanding and is ultimately useful for society. Future research and application will determine how well the four urbanization models (Figure 8.2) meet these goals.

Since urban regions probably manifest two or more of the urbanization patterns, evaluating combinations of the four core models seems promising. Thus combining concentric zones and satellite cities, or transportation corridors and satellite cities models, would be interesting. Even combining three options, e.g., dispersed sites with both transportation corridors and satellite cities might be informative. Landscape ecologists, urban geographers, and many others have much expertise to offer, from empirical measurement "on the ground" to modeling of diverse urbanization patterns. Identifying and evaluating the few basic patterns in a useful form for planners, ecologists, policymakers, and the public is an important step toward creating urban regions that sustain nature and us.

Several other dimensions of urbanization, from sea-level-rise effects to putting the model results into action, are introduced in Chapter 12. The core models and the guiding principles added to our repertoire in this chapter also remind us that, paraphrasing Isaac Newton, we mainly build on the shoulders of giants before us. Therefore, in the next chapter we turn to many principles already on the table for us to use.

9

Basic principles for molding land mosaics

An artist can translate a compelling inspiration into a painting or object that inspires the public, and even pleases the artist. In addition to inspiration and materials, skill is a key to success. Skill might be thought of as a set of principles, knowing what works and what doesn't – color mixtures, composition, types of lines, and much more. The artist has a palette of principles. When mixed with imagination and experimentation, they greatly increase the chance of a successful or inspiring result.

If one were designing wheels, using the known principles of wheel design greatly decreases the chance of producing square, oval, or one-spoked wheels. No matter how beautiful or well-made they are, such wheels do not work. If the land or an urban region is being planned or changed, we do not start from scratch. We use principles, subconsciously or specified. Water flows downward so streams are not designed flowing to hilltops. Trees require oxygen for their roots so we do not plant trees in water. People need security when asleep at night, so they are surrounded by shelter. Using known principles helps protect society from poor quality, and unethical, work.

Rather than simply ideas or hypotheses or even concepts, principles can be thought of as solid rigorous guidelines, a basis or foundation for planning and action. They do not apply everywhere anytime as we expect a universal law to do, but the often-considerable direct or indirect evidence supporting them is a basis for their widespread application (Dramstad *et al.* 1996, Forman 2004a).

Principles alone, however, lead to generic solutions. Monotonous, out-of-date, or lack-of-creativity might describe designs and plans using only principles on our palette. Instead, as for the artist whom we so admire, principles are mixed with imagination and inspiration to produce solutions for the land. Results are both dependable and creative.

This chapter is a palette or treasure chest of principles. All deal with land use, most with nature and people, and many with urban regions. They are not to be blindly followed. If your guidebook says that bears avoid the habitat type you are in and you see a bear moving rapidly toward you, it is wise to think beyond the guideline. Principles are to be creatively and intelligently used.

The bear example emphasizes that planning or action is also based on characteristics of the land. Land is not a blank canvas or a homogenous space. Spatial patterns, as well as flows and movements across them, are always present. The big challenge, and opportunity, is the integration of those existing land patterns with both principles and creativity. The goal is to improve the pattern and set of flows, and have improvements continue into the future.

Thus a set of *principles* useful for land-use planning and derived from a range of fields is presented. Overall these are statements of importance, of wide applicability, and with predictive ability. All have at least some empirical evidence, fit with indirect lines of evidence, and in some cases also have a known theoretical basis. Other scholars and planners, of course, would pinpoint a somewhat different list, including additional points. Indeed the reader can doubtless add to the list. Nevertheless, the bulk of the principles here seem to represent a consensus within each of the fields represented. Cutting-edge hypotheses and results are absent, as are narrowly focused principles with limited applicability. As always, both ongoing research and special attributes in a region dictate caution in applying or extrapolating a principle.

Not surprisingly, with a focus on natural systems and their uses in a region, landscape ecology is a particularly important contributor to the list. Yet principles are also drawn from transportation, community development, economics, conservation biology, water resources, and other fields.

Principles are conveniently placed into five broad categories, though clearly much overlap exists among the categories: (1) patch sizes, edges, and habitats; (2) natural processes, corridors, and networks; (3) transportation modes; (4) communities and development; and (5) land mosaics and landscape change.

Patch sizes, edges, and habitats

Principles in this first category focus on spatial pattern or structure of the land, especially relative to nature or greenspace. Consistent with the basic idea of nature conservation as a priority for society, rather than attempting to protect each species, the emphasis is on landscape patterns. The list of principles leans heavily on those presented in Schonewald-Cox and Bayless (1986), Salvesen (1994), Forman (1995, 2004a), Dramstad *et al.* (1996), Mitsch and Gosselink (2000), Farina (2005), Dale and Haeuber (2001), Opdam *et al.* (2002), Gutzwiller (2002),

France (2003), Lindenmayer and Burgman (2005), Groom *et al.* (2006), Wiens and Moss (2005), Perlman and Milder (2005), Fischer *et al.* (2006), and Lindenmayer and Fischer (2006). Chapter 4 elaborates on some of the principles, even suggesting others.

These principles are listed in four groupings: (1) patch size and edge; (2) natural habitats for conservation; (3) species-focused conservation; and (4) wetlands.

Patch size and edge

(A) *Large-patch benefits.* Large patches of natural vegetation are the only structures in a landscape that protect aquifers and interconnected stream networks, sustain viable populations of many interior species, provide core habitat and escape cover for most large-home-range vertebrates, and support near-natural disturbance regimes and plants dependent on them.

(B) *Edge width of a natural community.* Edge width, which largely results from penetration of wind, solar energy, and human influence into a natural community, is the distance with significant effects on sensitive ecological variables, such as desiccation, seedling mortality, herbaceous species, and upper soil layer conditions.

(C) *Edge and interior species.* A more convoluted natural-vegetation patch, or one that has been subdivided into two smaller patches, will have a higher proportion of edge habitat with slightly more generalist edge species, but will contain significantly fewer and smaller populations of interior species, including those of conservation importance.

(D) *Small-patch benefits.* Small natural-vegetation patches scattered across a less-suitable matrix act as stepping stones enhancing the movement of some species, provide some protection for widely scattered uncommon species, and, if near a large patch, may enhance species richness and movement associated with the large patch.

(E) *Populations in small patches.* Small patches, especially if isolated, tend to have smaller populations, which fluctuate more over time and have more inbreeding and resulting genetic deficiencies, and therefore a greater chance of local extinction or disappearance.

(F) *Human impacts and protected areas.* Closing spur roads and roads that bisect the interior of a large protected patch, and concentrating recreational opportunities and facilities for people in the edge portion of a protected area are effective ways to protect resources, especially in the interior of a large protected patch.

(G) *Boundary characteristics.* Boundaries or edges of a habitat, including their three-dimensional structure, distinctive microclimate and soil, and high vegetational density and species richness, affect adjacent habitats by functioning as a source of effects and as a filter of movements between the habitats.

(H) *Degradation of a natural community or ecosystem.* Degradation by human activity reduces vertical and horizontal structure, such as foliage layers, tree holes, vegetation gradients, and soil horizons, and reduces functional interactions and flows, including food webs, water flows, and mineral nutrient cycles.

Natural habitats for conservation

(A) *Number of large patches.* Consistent with risk-spreading theory, if each large patch of a particular habitat type contains almost all of its characteristic species in a landscape, then two or three large patches are probably sufficient to sustain almost all the species, but if each patch has a limited portion of the characteristic species present, four or five large patches are probably required.

(B) *Especially valuable patches.* Natural vegetation patches that play a particularly important role in the overall system (such as a key link in the landscape pattern), or contain unusual or distinctive characteristics (such as an important aquifer or rare habitat), are especially valuable for minimizing degradation.

(C) *Economically productive areas.* Remnant natural habitats in particularly productive areas especially merit habitat expansion, because they tend to be rare and to contain many rare species that thrive on the rich environmental conditions.

(D) *Habitat diversity.* Increasing the number of habitat types, primarily by including more substrate and microclimatic conditions or secondarily by maintaining more successional stages (e.g., fallow fields, shrubby areas), increases the number of native species present.

(E) *Tree holes and dead wood.* Dead wood, both standing and fallen, and cavities in tree trunks tend to be scarce in built areas, yet are especially important for biodiversity benefits.

(F) *Rare and representative habitats.* By protecting reasonable numbers and sizes of rare and representative habitats, nature (including the bulk of the native species present) should persist long term.

(G) *A small isolated habitat.* To protect a small isolated habitat long term typically requires the presence of an important role played by the habitat within a larger landscape pattern, and may also require widespread public recognition.

(H) *Ecology, cost, and threat.* Successful long-term land protection particularly focuses on location of the land relative to other protected lands, plus three characteristics subject to rapid change: (1) present ecological attributes of the land; (2) land cost and subsequent management cost; and (3) threats (urgency) to the land.

Species-focused conservation

(A) *Species of small isolated habitats.* To provide some long-term protection for species of dispersed small distinct habitats requires protection of extensive heterogeneous areas, or of numerous small sites, or of several large patches with enough connections across the landscape that most species distributions will be included in the large patches.

(B) *Species "perception" and conservation priority.* Animals and plants "perceive" and respond to different-sized structures and patterns, and thus successful conservation focuses on species especially sensitive to large structures and patterns, which are most likely to be lost or degraded by human activities in the landscape.

(C) *Keystone species.* Landscape patterns that protect keystone species (those with a disproportionately large effect on ecosystem function relative to their abundance or biomass), particularly predators, are likely to be especially effective in protecting biodiversity.

(D) *Species extinction proneness.* A landscape pattern that enhances the following species types – low mobility animal, large body size, low reproductive rate, top of food chain, large home-range size, hunted species, small population size, habitat specialist, and strong dependence on another species – reduces the chance of species loss.

(E) *Invasive species.* If an invasive non-native or feral species degrades a natural habitat, and ecological succession and other natural processes are unlikely to be an effective control, then carefully researched human control of the species is normally appropriate to restore the habitat.

Wetlands

(A) *Hydrologic functions of wetlands.* When not "full" of water, wetlands act as sponges slowing down and absorbing water flows, and then slowly releasing water through evaporation to air, percolation into ground, and runoff into surface water-bodies, that effectively reduces downstream peak flows and flooding (Figure 9.1).

(B) *Pollutants and wetlands.* Particulate pollutants settle out in wetlands, dissolved substances are absorbed by plant roots, diverse pollutants are filtered as water moves through soil, and some pollutants are broken down by microorganisms, that together results in cleaner water flowing out of a wetland.

Figure 9.1 Tidal wetland and river spanned by multilane highway bridge which facilitates wildlife and floodwater passage. Note: habitat heterogeneity in the floodplain; small meandering channel on mudflat; picnic recreation area on left; and continuous woody vegetation corridor along field that reduces agricultural runoff and enhances wildlife movement. Dover/Smyrna, Delaware, USA. Photo courtesy of US Federal Highway Administration.

(C) *Plants in wetlands.* Because the water table level is close to the irregular soil surface in a wetland, considerable spatial microheterogeneity and temporal change in water conditions are the norm, often producing a high diversity and biomass of adaptable, seasonally changing species.

(D) *Wetland complexes.* A connected cluster or complex of wetlands normally provides the highest wetland biodiversity and stability.

(E) *Ephemeral ponds.* Ephemeral ponds (or vernal pools) which dry out for a period most years often contain a concentration of both rare plants that thrive with alternating inundation and dry soil, and rare animals which benefit from the absence of fish, burrow deeply into the soil during dry periods, or seasonally migrate some distance from and to the pond.

(F) *Wetland surroundings.* Natural vegetation surrounding a wetland or ephemeral pond reduces sediment and other pollutant inputs, and is intensively used by many wetland animals, which also tend to move longer distances in the direction of other wetlands and suitable habitat.

(G) *Wetland restoration and creation.* Restoration is typically more successful than wetland creation, and establishing the right hydrologic conditions and flows is normally more important for the formation and stability of wetlands than are soil conditions and vegetation, that will develop naturally over time.

(H) *Wetland as pollutant filter.* Wetlands tend to be effective filters for water-borne suspended sediment, phosphorus, and biological-oxygen-demand (BOD), but less so for bacteria and nitrogen, unless the water flows a long distance through a wetland.

(I) *Rare species in wetlands.* Because wetland removal by drainage and filling has been so pervasive in urban regions, the wetlands remaining typically have among the highest concentrations of rare species in the region.

Natural processes, corridors, and networks

This second set of principles emphasizes function, the flows and movements across space, that in effect describes how the land or region works. Natural systems are the focus. The following references are especially useful for these principles: Forman (1995, 2004a), Dramstad et al. (1996), Harris et al. (1996), Ludwig et al. (1997), Beier and Noss (1998), Burel and Baudry (1999), Bennett (2003), Turner et al. (2001), Wiens (2002), Groom et al. (2006), Lindenmayer and Burgman (2005), Hilty et al. (2006), and Lindenmayer and Fischer (2006). Chapter 4 also contains insight on this subject.

Four groupings of principles are present: (1) natural processes and species movement; (2) water flows; (3) natural corridors and the matrix; and (4) natural networks.

Natural processes and species movement

(A) *Form and function.* Compact forms effectively conserve internal resources, convoluted forms enhance interactions with the surroundings, and network forms serve as an internal transport system, so that a natural vegetation patch with a rounded core, some curvilinear boundaries, and a few long lobes or attached corridors provides a range of ecological benefits.

(B) *Interactions between patches.* Species interactions (movements) are greatest between a small patch or site and its adjacent land uses, somewhat lower with nearby patches of the same type, and lowest with distant patches of a different type.

(C) *Metapopulation arrangement.* Human activities in the urban region often subdivide a large natural population into spatially separate small populations with few individuals moving among them (a metapopulation), in which case a few large natural patches, each surrounded by small patches, is an excellent design for sustaining metapopulations.

(D) *Metapopulation dynamics.* Species disperse outward from a large patch, providing genetic variation and reducing local extinction in nearby small patches, whereas species that disappear from a small patch are less likely to return or recolonize if the patch is isolated or surrounded by an inhospitable matrix.

(E) *Movement among small patches.* For a species that inhabits and moves among a few small patches, loss of a patch, especially a central one, tends to reduce population size, movement, and stability.

(F) *Straight and convoluted boundaries.* A straight boundary tends to have more species movement along it, whereas a convoluted boundary with lobes and coves provides diverse wildlife habitat and facilitates boundary crossing between adjacent habitats.

Water flows

(A) *Surface runoff.* Rainwater washing surfaces and soils of a land mosaic carries dissolved chemicals, erodes surface particles containing chemicals, and rapidly flows as stormwater into and along channels to cause a pulse of flooding, and to deposit its contents in gullies, streams, lakes, and other water-bodies (Figure 9.1).

(B) *Groundwater flows.* Surfacewater carries dissolved chemicals down into the ground where they may accumulate and contaminate the typically slow-moving water of an aquifer, or groundwater may be partially cleaned by flowing through soil or wetlands to water bodies on the surface such as streams and lakes.

(C) *Stream corridor.* A ("blue-green") ribbon of dense natural vegetation that covers the floodplain, both hillslopes, and a strip of interior habitat on both adjoining upland areas will normally provide protection against erosion, dissolved mineral nutrients, and toxic chemicals from the matrix, especially if the vegetation widens to surround entering intermittent channels.

(D) *Vegetation along small channels.* Vegetation protecting intermittent channels and small (first-order) streams is especially important for minimizing downstream peak flows and flooding.

(E) *Floodplain or riparian vegetation.* Dense floodplain vegetation, especially shrub cover, provides friction to reduce downstream flooding, provides shade, dead leaves and wood to enhance fish and other aquatic organisms, and increases the rich floodplain habitat diversity of wetland depressions, streambanks, sandy ridges, and surface microheterogeneity.

(F)   *River-ladder pattern.* A "river ladder" to protect rivers has vegetation strips along both sides of a floodplain to facilitate wildlife movement and protect hillslopes and adjacent upland, plus a sequence of large vegetation patches crossing the floodplain that reduce flooding, trap sediment, contribute wood for downriver fish habitat, provide organic matter for aquatic food chains, and maintain diverse habitats with rare floodplain species.

(G)   *Drainage basin and stream corridor.* The hydrologic, physical, chemical, and biological characteristics of a stream/river can be modified or mitigated by the riparian or stream corridor, but are much more affected and effectively managed by the types and spatial arrangement of land use across the watershed or drainage basin.

(H)   *Aquifer water.* Aquifer groundwater, which (except in limestone areas) moves very slowly and has little capacity to remove pollutants, is mainly kept clean by a complete cover of natural vegetation, particularly over its upslope portion.

Natural corridors and the matrix

(A)   *Corridor functions and their control.* Width and connectivity are the primary controls on all five key roles or functions of natural-vegetation strips or corridors, i.e., conduit, filter (or barrier), source, sink, and habitat.

(B)   *Small patches attached to corridors.* Small patches attached to natural corridors and networks provide "rest stops" for wildlife movement that, especially on long routes, typically increase the chance of a species reaching a destination.

(C)   *Gap in a corridor.* The ability of an animal moving along a natural-vegetation corridor to cross a gap or break in the corridor especially improves as gap length relative to the spatial scale of species movement shortens, and with more suitable conditions in and around the gap.

(D)   *Stepping stones between large patches.* For species movement between two large natural patches, a row of stepping stones (small patches) or a poor-quality corridor is normally better than no corridor, but a cluster of stepping stones with an overall linear alignment provides alternative routes and is likely to be more effective.

(E)   *Habitat contrast.* Greater habitat contrast or difference between a patch and a corridor or matrix decreases movement of species between the patch and an adjoining corridor or matrix, and hence across a landscape.

(F)   *Matrix heterogeneity.* Microhabitat heterogeneity increases the total species pool of the matrix and its role as a source of species, and if heterogeneity is arranged as a (gradual) gradient, rather than being patchy, species movement is either greater or less depending on gradient orientation relative to direction of movement.

Natural networks

(A) *Major natural-vegetation network.* The primary network (emerald network) of large natural patches and connecting corridors helps maintain distinct sections across a landscape, preventing coalescence of development and promoting a sense of community, local culture, and care for the land.

(B) *Loops in a network.* Loops or circuits in a network provide alternative routes for movement, thus reducing the effects of gaps and less suitable spots, and increasing the chance of successfully reaching a destination.

(C) *Landscape connectivity.* Most species evolved in highly connected heterogeneous natural landscapes, have had relatively little time to adapt to human fragmented ones, and occur in greater numbers (species richness) in more connected areas.

(D) *Species dispersal.* Since species disperse different distances and directions, a natural corridor and patch network with a relatively high average number of linkages per patch provides good dispersal opportunities which enhance the persistence of most species.

Transportation modes

This third set of principles involves highways and roads, commuter-rail lines, and walking. Transportation is a core spatial attribute and plays a major functional role in the urban region. It is a key factor in economic investment and development, as well as natural systems and their use. The following references are particularly useful for the principles here: National Research Council (1997), Warren (1998), Cervero (1998), Forman and Alexander (1998), Forman and Deblinger (2000), Ravetz (2000), Bullard *et al.* (2000), Simmonds and Hack (2000), Calthorpe and Fulton (2001), AASHTO (2001), Benfield *et al.* (2001), Willis *et al.* (2001), Forman *et al.* (2003), Forman (2004b), Dittmar and Ohland (2004), Handy (2005), Erickson (2006), Forman (2006), and Moore (2007). Also see Chapter 2.

Four groupings are addressed: (a) highways; (b) commuter-rail lines and communities; (c) roads in communities; and (d) walking and park systems.

Highways

(A) *Highway as source of effects.* Wider and especially busier highways, as concentrated linear sources of ecological effects, increasingly alter local hydrology, wetlands and streams, block animal movement across the landscape, subdivide natural populations into smaller populations, road-kill animals, and disperse air pollutants into the environment (Figure 9.1).

(B) *Degradation zones by highway.* Increased vehicular traffic on highways creates wider adjacent zones of degraded animal communities (presumably due to traffic noise), and wider highways (often with more traffic) usually are greater sources of non-native species, eroded earth material, stormwater contaminants, and atmospheric pollutants.

(C) *Highway protection of the matrix.* A more concentrated, safe, and efficient transportation system to access resources, homes, and other human land uses is valuable for reducing dispersed human impacts on nature and natural systems across the landscape.

(D) *Highway network.* Busier and wider highway corridors increasingly reduce landscape connectivity for wildlife and subdivide an urban region into sections, with a mesh size normally suitable for relatively separate small populations of large animals.

(E) *Perforated highway corridor.* Increasingly perforating a transportation corridor with passages, from tiny wildlife tunnels to culverts, underpasses, and overpasses, reduces habitat fragmentation by providing for relatively natural movements and flows of wildlife and water.

(F) *Closing roads.* Progressively closing spur roads and low-usage roads in and by medium-to-large natural patches is an especially effective way to create large natural patches and their many important benefits for nature and society.

(G) *Adding radial-route capacity.* Adding transportation capacity on a city's radial route stimulates growth and development in that direction.

(H) *Adding a ring road.* Adding an outer ring road provides flexibility in movement for suburban (peri-urban) residents and catalyzes growth and development, with associated habitat loss, over a broad outward zone.

(I) *Trucking center.* A truck (lorry) transportation terminal near the metro-area border facilitates the transfer of manufactured goods and agricultural products for long-distance trucks, as well as small-truck movement serving local farms, industries, markets and restaurants, in effect providing economic efficiency and better traffic flows on congested urban streets.

Commuter-rail lines and communities

(A) *Commuter rail lines.* Light or heavy rail lines and streetcars/trollies that extend outward, offering convenient service beyond the metro area, provide greater modal (transportation types) flexibility for suburban residents and help limit vehicular traffic.

(B) *Transit-oriented development.* TOD that meshes mixed-use residential-shopping areas with local natural ecosystems within 800 m (half-mile) of a station on a commuter transit line has a higher proportion of people commuting to work on public transport, and also may have more walking, bicycling, and local shopping, a tighter community, and a greater sense of place by residents.

Roads in communities

(A) *Traffic calming.* Traffic-calming techniques that slow vehicle movement increasingly provide safer, more convenient walking opportunities for children and the elderly, and enhance a sense of community in neighborhoods.

(B) *Accessibility and local spaces.* Road infrastructure which effectively provides for both accessibility and local community spaces and private spaces successfully addresses both broader social goals and narrower neighborhood and individual goals.

Walking and park systems

(A) *Park system.* Providing routes for movement of people and/or species among parks changes a group of parks into a park system, with consequent benefits to both nature and people.

(B) *Greenspaces and neighborhoods.* An effective urban park system has greenspaces conveniently walkable for residents of all neighborhoods.

(C) *Sustainable park system.* To establish a sustainable park system, each park and each connection is important, and both government and the public understand how the interdependent pieces fit together to work as a whole.

Communities and development

In this fourth area of principles, the focus is a community, an aggregation of interacting residents in a city, town, or village. Development emphasizes the spread of built areas, including economic investment across the land. Both communities in place and the process of development strongly interact with natural systems. In contrast to the preceding section on human movement patterns, the social and economic focus is on where people live.

Principles here are largely extracted from Yaro et al. (1990), Sukopp and Hejny (1990), Bartuska (1994), Campbell (1996), Seddon (1997), Warren (1998), Donahue

(1999), Atkinson *et al.* (1999), Ravetz (2000), Beatley (2000), Jacobi *et al.* (2000), Warner (2001), Willis *et al.* (2001), Macionis and Parillo (2001), Benfield *et al.* (2001), Peiser (2001), White (2002), Grimm *et al.* (2003), LeGates and Stout (2003), Campbell and Fainstein (2003), Nassauer (2005), Handy (2005), Kellert (2005), Hersperger (2006), Clark (2006), Moore (2007), and Robert Yaro (personal communication). The emphasis is much more on land planning than on management of existing land (Atkinson *et al.* 1999, Willis *et al.* 2001, White 2002). Also see Chapter 2.

Three subgroupings are useful for this topic: (a) locating development; (b) environment and community; and (c) social dimensions and sense of community.

Locating development

(A) *Development and low-ecological-value areas.* Guiding potential growth and development to areas of low ecological value is a major step in protecting and sustaining natural systems.

(B) *Concentrating or dispersing development.* Concentrating rather than dispersing development greatly increases the protection of natural systems and reduces the dependence on transportation infrastructure and vehicular usage (Figure 9.2).

(C) *Coalescence of communities.* Preventing the coalescence of adjoining communities, e.g., with greenspace strips, helps maintain the identity and distinctiveness of each community.

(D) *Mixed-use communities.* Intermixing residential, commercial, and light-industry areas in sections of a community reduces vehicular travel, but causes more nearby land-use conflicts than in single-use communities.

(E) *Edge nodes of industry and employment.* Concentrating light industry (and sometimes medium industry) in nodes on the edge of residential/commercial towns and small cities helps reduce both vehicular travel and land-use conflicts.

(F) *Heavy industry centers.* Aggregating compatible heavy industries on a site with efficient water, power, and waste-disposal plus convenient public transport for diverse nearby employees, away from major rivers/streams, and downwind of population centers and valuable nature, minimizes environmental problems and maximizes benefits.

(G) *Land prices.* Overall, land prices decrease with distance from a city's central business district, a pattern mainly modified by geomorphology and by major nodes of public or private investment.

(H) *Radial transportation corridors.* Radial transportation corridors are major catalysts of commercial and residential expansion, either directly as

Figure 9.2 Compact urban development with adjacent viable semi-natural land and agricultural land. Richmond, Virginia, USA. Photo courtesy of USDA-Soil Conservation Service.

strip development, or indirectly as nodal growth along a transportation corridor begins, elongates, and coalesces.

(I) *Strategic position.* The strategic position of a community conveys a commercial or other advantage over other places, though in time any advantage reflects the balance of changes in the community relative to those elsewhere.

(J) *Hazardous areas.* Establishing protected natural lands on high-hazard-risk areas helps avoid the social and economic disruptions of community "disasters."

(K) *Compact development*. Compact development enriches the sphere of an individual's social, cultural, employment, and other opportunities in a small area, reducing vehicular travel, and providing economic support for public transport and walking/biking paths.

(L) *Development density*. Residential development at sufficient density helps support public transport conveniently accessible for both residential and employment locations.

(M) *Geometry of a node (πr^2 problem)*. Since area is scarce around the center of a circle and increases rapidly moving outward, a land price gradient tends to produce concentric land-use zones, which may be broken by convenient transportation radii and planned slices or nodes of different land use.

(N) *Infill*. Infill development on greenspaces in a built area is often beneficial for creating compact neighborhoods, but only up to the point where quality parks are too far apart for most residents to walk, and stepping stones too far apart for effective species movement across the built area.

Environment and community

(A) *Metabolism/ecosystem/machine analogy*. Using the structure and flows of an organism, ecosystem, or machine to understand a city emphasizes the importance of limited diverse inputs and outputs, and maintaining a diverse, but not too complex, structure within the city, both of which provide stability and adaptability for the inevitable big surprises ahead.

(B) *Human–environment relationships*. In addition to social needs and economic opportunity, human–environment relationships are at the core of a community and are sustained by an effective mix of greenspaces, built areas, infrastructure, and institutions.

(C) *Environmental management*. Management of urban environmental resources and problems that places short-term crisis-prevention measures as part of long-term solutions for a larger potential future community is likely to save costs, maintain public support, and establish a more sustainable future.

(D) *Impermeable surfaces*. Limiting the amount of impermeable-surface area, especially in suburbs, reduces rapid-runoff peak-flow flooding, recharges groundwater, reduces pollutant levels reaching water bodies, and improves local streams and fish populations.

(E) *Drainage connection*. Drainage connection area (impermeable surface directly connected by pipes or ditches to water bodies) is reduced by channeling stormwater into vegetated depressions and drainage basins, which help reduce flooding, filter water pollutants through the soil, and improve water quality and fish populations in water bodies.

(F) *Wetland uses*. Multi-use wetlands with adequate water flow and attractive paths or boardwalks help provide recreation, biodiversity, aesthetics, flood control, and pollutant absorption, and can separate or center neighborhoods.

Social dimensions and sense of community

(A) *A central park*. Parks in the center of communities normally are intensively used, serve as meeting places, receive considerable maintenance, and have severely degraded nature.

(B) *A linear edge park*. An edge park along the border of a community provides amenities for the existing community and for present or future outside communities, and, with limited human use, provides natural habitat and connectivity for movement of some species.

(C) *Market-gardens*. Market-gardening (truck farming) near a city provides fresh fruits and vegetables at low transport cost to city markets and restaurants, plus diverse environmental benefits on the city's outskirts.

(D) *Human habitat*. For planning purposes, a good human habitat is a community offering a choice in the diversity of frequently needed and used places (e.g., grocery, school, park, eatery) located in relative proximity to its residents.

(E) *Neighborhood units*. Neighborhoods serve as the basic social and planning units of a larger community or district, and several neighborhoods connected to a cultural and/or shopping center are likely to sustain the larger community.

(F) *Urban district*. An aggregation of interacting neighborhoods with a distinctive identity forms a district (or urban "village"), a place that residents identify with and that the broader city or metropolitan area uses for identification and planning.

(G) *Green, profitable, and fair*. Combining environmental protection, economic growth/efficiency, and social justice/economic opportunity/income equality as equal parts under the rubric of sustainable development still seems to be utopian, marketing, impossible, or a ruse, yet balancing such big human and environmental objectives is attainable and should be the norm.

(H) *Aesthetics and basics*. Adding aesthetic forms, after providing the basics of water, neighborhoods, jobs, natural areas, etc. for a community, enhances a sense of place and stimulates people to actively care for it.

(I) *Sense of place.* The intertwining of built structures and greenspaces that persist over time creates a place that people care for and remember.

(J) *Community gardens (allotments).* Joining neighbors digging in the soil and growing their own plants and foods on tiny adjoining plots enhances an understanding of nature and creates valuable social bonds that strengthen a community.

(K) *Commuter-station areas for urban residents.* Community gardens, bicycle parks, walking paths, and recreation areas centered around commuter-rail stations provide important accessible resources and values for urban residents concentrated in high-density metro areas.

Land mosaics and landscape change

This fifth and last category of principles highlights the big picture. Land mosaics emphasize the structure or spatial pattern of a landscape or urban region, including how nature and people are arranged. Land change then focuses on how the pattern is altered or changes, plus the associated functional changes over time. Change may be catalyzed by overall planning or decision, or produced by the multitude of little steps taken by people (Odum 1982), or caused by natural systems. For these subjects, Pickett and White (1985), Zonneveld and Forman (1990), Forman (1995, 2004a), Dramstad *et al.* (1996), Ludwig *et al.* (1997), Losada *et al.* (1998), Dale and Haeuber (2001), Ingegnoli (2002), Gutzwiller (2002), Foster and Aber (2004), Chen *et al.* (2004), Lindenmayer and Fischer (2006), Erickson (2006), and Moore (2007) are particularly useful. Also see Chapter 3.

The two subgroupings are: (a) land mosaics; and (b) landscape change.

Land mosaics

(A) *Structure–function–change feedbacks.* Landscape structure or pattern controls landscape function (how the area works), which alters structure, in turn causing function to change.

(B) *Spatial scales.* Ecological and human conditions in an area (such as a landscape) are strongly affected by patterns and processes at three scales: the broader scale (e.g., region); the finer scale (e.g., large patches within the landscape); and surrounding areas (e.g., competing or collaborating) at the same scale.

(C) *Hierarchical structure.* A spatial hierarchy of habitat sizes, stream orders, and population sizes controls the amounts and directions of flows and movements across a landscape, and patterns and processes of a particular type tend to differ at different spatial scales.

(D)   *Grain size of the mosaic.* A coarse-grained landscape (mainly composed of large patches) that contains fine-grained areas (mainly small patches) is better than either type alone, because it effectively provides for many large-patch benefits, multi-habitat species including humans, and a wide range of habitats and natural resources (Figure 9.2).

(E)   *Mosaic pattern and multi-habitat species.* Species (and people) that regularly use different habitats or land uses are favored by convergency points (junctions where three or more habitats converge), adjacencies (different combinations of adjoining habitat types), and habitat interspersion (habitat types scattered rather than aggregated).

(F)   *Environmental gradients and patchiness.* Environmental gradients with gradual ecological change over space are sometimes evident, though patchiness with distinct boundaries predominates on land, because of patchy substrates and especially human activities that typically sharpen boundaries.

(G)   *Nature and a grid.* A regular grid, such as of roads and strip development, may be the ecologically worst way to distribute a small amount of built area over a natural landscape, since the grid leaves only small natural patches, truncates connectivity, and removes much of the irregularity and heterogeneity characteristic of nature's species-rich communities.

(H)   *Key variables of urban areas.* Human population density and spatial proximity are considered to be the two leading variables, with functionality the third, providing understanding and predictive ability for most human patterns and issues in urban areas.

Landscape change

(A)   *Ecologically optimum change.* The optimum way to change a large natural landscape to a less ecologically suitable one is to progressively remove vegetation in strips from two adjacent sides of the landscape, maintain a few large green patches in the middle phase, and then sequentially remove the patches.

(B)   *Specific changes within an optimum sequence.* Determining an optimum spatial sequence for a changing landscape permits one to pinpoint at any stage the best and worst locations for a specific change, either deleterious or beneficial.

(C)   *Spatial processes.* With the spread of human activities, natural areas may be perforated, dissected, fragmented, shrunk, and/or eliminated with quite different ecological consequences, even though habitat loss and isolation normally increase with all of the processes.

(D) *Change in mosaic pattern.* Ongoing human activities and natural disturbances keep the structure and habitat diversity of a land mosaic changing over time, as land uses and successional-stage habitats "move around," even though natural resources of the whole landscape may be in a degradation, meta-stable, or restoration trajectory.

(E) *Greenspace in a changing context.* All greenspaces change over time from interactions with adjacent and more-distant land uses, with the intensity or rate of flows/movements decreasing with distance and increasing with direction of incoming wind/water flows, animal locomotion, and human influences.

(F) *Worst urbanization.* Regional urbanization in dispersed sites surrounding a metropolitan area, and to a lesser extent along transportation corridors, appears to cause the most extensive nature-and-human resource degradation (Chapter 8).

(G) *Best urbanization.* Urbanization focused around satellite cities, which causes the least overall resource degradation, appears to be the best regional development pattern, though factors specific to a region may indicate a preference for combining satellite-city development with concentric-zone development adjacent to a metropolitan area (Chapter 8).

(H) *Cumulative effects.* Cumulative effects represent the combination of spatially separate effects, previous effects at different times, and different types of effects, and therefore a group of different and dispersed solutions is normally required to significantly reduce or mitigate cumulative effects.

(I) *Time lags.* Time lags reflecting the inertia or resilience of nature mean that some ecological responses (such as biological diversity and populations of long-lived species) are delayed after a change, that some ecological conditions today reflect earlier patterns, that mitigation may be effective well after landscape degradation, and that a successful response may be delayed after mitigation.

(J) *Plan/design for long term.* A land-use plan which provides an adaptable pattern to anticipate and respond to changes and which outlines broad land-use areas or zones, with only spots designed in detail, is more likely to be a successful long-term plan.

(K) *Region and local.* Planning regionally for broad-scale patterns (e.g., large greenspaces, highways) and then planning locally (e.g., neighborhoods, aesthetics) to effectively mesh with them is likely to successfully address both regional and local needs.

(L)   *Communities and history.* Cities and towns are a product of historical development, yet they have also helped shape that history.

In conclusion, a treasure chest of principles has been opened. Many are or will become second nature to practitioners and scholars dealing with land use and urban regions. The list is also a handbook to be kept handy for solving problems. In effect, the principles are convenient handles for molding better urban regions where nature and people both thrive long term.

Such a cornucopia of riches calls out for a few governing principles or broad paradigms, from which the detailed statements or principles follow. Perhaps the patch–corridor–matrix model or pattern-process paradigm illustrates one of the broad paradigms (Forman 1995, Turner *et al.* 2001, Robert McDonald, personal communication). Many of the principles, at least indirectly, follow from that. Articulating the few broad paradigms covering all principles awaits an exceptionally creative mind. At the other end of the conceptual scale, some of the principles articulated follow from more detailed or basic theories, such as central place and hierarchy theory (O'Neill *et al.* 1986, Hall 2002).

Finally, experts in specific fields can and hopefully will delineate more, better, and fuller lists of principles in those fields. The value of the preceding treasure chest is to see the principles from different fields listed together, and to see some integrated principles that cross fields. Consider this list to be a palette, much in need for direct use today. But also consider it a work in progress, readily amenable to enhancement and enrichment on into the future.

10

The Barcelona Region's land mosaic

Suppose you were faced with developing a regional plan that highlights natural systems and their human uses for one of the world's great cities. You have never been to the city or its surroundings, though you once lived in the broader geographic region. What would you do? Here is my story, and especially its result.

Barcelona struck me as a vibrant livable place with a cutting-edge can-do attitude. Best known are the amazing multicolored organic structures created more than 80 years ago across the city by Antoni Gaudi. I was inspired by his Parque Guell, especially the stunning evocative mosaics created with broken pieces of brightly colored ceramics (Color Figure 40). They are as magical today as when he did them. Any land mosaic I propose for the Greater Barcelona Region should be as inspiring, valuable, and long-lasting as those ceramic mosaic masterpieces in tiny spaces.

In essence, a 150-page conceptual plan (including 28 maps) was prepared over a 15-month period for the city's Mayor and Chief Architect (Forman 2004a). The project objective was to evaluate and highlight the importance of the urban region (rather than the city), the major natural systems therein, and the diverse human uses of these natural systems. This challenge was addressed by the development of a "land mosaic for natural systems and people" based on landscape ecology and other principles. I extensively visited the region, talked with diverse knowledgeable experts, and accumulated shelves of valuable information, maps, and literature. The resulting report listed important assumptions, stated basic principles, outlined a vision, portrayed simple spatial models, applied the models to the region, and identified options. Major themes were highlighted in detail. And three plan options for the Greater Barcelona Region (GB Region) as a whole were described, mapped, and compared.

Perspective and approach

Now let us examine more closely the project and see the major results and recommendations in context. The background perspective for the project was very similar to that presented in the opening section of this book.

Land is home and heritage, also capital and investment. Both types are scarce in an urban region, yet land is still being "wasted" there. A dense population depends daily and fundamentally on natural systems that are out of sight, out of the city. But like a hungry giant, uncontrolled urban expansion devours the closest and best resources. Such regions are dynamic mosaics with human pieces expanding and natural pieces shrinking, leaving the fundamental human dependence on nature's resources riskier, less sustainable.

Flows and changing patterns across the land are central foundations for planning an urban region. Surface water flowing in streams and rivers supports many human needs, from clean drinking water to recreation, wastewater treatment, and aesthetics. Groundwater flows create "underground reservoirs" that support wells, agriculture, and diverse natural plant communities. Wildlife moves across the land, a key value for recreation, even human culture. At the same time, unpolluted water becomes scarce and expensive. Traffic jams proliferate. Highways subdivide nature into pieces. Appealing recreational and tourist sites degrade. Impermeable surfaces spread and flood peaks get higher. Productive agriculture and family farms shrink. All so familiar. People of the region, long dependent on local resources and benefits of natural systems, must increasingly depend on more distant, more expensive resources.

In Barcelona we stand on a threshold at the onset of the century. Most of the important agricultural areas in the region could shrivel up, or alternatively continue as a key resource in the land. Cities, towns and villages could merge, for example in the Central Valles area, into a single huge amorphous urbanized zone, or alternatively, the municipalities could maintain, even enhance, their identities and distinctiveness. The Llobregat River delta right next to the city could be all urbanized, or could be a distinctive, striking area celebrating its unique resource, abundant clean water. Towns and small cities could drown in traffic, or could have convenient walkable small-to-medium industries and neighborhood parks around their edges. And on and on. A rich set of natural resources for the people could still be in Barcelona's future, but at the present rate and trajectory, that set will be noticeably diminished in a decade or two.

Investing in natural systems also pays economic dividends. Thus revamping floodplain design, targeting a handful of pollution sources, maintaining large agricultural landscapes, and concentrating rather than dispersing development can significantly reduce costs, e.g., of flood damage, increasingly scarce water supply, infrastructure and servicing, and market-gardening products. More

broadly, planning that heads off crises and creates positive trajectories and legacies for the region is good economics.

A land-use framework that spatially arranges nature and people so they both
thrive long term is the focus (Forman 2002a). In contrast to legal and regulatory
approaches that can and do change "overnight," this spatial approach tends to
provide a longer-term future for the region. Indeed, when the public sees, uses,
and understands the value of certain land and its arrangement, degrading or
transforming it is less likely. Also, rather than doing separate plans for each
sector (e.g., transportation, housing, water) and then trying to mesh them, land
areas and natural systems, often with several uses for society, are the fundamental pieces to fit together to form the completed puzzle. Finally, flexibility and
adaptability to provide stability in the face of big changes and surprises ahead
remain an underlying planning thread.

Urban planning traditionally enhances the quality of people's life and promotes intelligent growth, whereas conservation planning protects the natural
systems and nature that people depend on, use, and value (Noss and Cooperider
1994, Dailey and Ellison 2002, Opdam et al. 2002). No model or case study was
found that sustains the diverse natural resources and nature in the region
around a major city. A new strategic approach is needed to mesh both halves,
people and nature, and create a whole. The Barcelona project thus targets and
highlights the gaping hole or weakest link in current urban-region thinking,
i.e., nature and diverse natural systems for people.

In essence, the *objective of the planning project* is to outline promising spatial
arrangements and solutions that enhance natural systems and associated human
land uses for the long-term future of the Greater Barcelona Region.

Land mosaic with distinctive features, the Greater Barcelona Region

Land-mosaic theory and principles focus on the spatial arrangement of
land uses in large heterogeneous areas such as landscapes, regions, or the area
seen from an airplane window or in an aerial photograph, exactly the right spatial scale for effective planning (Chapter 1). The landscape or region exhibits
three broad characteristics: structure (the spatial pattern or arrangement of
land uses present), function (the movement or flows of water, materials, species,
and people through the pattern), and change (the dynamics or transformation
of pattern over time) (Forman 1995, Farina 2005, Gutzwiller 2002). The land
mosaic or structural pattern is conveniently reduced to only three types of
elements: patches, corridors, and a background matrix. Patches are large or
small, dispersed or clustered, and so on. Corridors are narrow or wide, continuous or disconnected, etc. The matrix is single or subdivided, perforated or
dissected, and so forth. Adding a housing development, a nature reserve, or a
highway, for example, changes the mosaic pattern, with movements and flows

thereby altered in generally predictable ways. The patch–corridor–matrix model thus provides a useful flexible handle for land-use planning.

A land-use map portrays the big patches, corridors, and matrix of the Greater Barcelona Region (Acebillo and Folch 2000). The flows and movements of water, materials, species, and people across this mosaic match the activity of a bee's nest. Looking ahead, existing *trends* strongly suggest the following changes for the Greater Barcelona Region 50 years from now, and in most cases 10 years hence: (1) more people, buildings, sprawl, wasted precious land, roads, traffic, and coalescence of municipalities; (2) less clean water, parkland, agricultural land and production, and area for natural systems and nature; (3) also, hotter drier climate, spread of second homes, more buildings on steep slopes and on skylines, near-disappearance of clean stretches of streams and rivers, and reduced biodiversity (Barraco *et al.* 1999, Folch 2000, Acebillo and Folch 2000, Prat *et al.* 2002, Busquets 2005, Rowe 2006). Pulses rather than linear trajectories are likely to occur: immigration and population growth spurts, climate warming effects, mega-floods, serious economic drops, and less predictable surprises.

The current land mosaic of the GBRegion contains several distinctive and important *attributes on which to base a plan*. These include (Roda *et al.* 1999, Acebillo and Folch 2000, Prat *et al.* 2002): (1) an impressive set of large protected natural areas on mountains; (2) high visual quality on hillslopes and skylines with little urbanization; (3) the Llobregat and Tordera river floodplains/deltas as major sources of abundant clean water in a relatively dry climate; (4) wetlands, coastal vegetation, and marine littoral ecosystems as traces of a rich heritage reduced to a few threatened locations; (5) outward spreading towns ready to coalesce in some areas; (6) separate houses on large lots, a sprawl pattern ready to explode across the region; and (7) local rail lines and train stations apparently less abundant than in comparable areas of Europe, suggesting vehicular traffic growth as a major threat.

Approach and methods

The overall planning approach is conceptual rather than quantitative. The land-use focus is on nature itself and on nature–human interactions, but not on society itself. Planning solutions are integrated, and high-to-low priorities are generally evident rather than presented in a ranked list. The target is a broad-scale regional plan with maps, not detailed fine-scale maps, site analyses, and specifications. To attract and engage decision-makers and potentially gain public support, the approach strives to be both visionary and feasible.

For this relatively large Greater Barcelona Region and with limited time, I chose not to: (1) drown in details of extensive databases; (2) try evaluating the contents of numerous existing and proposed plans; and (3) make widespread observations on finer-scale secondary issues. This is not a traditional urban or regional plan in the sense of providing mainly for people and growth, since natural systems stand alongside people as major goals to mesh in the mosaic of land (Mata and Tarroja 2006). Nor is it a multi-stakeholder or committee-report plan that ends up, eventually, proposing an incremental or least-common-denominator change. Nor is this an action plan with a stepwise sequence of specific implementation recommendations, which would be closely dependent on traditions, laws, and regulations, as well as political and economic conditions. At each step the planning process is designed to be simple, lucid, and close to the land and people. For a plan, or parts of it, to get off the shelf and be implemented, it must be understandable by decision-makers as well as the public, who both must explain it, defend it, and generate excitement in it.

An existing area and semi-governmental organization, Barcelona Regional, has been effectively used in recent years for planning that involves the city and many surrounding municipalities (Acebillo and Folch 2000). Still, important urbanization and other processes occur outside this planning area. Thus the Greater Barcelona Region, with twice the area (i.e., 6500 km^2) and a radius of about 65 km (40 mi), was selected for this planning project (Forman 2004a). The area includes El Vendrell, Igualada, Calaf, Manresa, Vic Valley, and Tordera floodplain.

Later I identified 16 sections of the Greater Barcelona Region for more detailed maps and descriptions. Also solutions are frequently woven into the report for small places, such as gullies, streams, highways, and the edges of towns, which are widely repeated and have a large cumulative effect across the region. This is a long-term plan for the Greater Barcelona Region as a whole. Linkages with other regions are important. Also, plans limited to individual portions of the region would be different, but valuable, as long as they are readily compatible with solutions for the Greater Barcelona Region as a whole.

Six intensive visits of 3–6 days each at all seasons over a 16-month period permitted me to visit all portions of the Greater Barcelona Region (though time prevented a much-needed analysis of interactions with other regions) (Forman 2004a). I traveled with regional experts and consulted with many other helpful knowledgeable leaders. Numerous reports and published materials were provided and many other publications accumulated. In addition, Barcelona Regional produced, always in a time-efficient manner, a rich set of background information, images, and maps for this project.

Assumptions, principles, vision, spatial models, three plan options

Thirty-five important assumptions were stated, usually each in a single sentence (Forman 2004a). The plan rests on these. Also, 44 principles from landscape ecology and other fields were succinctly stated (also see Chapter 9). These range from habitat conservation and stream corridors to transportation, development, and landscape change.

The basic principles point to a vision, effectively a set of flexible land patterns to sustain natural systems and people at a high level in the GBRegion (Forman 2002a). Scattered glimpses of the vision resemble a "wish list" of desirable characteristics. The vision is not to perpetuate existing conditions, which will surely change markedly under any scenario. Rather, it is something to rest our ladder on, and climb toward.

The principles are meshed with the distinctive existing patterns and processes of the Greater Barcelona Region. The first step selects the "primary" principles, those offering the greatest overall benefits. The second step portrays the primary principles as simple spatial models or diagrams, which are applied or compared with specific patterns in the region to identify promising options. Usually two to four options for each principle are evaluated by simply listing the benefits and disadvantages. The third step, basically a best-judgment iterative process, then combines the preferred options (most benefits, least disadvantages).

These general steps lead to *three comprehensive spatial solutions* for the GBRegion, each of which should achieve the original stated objective (Forman 2004a). Also each solution is considered to lie in the envelope of feasibility. For convenience, the three options are labeled the "Most-promising Plan," the "Solid Plan," and the "Minimal Plan." Rather than being three alternatives with quite-different central themes, the three plans have common central themes and differ in how much or how strong proposed changes will be. Thus the three are presented flexibly, so decision-makers and the public can add or subtract pieces from any of the plans.

Finally, all three plans provide a mosaic of land uses for the whole urban region, as well as for each portion and piece within it. In this way one can easily see how changes or non-changes for the individual pieces fit into the big picture. Thus a trajectory of implementation could be established for the whole region. Or portions or pieces of the region could be addressed in the context of the big picture and its trajectory.

Nature, food, and water

Now we turn to the seven *primary themes* that emerged from this process and are incorporated in the plans and solutions for the Greater Barcelona Region. These represent the conceptual heart of the project. The first four: (1) emerald network, (2) major food areas for the future, (3) water for nature and us, and (4) streams, rivers, and blue-green ribbons, are presented in this section. The other three: (5) growth, development, and municipalities, (6) transportation and industry, and (7) nature and people in municipalities, emerge in the following section.

Emerald network

Large natural-vegetation areas or patches interconnected by vegetation corridors form the emerald network (Forman 2004a). This is the fundamental backbone of natural systems protection in a landscape or region, the *piece de resistance* of the Greater Barcelona Region (Color Figure 41). The network provides numerous values, from a permanent buffer against species extinction to a limit against endless urbanization, and is composed of emeralds and connections. Two solutions were outlined: (1) emeralds, the crown jewels of nature; and (2) connected land: five types of connections.

Emeralds, the crown jewels of nature

Large natural-vegetation areas or emeralds provide a group of benefits that cannot be provided in any other way, including: water quality protection for aquifers; connectivity of headwater streams; habitat to sustain populations of patch-interior species; a source of species dispersing through the matrix and to small patches; and ability to absorb or persist through natural disturbances over time (Forman 1995). Additional important benefits for society include flood control, adequate water for sewage treatment facilities, aesthetics, biodiversity, wood products, and recreational opportunities from family picnicking to bird watching, hunting, hiking, and youth education (*Parc de Collserola* 1997, Liddle 1997, Bacaria *et al.* 1999, Roda *et al.* 1999, Blondel and Aronson 1999, Atauri and de Lucio 2001, Grove and Rackham 2001).

Eight large protected natural areas currently exist in the Greater Barcelona Region (Garraf, Collserola, Montnegre-El Corredor, Montseny, Sant Llorenc del Munt, Montserrat, Serra de Castelltallat, and Serralada Transversal) (Roda *et al.* 1999, Acebillo and Folch 2000), and two additional ones are proposed. The Ancosa-Miralles emerald (west of the Penedes) combines and gives integrity to a cluster of five separate medium and small existing protected areas, and fills a void in the western quarter of the Region. The Serra de Rubio emerald (north of Igualada) connects headwater stream areas of the Anoia and Llobregat rivers, protects some European-Community-listed rare habitats, and helps link the isolated Castelltallat to Montserrat and Ancosa-Miralles. The combination of distinctive predominant vegetation and predominant rock surfaces emphasizes that all ten emeralds are ecologically important. Three emeralds, Garraf, Collserola, and Montnegre-El Corredor, are increasingly threatened by urbanization, and in Serralada Transversal watershed protection is needed for the critical reservoirs.

Fire management has special importance in the GBRegion because built areas alternate with large natural areas (Vallejo and Alloza 1998, Moreira *et al.* 2001, Forman 2004a). Fire management tends to focus on the large natural areas, where three overriding objectives often have to be spatially meshed. Protecting (1) forests growing wood products, (2) rare fire-adapted species, and (3) built areas (near natural areas) requires quite different strategies. Thus a carefully designed arrangement of fire-management techniques is required.

Connected land: five types of connections

The overall objective is to provide landscape connectivity to facilitate the movement of species as well as people (Saunders and Hobbs 1991, Forman 1995, Bennett 2003, Ahern 2002, Jongman and Pungetti 2004). Wide connecting areas are the most certain long-term way of accomplishing this in the face of human activities changing and spreading in unforeseen ways. Probably few species require corridors, but movement of numerous species is enhanced with them. Also the fauna is better off on connected than isolated patches. Where vegetation corridors cross major transportation routes it is important to construct passages, either underpasses or overpasses, for effective wildlife and people crossing (Rosell Pages and Velasco Rivas 1999, Forman *et al.* 2003, Iuell *et al.* 2003, Trocme *et al.* 2003). Walking paths in the corridors provide connectivity for hiking and recreation in the GBRegion. Serious hikers could walk across or even around the entire region. More importantly though, everyone will live not too far from a path for a relaxed brief stroll in nature.

Five types of connections, all providing for species movement and a walking path between emeralds, offer flexibility (Color Figure 41) (Forman 2004a). Natural vegetation forms the predominant land use, though in some connections farmland or "people parks" are also quite appropriate:

(1) *Reconnection zone*: a wide vegetation connection of nearby protected areas that normally creates a single stronger integrated protected area.
(2) *Green ribbon*: a wide vegetation corridor that crosses overland and normally connects protected areas.
(3) *Blue-green ribbon*: a wide vegetation corridor, usually covering floodplain and adjacent slopes, that protects a stream and may connect to a protected area.
(4) *Ribbon of pearls*: a wide vegetation corridor with attached small natural-vegetation patches (pearls) as "rest stops" to enhance wildlife movement along its length.

(5) *String of pearls*: a narrow linear vegetation-lined walkway with attached small natural-vegetation patches as "rest stops" to enhance wildlife movement.

Major food areas for the future

The *values of agriculture* in a urban region are striking. In the Greater Barcelona Region it provides food products for today's residents, as well as long-term flexibility and stability for the region through periods of major change. Urban agriculture benefits include the historical symbolism or heritage of farmland, the active roles of farm families, the educational dimensions of farms, aesthetics and rural character of landscapes, enhanced game populations, important wildlife species and biodiversity, and the ethics of protecting prime food-producing areas in a world with growing hunger. Protecting the best soils is a priority – a nation stands on its soils. Three solutions were outlined: (1) the large productive agricultural landscapes; (2) agriculture-nature parks; and (3) concentrated greenhouses.

The large productive agricultural landscapes

Large agricultural landscapes are much better than the same area in small pieces. Farm operations are more efficient, negative impacts from other land uses and people are less, large open areas support key open-land wildlife species, and long-term protection is easier and less expensive. The large agricultural landscapes of the GBRegion – vineyard area (Penedes), grain area (Calaf Valley, including today's fields-and-woodland around it), livestock and grain area (Vic Valley including western portion to Prats de Llucanes), small-market and family-food-garden areas (Tordera floodplain and Llobregat floodplain and delta), and concentrated greenhouse areas (certain Maresma valley bottoms) – provide highly diverse products and income (Color Figure 42) (Acebillo and Folch 2000, Forman 2004a). As markets and other factors change over time, the prime soils and local farming cultures of large agricultural landscapes maintain the flexibility and stability critical to the region.

Each major agricultural landscape has significant problems, but the basic viability of the Vic agricultural land and the Lower Llobregat floodplain and delta are now threatened. Spread of Vic and nearby towns, plus fragmentation of the land into pieces, threatens the Vic Valley. Urbanization on the Lower Llobregat floodplain would eliminate significant productive land, and probably accelerate contraction and loss of the key water-related resources of the delta. Currently the Penedes land is of exceptional economic importance, and the Llobregat floodplain/delta is of combined major ecological and economic significance. In such circumstances, every hectare counts.

Numerous small farming areas and isolated fields across the region provide stability for its future, but, since they exist at a finer scale, are not included in this regional perspective and plan. Finally, within an agricultural landscape, stream corridors, woods, hedgerows, and scattered shrubs and trees are of particular importance for nature and wildlife.

Agriculture–nature parks

The present set of protected natural areas is overwhelmingly on mountains or hills and designed for forest-dependent values. Therefore the rich nature and biodiversity dependent on farmland, especially smaller fields and less intensive agricultural practices (e.g., in the rapidly urbanizing Valles) have almost no protection in the GBRegion. For example, a significant number of the European-Community-listed rare migratory birds in the region are concentrated in successional habitats of valley farmland (Pino *et al.* 2000).

Agriculture–nature parks that combine active farming and nature protection appear to be the optimum long-term solution (Color Figure 42) (Forman 2004a). Some portion of the farm fields would be maintained in designated successional habitat types. Agriculture–nature parks are most appropriately located in valleys adjacent to large protected natural areas or emeralds, where farm families and the management expertise of conservation-park personnel could be combined to enhance farm production and nature conservation on the parks. Seven proposed agriculture–nature parks are dispersed across the region: (1) southern boundary of Castelltallat; (2) valley east of Sant Quirze projecting into Serralada Transversal; (3) valley east of Manlleu projecting into Serralada Transversal; (4) by Tordera River north of Montnegre-El Corredor; (5) Eastern Valles on south side of Montseny; (6) southwest of Igualada on northeast side of an expanded Ancosa-Miralles protected area; and (7) southeast side of Penedes along an expanded northwest side of the Garraf. Over time the agriculture–nature parks will become major, but different, food-producing areas, further enhancing the economics and stability of the Greater Barcelona Region.

In the important Lower Llobregat floodplain and delta a remarkable "agricultural park" exists next to some protected wetlands, coastal vegetation, and undeveloped (without nearby buildings) coastline (Boada and Capdevila 2000, Acebillo and Folch 2000). Market-gardening (truck farming) and family-food gardens predominate as the region's primary orchard area for cherries, apples, etc., and a major producer of artichokes and vegetables. A rather similar area exists in the Lower Tordera River floodplain. Assuming a significant expansion of protected wetland, coastal vegetation, and undeveloped coastline in each floodplain/delta area, both areas seem appropriate as agriculture parks.

Concentrated greenhouses

The major clusters of greenhouses in certain valleys of the Maresma together represent an important area of food and flower production. Two relatively simple changes would greatly increase the value of these for nature and people. Scattered shrubs, trees, and tiny woods, plus vegetation along gullies would provide habitats and feeding areas for birds and other wildlife. The gain to society would be still greater with the addition of small recreational facilities. Two problem areas apparently need evaluation and solution: greenhouses scattered on surrounding steep slopes, and excessive fertilizer used in the plant-production process.

Water for nature and us

Water scarcity and flooding pose a dilemma. Could water shortage be the Achilles heel of Barcelona's growth and influence? Is an August 2002 flooding of nations and major cities of Northern and Central Europe a harbinger of the future for the GBRegion? Although no single major solution exists, a package of solutions looks promising (Forman 2004a). Each solution in the package addresses two or more specific water-related issues. Engineers and hydrologists combined with aquatic biologists and landscape ecologists are central to the solutions. Four solutions were outlined: (1) too little water: scarcity; (2) too much water: floods; (3) two flows: stormwater and sewage; and (4) wetlands.

Too little water: scarcity

Most global-climate-change experts agree that in this region in the 2020s annual precipitation will be about 10–15 % lower, and 20 % lower by the 2050s (McCarthy *et al.* 2001). During this period, exactly the time period considered in the present plan, increases are likely in: frequency of intense precipitation events; flood hazard; risk of water shortage; and summer drought risk. That suggests both less available water and more flooding ahead.

If it is likely that no new clean fresh water will be discovered and the sky will provide less, two strategies to get large amounts of water within the GBRegion seem possible (ignoring expensive desalinization and bringing water from other regions). Reduce demand, and clean up dirty water. Reducing demand through water-conservation measures is mainly a finer-scale approach focusing on individuals, industries, municipalities, and government. Here the focus is on land use for the long term.

There is lots of water around the GBRegion flowing in streams and rivers. But almost all of it looks and is dirty. Hardly anyone sees fishermen there, for good reason. Hardly anyone would drink out of these never-ending flows of stream and river water. So, cleaning up the dirty water makes a huge supply of clean water available right around us, even into the future (Prat *et al.* 2002, Forman 2004a).

Solutions in the water-scarcity package include stream corridors and riparian vegetation that recharge groundwater, filtering water though the ground to clean it, protecting aquifers, creating a stormwater drainage system so sewage treatment plants can work and clean better, stormwater-created wetlands to clean water, and more (Color Figure 43) (Rieradevall and Cambra 1994, Decamps and Decamps 2001, Prat *et al.* 2002). Also, where do the wastewater pipes from rural homes, villages, and towns go? With a better system for these numerous local sites, much more clean water would be available throughout the GBRegion.

Providing more clean water for residents also benefits natural systems. Riparian vegetation on floodplains, particularly shrub cover on upper portions of stream systems, will provide rich habitat for wildlife and greatly enhance biodiversity. Rejuvenated streams and rivers, native fish, fish-eating herons, other species, and yes, fishermen, will be widespread.

Aquifers with clean water can be well protected with extensive natural vegetation. Pollutants from urbanization and industry basically accumulate in the typically slow-moving aquifer water, thus degrading the water source. To reduce saltwater intrusion into the Llobregat delta (and elsewhere along the coast) (Acebillo and Folch 2000), the most important aquifer in the region, requires minimizing the input of pollutants, minimizing the pumping out of water in both the lower floodplain and the delta, and increasing the normal flow of river water.

Too much water: floods

Global modelers highlight frequent intense-precipitation events and greater flood hazard ahead in the GBRegion. But with roads, parking lots, and buildings, urbanization adds extensive impermeable surfaces. Perhaps worse, pipes and drainage channels often carry the rainwater right to gullies, streams, and rivers. The inevitable result is increased pulse flooding, where water levels rise quickly, reach higher levels causing damage, and drop quickly leaving little water available in the channel (Decamps and Decamps 2001, Forman *et al.* 2003). On this trajectory the "hundred-year flood" may come twice a decade.

The flood-reduction package of solutions contains (Color Figure 43) (Forman 2004a): (1) emerald-network protection against slope erosion and runoff; (2) stream corridors and riparian vegetation that recharge groundwater and aquifers; (3) a stormwater drainage system with pipes that create wetlands (or sponges); (4) disconnecting impermeable surfaces around built areas; and (5) small basin parks along floodplains to hold some floodwater from and for the surrounding neighborhood. The solutions that also address water scarcity are especially cost effective.

Although flood reduction is a prime societal goal, once again natural systems and nature will be big winners with these solutions. Establishing and maintaining woody vegetation along stream and riverbanks provides a multitude of benefits (Decamps and Decamps 2001). Recharging groundwater keeps streams flowing and can support wetland vegetation, an extremely rich source of wildlife and biodiversity. Small neighborhood basin parks along lower portions of a stream system would provide seasonal wetlands for wildlife.

Two flows: stormwater and sewage

Numerous sewage-treatment facilities are spread across the region to serve cities and towns (Acebillo and Folch 2000). Currently stormwater runoff and human sewage enter and flow in the same piping system to a secondary sewage-treatment facility. During rainstorms, especially heavy ones, the systems apparently are often overloaded so overflow raw sewage directly enters a stream or river. Villages with less than about 2000 people have some home septic systems, but mostly wastes seem to be piped to a nearby gully or stream (Narcis Prat, personal communication). From a public-health perspective, treatment of human sewage is the most effective way to stop the spread of E. coli and many other illnesses and diseases. Also, stormwater washes a range of toxic substances from urban and highway sources into sewage-treatment facilities.

Separate drainage systems for stormwater and sewage would provide numerous ecological and human benefits (Color Figure 43) (Forman 2004a). By removing stormwater flows from the sewage-treatment facility, the treatment of human sewage would be more effective. During rainstorms, overloads and overflows of raw sewage would be much less frequent. Disease spread would be diminished and public-health authorities happier. Streams and rivers would look less like a sewer and would smell better. They would become places to walk along and enjoy, almost as if a new linear park suddenly appeared by every major town and city. Yet the ecological gains would be more extensive. With effective secondary treatment of human sewage, the aquatic ecosystems in almost every stream and river would improve. Native fish, herons, and fishermen could thrive.

The stormwater story is more interesting. Many stormwater drainage pipes, rather than leading to sewage-treatment facilities, or even directly to a stream, could simply lead to a depression (*stormwater depression*) in the ground, tiny or large, which may quickly become a wetland (Forman 2004a). Floodplains with a high watertable are good places for wetlands. The depression temporarily holds water from a storm; some evaporation occurs while water is slowly absorbed, sponge-like, into the ground. The water level in the depression tends to slowly drop, and in a seasonal wetland may remain below ground level for a prolonged period most years. The fluctuating water levels support wetland plants and wildlife (Boada and Capdevila 2000), including rare ones in the region. Typically some of the water absorbed in the wetland moves through the soil which filters out many toxic substances that currently flow through sewage-treatment facilities into the stream. So, the fish and other aquatic organisms "like" this separate stormwater system and its many wetlands dispersed across the GBRegion. A long-lost heritage returns.

Apparently a few neighborhoods currently have dual drainage systems. Several areas are identified as highest priorities to begin the process of separating stormwater and sewage flows: (1) all new urbanization projects; (2) all towns in the Tordera River Valley; (3) municipalities along the Llobregat River from Sant Vincenq del Castellet to El Prat; and (4) municipalities upstream of the El Foix and Ter River reservoirs. This is an extremely effective way to accomplish multiple ecological and human goals, and see results quickly.

Wetlands

Long ago wetlands were doubtless scattered across the region, especially on floodplains and near the bases of mountains. Today a small handful remains mainly in the Lower Tordera floodplain and in the Llobregat delta (Color Figure 43) (Acebillo and Folch 2000). Wetlands include open ponds and marshes where groundwater is at or near the surface, but most restored wetlands would be shrubby or wooded swamps where the watertable is somewhat lower. Open areas have more water birds and wooded areas more songbirds, both apparently appreciated by the public.

The most important factor in successful wetland restoration is to get the hydrology right (Mitsch and Gosselink 2000, France 2002, Forman *et al.* 2003). For an extended period during the year, the input of water to a depression (e.g., with clay in the bottom) needs to roughly equal water output due to percolation downward, runoff from the surface, and evaporation upward. On most floodplains, which normally have year-round groundwater, wetland species will quickly arrive and visible success will be obvious. The only places where large wetlands are feasible are the Lower Llobregat and Tordera floodplain/deltas. Small and tiny wetlands could be produced throughout the region, for instance, as depressions at the ends of stormwater drainage pipes and as basin parks by streams and rivers.

Streams, rivers, and blue-green ribbons

Blue-green ribbons (stream corridors) are wide natural-vegetation strips that protect a stream or river, provide for wildlife movement, and include a walkway (Color Figure 43) (Binford and Karty 2006, Decamps and Decamps 2001, Forman 2004a). They provide multiple goals for people and nature, including connectivity for movement by both aquatic organisms and terrestrial species along a valley. To be most effective the natural-vegetation ribbon covers the floodplain, both hill-slopes, and adjacent strips of well-drained upland. Some high-quality streams (e.g., Riera Sorreigs northwest of Vic) are presently protected for part of their length by a stream corridor. Four major issues are addressed: (1) water quality; (2) floodplain riparian vegetation; (3) industries, streams, and rivers; and (4) four rivers.

Water quality

Four *types of water* may be recognized: (1) looks and is dirty; (2) looks clean but isn't (it's not safe for swimming or eating fish); (3) swimmable and fishable (it's safe to swim and eat the fish, but don't drink a drop); and (4) drinkable (potable). An aquatic biologist can easily describe these types rigorously from the aquatic life present (Rieradevall and Cambra 1994, Prat *et al.* 2002). Almost all the streams and rivers in the Greater Barcelona Region have the first type of water; they mostly look and are dirty. They are unswimmable, unfishable, and undrinkable. At present, drinkable water basically comes from wells in uncontaminated groundwater or from surfacewater that goes through an expensive water-treatment facility.

Mineral nutrients, with nitrogen and phosphorus of prime concern in the region, are introduced into streams and rivers in human sewage, livestock (particularly pig) waste, and greenhouse fertilizer use (Color Figure 43). Nitrate levels are so high in groundwater north of Vic that the public water supply is apparently unsafe to drink in 20 to 30 towns. (A few towns near the Tordera River seem to have the same problem). Both surface runoff and groundwater flows carry nitrogen and phosphorus to the Ter River and its tributaries, which thereby become eutrophicated (filled with green floating algae because of the enrichment). The river then carries the nutrients, which eutrophicate the all-important reservoirs. Sodium and chloride from salt mines and mine-waste piles in the Cardona and Navas valleys reduce water quality somewhat down the entire Llobregat to its mouth. Also, industrial pollutants are probably a widespread water problem in the region.

Floodplain riparian vegetation

Typically 80 to 95 % of the water entering a stream system enters in the small upper tributaries, whereas flood hazard is primarily in the lower portion. Water that enters the upper channels is either absorbed into the ground to recharge the surrounding groundwater, or it rushes downstream. *Riparian vegetation* along the gullies is a major determinant of how much enters the groundwater, and how much becomes downstream flood hazard (Decamps and Decamps 2001). Trees with scattered shrubs and debris from previous floods, and even herbaceous vegetation, provide friction against flowing water, which increases absorption into the ground. However, a good shrub cover (e.g., of *Salix* willow) along gullies and streams in the upper portion of a stream system is an especially effective way to reduce flood hazard downstream. Poplar (*Populus*) for wood production, if combined with shrub cover, can provide effective friction. In the GBRegion two factors tend to remove valuable shrub cover, browsing by goats, sheep, etc. and family-food gardening, especially in the long stretches between towns and villages. Both lead to greater downstream flooding.

The lower portion of a stream system receives the water from upstream and is the key flood hazard area (Decamps and Decamps 2001). Floodwaters may cover the floodplain valley and cause property damage including bridge washout. Rich wildlife habitat and biodiversity here are enhanced by: (1) shrub cover; (2) uprooted floating plants being widely deposited across the lower floodplain; and (3) diverse surface micro-topography that creates variations in groundwater and surfacewater conditions present. Important recreation benefits can vary from walking paths near small streams to promenades and "people parks" by rivers.

Industries, streams, and rivers

Traditionally industries were established along rivers and major streams, around which villages and towns grew. Today electric power, fuel supplies, and water supplies reach across the land and support modern industry. The gradual removal of old industries from the streams and rivers of the Greater Barcelona Region would produce an enormous cumulative benefit to natural systems and people (Color Figure 43). (Heavy industry [discussed later] and industrial sites of exceptional architectural/cultural heritage value of course are special cases.) Without the existing almost-non-stop stair-stepped sequence of small dams and elongated ponds, and without the associated need for industrial evaporative-cooling, less direct evaporation from the stream surface would occur. A more-natural rapidly flowing and winding river with stretches of splashing riffles that oxygenate the water would appear. Aquatic habitat heterogeneity would increase, and more normal-period water flows in the lower portions of rivers would be restored.

Furthermore, without the input of industrial byproduct wastes, such as metals and toxic organic substances, stream and river water quality should greatly improve. In principle, light, medium, and heavy industries clean their water, e.g., in waste-treatment ponds, and pour reasonably clean water back into the stream or into a sewage-treatment facility. Older industries, built before basic pollution standards, are doubtless less effective, on average, in maintaining clean streams and rivers. Finally, old industries by towns are often in prime riverside locations readily converted into wonderful parks and places for residents.

Four rivers

Among the streams and rivers of the GBRegion four are especially important, the Ter, Tordera, Besos, and Llobregat (Color Figure 43) (Acebillo and Folch 2000, Forman 2004a). Various tributaries of the four, as well as the Foix River, also have regionally important water flows and other characteristics. Here the contrasting rivers are briefly introduced:

(A) The Ter River system mainly north of Vic drains a large agricultural landscape with abundant livestock, particularly pigs. The Ter then flows eastward into three reservoirs in the Serralada Transversal which are a major water supply for the Barcelona area. Phosphorus, and particularly nitrogen, from livestock wastes heavily pollute the river and cause eutrophication problems in the reservoirs.

(B) In contrast, the Tordera River system is by far the most natural one in the GBRegion. It originates in mountains and in the Girona area, flows through some farmland, woodland, and town areas to a wide lower floodplain delta covered with market-gardening, plus scattered buildings and small wetlands.

(C) The Besos River system drains much of the Valles, formerly covered with small farms and now being rapidly urbanized and fragmented. The Lower Besos is channelized between transportation corridors and passes through Barcelona to its mouth. Although floods occur, the Lower Besos is commonly a small channel of water mainly emanating from sewage-treatment plants. A specific plan for the Valles and Besos is needed.

(D) Finally, the Llobregat, as the largest river system, drains a highly heterogeneous forested, agricultural, and built area representing nearly half of the GBRegion. The Lower Llobregat is a wide urbanization-lined floodplain, normally with rather little river water, that ends in an impressive delta. The delta is covered with market-gardening and family-food gardening, plus the Barcelona Airport, a town, and transportation corridors. The coastal strip includes limited amounts of coastal vegetation, undeveloped coastline, development, and highly significant wetlands. Overall, the Llobregat River is degraded by reduced normal water flow, pollutants from nearby industries and development, overloaded sewage-treatment plants, and structures encroaching on the floodplain.

Built areas and systems

Three major groupings of issues are addressed: growth, development, and municipalities; transportation and industry; and nature and people in municipalities.

Growth, development, and municipalities

This section pinpoints particularly suitable places for growth, inappropriate places for growth, and beneficial forms of growth. Solutions emerge in five areas: (1) key areas for growth and development; (2) green-net areas; (3) municipalities for limited growth; (4) strengthening floodplains, slopes, major food areas, and emerald network; and (5) rest of the region.

Key areas for growth and development

One of the most important strategies for maintaining natural systems and their resources for people long term is to focus growth and development in areas where environmental damage will be low (Forman 2004a). Then specific locations can be selected based on transportation, access to resources, and other important socioeconomic factors (Folch 2000). Five satellite cities around the Greater Barcelona Region and two areas near Barcelona are thus highlighted as promising for growth and development (Color Figure 44): Igualada, Manresa, Mataro/Argentona, El Vendrell, Vic, "Llobregat West", and the Lower Anoia area. Economic bases of the seven areas highlighted for growth might include the following:

(A) *Igualada*: industry; commercial services for the Calaf Valley and nearby Terragona area to the west; some nature tourism focused on three surrounding emeralds; (also see below for the nearby Lower Anoia area).

(B) *Manresa*: industry; a transportation hub; agricultural-products center for valleys around Calaf, Berga, and Solsona; some nature tourism focused on four surrounding emeralds.

(C) *Mataro-Argentona*: industry; residential commuters for the Barcelona metropolitan area; coastal tourism; some of the targeted nature tourism for the Tordera River watershed.

(D) *El Vendrell*: industry; transportation hub; commercial/service center for coastal tourist towns along the Southwest Coast; commercial support for the Penedes agricultural area.

(E) *Vic*: industry; commercial center for the Vic Valley and agricultural area to the west; transportation hub; and nature tourism focused on the Serralada Transversal. Because of the threat of loss and fragmentation of regionally significant agricultural land, moderate and compact growth rather than major or dispersed growth is especially important.

(F)   *Llobregat West* (general area around Gava-Sant Boi-Torrelles and perhaps northward): residential commuters for the Barcelona metropolitan area.

(G)   *Lower Anoia area* (general area around Piera and surrounding towns): a promising area for a heavy-industry center and a trucking transportation center, strategically located near Barcelona and Igualada.

Large adjacent green areas and the sea help provide a human scale (and a linkage to the land) for the City of Barcelona as an extremely appealing urban area (Acebillo and Folch 2000). Urban models and experience emphasize that growing cities do not have to expand contiguously into adjacent areas. Indeed, most of the most appealing, livable, and successful cities maintain large open-space green areas adjacent to the city to provide rich benefits to society.

Although Collserola and the Lower Llobregat floodplain and delta are far too important for the future of the GBRegion to be used for development (Parc de Collserola 1997), the plan includes no major restriction on growth in the Valles or the southwestern part of the Maresma (Forman 2004a). Assuming growth of the Barcelona area will occur, densification of certain existing built areas, along with limited residential expansion in the Llobregat West area, should be considered. If industrial and transportation growth is to occur nearby, the Lower Anoia area should be considered. Where hilly terrain is present, such as in Llobregat West, exceptional environmental sensitivity and planning are required. Some particularly beneficial forms for growth are identified, including the early establishment of local integrated park systems, water-protection systems, and infrastructure/public-transport systems.

Green-net areas

Where several growing towns or cities are in proximity, they could simply coalesce into a large urbanized area like a large inefficient non-city, or, if adequate incentive and investment exists, could be formed into a major new planned city near an existing major city. A third option, whereby each town or city could grow in a limited manner and retain its distinctive identity, is recommended here. To achieve this, a *green net* of vegetation, farmland, or park-land strips is established near municipality boundaries (Forman 2004a). Strips can change, for example, from natural vegetation to parkland according to local wishes, but must remain unbuilt greenspace. A minimum strip width seems appropriate though adjacent municipalities could choose to have wider strips. In this way each individual town can expand within its enclosed area, but not merge with an adjacent town or lose its identity and distinctiveness.

Seven areas seem particularly appropriate as green-net areas (Color Figure 44):
(a) Western Valles (San Cugat-Terrassa-Sabadell area); (b) Central Valles (Parets-
Caldes-La Garriga-Grenollers area); (c) Eastern Valles (Cardedeu-Canoves-Sant
Antoni de Vilamajar area); (d) Lower Anoia (see above); (e) Llobregat West (Gava-
Sant Boi to Sant Andreu); (f) Southwest Coast (Sitges to Segur de Calafell); and
(g) Maresma Coast (Badalona to Malgrat de Mar).

Compact rather than dispersed growth is the other key strategy in a green-
net area. A green net mainly follows the outer boundaries of municipalities.
Compact growth channels development in and adjacent to the existing, usually
central urbanized area within a municipality. For example, a small municipality,
Mollet del Valles, has concentrated urbanization in one portion of the municipality, thus protecting natural systems, productive agricultural land, and cultural
resources, and helping create a strong sense of community.

Municipalities for limited growth

Because of the proximity of regionally significant natural resources,
urbanization near certain cities and towns would be damaging and inappropriate (Color Figure 44) (Forman 2004a). For example, to maintain the prime
soils and integrity of the large agricultural landscapes of the region, very limited growth adjacent to the present built "footprint" of Calaf, Vilafranca del
Penedes, Tordera, El Prat de Llobregat, and a few other towns is appropriate.
Several towns around the edges of critical agricultural-production areas could
grow, but not toward the production area. Limited growth, either in total or in
a particular direction, is also appropriate for municipalities that threaten the
natural resource values of emeralds and their interconnections, valuable stream
corridors, European-Community-listed habitats, water-supply reservoirs, and so
forth. In effect, limiting growth in key areas is just as important as focusing
growth in other areas, to provide for the future of the GBRegion.

Because of the exceptional resources and benefits provided, the Lower
Llobregat floodplain and delta represents the other area where very limited
or no growth is paramount (Forman 2004a). The floodplain and delta provide
the best aquifer for clean water in the region, a rich agricultural park bene-
fitting numerous residents and city markets, the most important wetlands in
the region, flood-buffering benefits, riverine wildlife and fish, a large open space
with vistas, clean air for city residents, and more. Today could be the last chance
to "permanently" establish those resources and benefits for future generations,
which are likely to need them even more. Catalunyans revere unique "patri-
mony places." The combined Lower Llobregat floodplain and delta is a flagship
one.

Strengthening floodplains, slopes, major food areas, and the emerald network

This section recognizes a set of diverse highly significant places in the Greater Barcelona Region where most buildings are inappropriate (Forman 2004a). The absence of buildings on floodplains means less flood damage. No buildings on steep slopes means fewer erosion and sedimentation problems, reduced costs of infrastructure construction and maintenance, less home damage from heavy rains or from fires sweeping up-slope, and less aesthetic degradation of view-sheds. Quantities of high-quality grapes only grow on the Penedes agricultural soil, which is much too valuable for growing houses. Similarly, most buildings in emeralds or in their connecting ribbons degrade key values for natural systems and wildlife.

To accomplish the important objectives for the Greater Barcelona Region, most structures on floodplains, steep slopes, agricultural landscapes, and the emerald network need to be gradually removed. (Exceptions include certain major historical/cultural structures, farm-related buildings, park-management buildings, bridges/viaducts, and some roads.) A wide range of approaches from incentives to regulations should be helpful in accomplishing success.

Rest of the region

Excluding the areas identified in the plan as being especially important for the future, the rest of the region remains as appropriate to accommodate some growth and development. Constraints on development over this extensive area are few and fairly obvious. Don't build in floodplains or on steep slopes, and avoid valuable habitats or damaging populations of rare species (normally places to maintain natural woody vegetation). Do build wisely after thinking both locally and regionally.

Transportation and industry

This section introduces several key dimensions of surface transportation and different types of industry, with solutions outlined in three areas: (1) traffic, public transit, and municipalities; (2) rail, trucking, highways, and wildlife; and (3) industry: large, medium, and small.

Traffic, public transit, and municipalities

Rail lines and numbers of people transported appear to be lower in the Greater Barcelona Region than in many comparable parts of Europe. Vehicular traffic is expected to grow in the vicinity of cities worldwide, but with fewer railroads, it could grow enormously in the GBRegion. Three strategies will help *limit vehicular traffic* growth: (1) reduce commuter traffic by strengthening the nice pattern of many municipalities that have light and medium industry on the edge of town, easily accessible for employees by walking and biking; (2) create walkable/bikable municipality and neighborhood people-parks on the edges of and near towns; (3) invest in public transport systems, such as light rail, small-bus systems, or modular-bus rapid transit, in the region. Attractive stations within walking distance of residential areas and with ample parking should reduce traffic.

Rail, trucking, highways, and wildlife

In a preceding section, seven areas in the GBRegion were targeted as especially promising for future growth (El Vendrell, Igualada, Manresa, Vic, Mataro/Argentona, Llobregat West, and the Lower Anoia) (Color Figure 44). If significant growth occurs in these areas, it is important to establish appropriate rail and highway infrastructure early, both for long distance and for public transport plus highway access to serve local residential, commercial, and industrial areas. A specific transportation plan that fits within, and is consistent with, the GBRegion land-mosaic plan is a priority.

A *truck distribution center* or hub on the outskirts of major cities permits long-distance trucks to load and unload goods, then taken by smaller vehicles on streets throughout the city. Also farmers deliver products to the trucking center, which are transported long distance or into the city. It may be useful to develop a trucking hub in the Lower Anoia for convenient access to the western side of Barcelona and the municipalities near the Llobregat (Color Figure 44) (Forman 2004a).

Busy highway corridors across the land are effectively wide strips of concentrated ecological impacts (Rosell Pages and Velasco Rivas 1999, Forman *et al.* 2003, Iuell *et al.* 2003, Trocme *el al.* 2003). Moreover, highway corridors function as major barriers that subdivide the land both for people and wildlife. Landscape connectivity for larger animals and for people to cross highways requires *wildlife underpasses or overpasses*. (Seven massive overpasses [false tunnels] were observed in the GBRegion.) If designed properly, such wildlife overpasses (and underpasses) are quite effective for crossing by boar, badger, geneta, roe deer, larger deer species, as well as people (Rosell Pages and Velasco Rivas 1999). Twelve locations are pinpointed as especially strategic for new highway crossings for wildlife and people (Color Figure 41). Where a green corridor connecting emeralds crosses a busy highway, a minimum of two underpasses or overpasses is appropriate to provide flexibility. Effective, even innovative, designs of wildlife passages over or under a highway require the combined expertise of animal behavior and engineering.

Industry: large, medium, and small

Light and medium industry, as a major economic force for the GBRegion, is widely dispersed across the land, commonly on the outskirts of municipalities, and thus conveniently located for employment. In contrast, older medium industry is often aligned along streams, and older heavy industry along rivers, one of the key reasons aquatic ecosystems are so degraded. However, today fuel, electricity, water, and space for waste treatment, as key resources for industry, are widely distributed across the land. Relocation of older stream-based medium industries to the edges of appropriate towns and cities would greatly improve natural systems in the GBRegion.

Heavy industry is also very important to the economy of the region, and the most promising approach appears to be relocation to one or more new *heavy-industry center*. Here modern energy-efficient, water-efficient, and minimal-pollution conditions could be established. Some waste-treatment needs of industries are compatible and can be pooled in one area. Two types of heavy industry would be especially appropriate for such a center: (1) existing industry from along major streams and rivers, especially the Tordera and Llobregat (Color Figure 44); and (2) attractive new clean industry that supports new jobs and economic growth.

The optimum location for a heavy-industry center appears to be somewhere in the Lower Anoia area (general area around Piera and surrounding towns) on relatively flat terrain generally away from streams and rivers (Color Figure 44) (Forman 2004a). Good highway access and rail lines (also to the port if essential) could be efficiently developed for transport of materials and goods. Also, convenient public transport and highway access to different municipalities should provide for diverse dependable employment. An alternative strategy would be to provide two or more heavy-industry centers, for example, on the north sides of El Vendrell and/or Manresa.

One existing stretch of the Llobregat River valley from north of Martorell to south of Sant Andreu de la Barca with considerable heavy industry is of particular concern. Apparently the most severe air pollution in the region, especially particulates, NOX, and SOX, is centered here (Acebillo and Folch 2000). Preponderant winds are from the northwest, west and southwest, so pollutants are carried directly downwind to the densest population in the Barcelona metropolitan area and the Valles. River water quality is extremely poor from this area downriver in the Lower Llobregat floodplain. Riverside locations vacated by industry would provide great opportunities to enhance the Lower Llobregat area, and also for flood reduction, the development of attractive local people parks, and perhaps new minimal-polluting development.

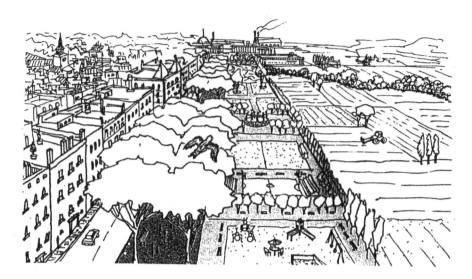

Figure 10.1 Edge park concept for a town or small city in the urban-region ring. For the numerous outward-expanding towns often present in an urban region, generic solutions such as illustrated may be tailored to each town's distinctiveness. The edge park design (here an allee and series of small "outdoor rooms") highlights four values: (a) greenspace amenity for the existing adjoining neighborhood; (b) amenity for a future adjoining neighborhood that develops on the outside; (c) corridor for movement of some species; and (d) connection to a stream corridor which commonly slices across the land and into a town, thus facilitating species movement and a potential walking-trail system. Also note the compact town development and the node of small and medium industry at the edge of town, providing ready access to employment and limiting commuter-vehicle traffic. Adapted from Taco I. Matthews drawing (Forman 2004a).

Nature and people in municipalities

Two quite different targets are addressed here. First, for the numerous towns and small cities present, are attributes which, when multiplied many-fold, may have a major cumulative effect across the region. Second outlines the solution for the most controversial spot in the region, the lower Llobregat floodplain delta right next to the city.

Towns and small cities

Towns and small cities in the region often show a rather nice logic. A historic distinctive residential and shopping core is surrounded by newer multi-story residential and commercial areas, and on the edge of the built-up area, small and medium industry of recent decades is aggregated in spots, and agriculture (sometimes combined with forest) surrounds the area (Figure 10.1) (Forman 2004a). Highways pass by and/or enter the town. That generalized pattern provides nearby employment and is convenient for residents.

The Greater Barcelona Region contains numerous towns and small cities, so solutions at their scale (some introduced earlier) for people and nature have a large cumulative effect: (1) the green net around nearby growing towns is composed of relatively narrow vegetation, farmland, and/or parkland strips; (2) compact, rather than dispersed, growth within the area of a municipality protects natural systems, agriculture and cultural resources, and helps create a sense of community; (3) a ribbon connecting large emeralds contains a walking path accessible to local residents; (4) a green corridor may have a pearl as a small patch of vegetation next to which a people park could be created for the municipality; (5) linear parks along the edge of towns would enhance adjacent neighborhoods by providing shaded walkways and benches, playgrounds, and small areas for active sports (Figure 10.1); (6) convenient walkability, bikability, and public transportation will help reduce traffic congestion and effects; (7) separation of human sewage pipes from stormwater drainage pipes in all future development provides many local water-quality benefits; (8) stormwater drainage pipes and other opportunities exist to create tiny or small wetlands in a municipality to greatly benefit wildlife and nature appreciation for residents.

The Llobregat and us

The Llobregat River is the great river of history and of the Greater Barcelona Region. Today one of the impressive cities of the world lies next to the river valley, and almost all travelers in and out cross it. Yet the river and its floodplain lie nearly invisible to travelers and residents alike.

Imagine creating a world-class *great park* that brings the river valley alive for all to see and enjoy right next to the impressive city (Color Figure 44; Forman 2004a). Attractive overpasses passing over transportation corridors to a glorious magnet on the floodplain overlooking the river, with its cleaner and greater flowing water. Promenades, picnic areas, ball fields, children's playgrounds, nature cafés. Family-food gardens, small-market farms, community-group gardens. Wetlands, biodiversity, abundant visible birds, migration stops. Highlights of history, culture, and heritage. Loop roads, bike routes, jogging and walking trails. Water features, reflection ponds, high points, stunning inspirational views. Everything in the park designed to let the occasional big floodwater pass by. With such a vision, the surrounding area could expect an economic boon. Real-estate value would skyrocket in places. Most family food-gardens and market-gardening would continue as a productive setting for the park. Overall quality of life would noticeably rise.

Two of my experiences there are shared. First, I saw the exact spot where Barcelona gets 40 % of its water, and despite the adjacent water-treatment facility, I will continue drinking bottled water. Second, nearby I was treated to a view of a dozen species of beautiful water birds in the Llobregat, with three flamingos standing tall. Unforgettable.

Finally, a symbol or flagship identity for the great park is useful. For example, the ships of history tied the city and its river to the Mediterranean and the world. The park could highlight huge concrete-and-steel Greek, Roman, Spanish, Columbus', and Catalan ships facing up the Llobregat, as if "at anchor on the floodplain," and for the public to walk up on. Clearly visible to Montjuic, Collserola, and Llobregat West, as well as to highway motorists, train travelers, and airport travelers, even at night, the Great Park "BarcaBarcos" would grab attention and be memorable.

Three plan options for the region

Three plans are presented for the GBRegion as a whole (Forman 2004a). A "Solid Plan" is considered to be the least ambitious plan that warrants reasonable confidence of meeting the objectives at the beginning of this report. The "Most-promising Plan" is stronger (yet within a feasibility constraint), contains more flexibility and stability, and seems relatively certain to attain the objectives. Finally, the "Minimal Plan" is, as suggested, considered to be the minimal solution that may attain the objectives, but success is less certain. The Solid Plan is summarized in some detail, and then the three plans with solutions are briefly compared.

The Solid Plan and a comparison with alternatives

The following are 15 major dimensions of the Solid Plan:

(1) Ten large emeralds widely distributed across the GBRegion are the crown jewels of natural systems and nature, providing numerous resources to people, from water supply to flood control, biodiversity, and heritage values.

(2) The emeralds are interconnected with several types of connections, thus providing for wildlife movement, people walking/hiking routes, and stability for biodiversity through time and surprises.

(3) Six large different food-producing areas are protected and provide stability for the future.

(4) Seven agriculture–nature parks are established to maintain somewhat different farming in smaller areas, add regional stability, and maintain important successional habitats that are progressively disappearing in the region.

(5) More clean water is available in the Ter and Foix reservoirs and in the Tordera and Llobregat rivers, due to steep-hillslope protection, strategic stream-corridor protection, expansion of riparian vegetation that recharges groundwater/aquifers along streams, removal of some industry along rivers, and separation of stormwater and sewage drainage systems in strategic locations.

(6) Much-restored aquatic ecosystems, with associated fish, herons, and fishermen, in rivers and streams are present for the same reasons.

(7) Flooding danger is reduced with steep-hillslope protection, strategic stream-corridor protection, increased riparian vegetation in and along key streams, and basin parks along lower portions of the Llobregat, Tordera, and gullies in the Maresma.

(8) A large increase in wetlands and associated biodiversity is accomplished in the Tordera and Llobregat floodplains/deltas (due to land protection for wetland restoration), and in local valleys across the region (due to outflows from stormwater drainage systems).

(9) Five satellite cities (El Vendrell, Igualada, Manresa, Vic, Mataro/ Argentona) are targeted for potential future growth where ecological impacts would be relatively low.

(10) The Llobregat West area is targeted for potential future expansion for the Barcelona metropolitan area, where ecological impacts would be comparatively low.

(11) The Lower Anoia is highlighted for a potential heavy-industry center (especially for relocation of industry from along the Llobregat and Tordera rivers, and to attract new industry) and perhaps a trucking transportation center, where ecological impacts would be comparatively low and socioeconomic benefits great.

(12) Six green-net areas are established to maintain the distinctive identities of municipalities threatening to coalesce and to provide nearby nature (Lower Anoia, Western Valles, Central Valles, Eastern Valles, Maresma Coast, and Southwest Coast).

(13) Moderation of traffic growth is addressed with people parks and light- and medium-industry on municipality edges, plus expansion of convenient rail lines and stations.

(14) Several critical locations are improved by the gradual removal of misplaced buildings that degrade natural systems and associated human uses.

(15)   A major opening of Barcelona and neighboring municipalities to a magnificent Great Park in the impressive Llobregat River floodplain is established.

The Solid Plan, Most-promising Plan, and Minimal Plan all provide numerous benefits to natural systems and people, and are considered likely to achieve the stated objectives (Forman 2004a). Yet major differences exist (Table 10.1).

Overall, compared with the Solid Plan, the Most-promising Plan adds a range of major benefits for the GBRegion, especially: a stronger emerald network; retention of more agriculture and nature in the Valles; stronger protection of natural systems and nature in the Tordera River watershed; enhanced flood control; greater protection of the Foix and the Ter reservoirs; and enhancement of the Llobregat River. The Minimal Plan may attain the objectives stated at the outset of the report, but relative to the Solid Plan, the primary losses are: eliminates two of the ten emeralds; weakens several of the connections between emeralds; leaves the southwestern portion of the GBRegion poorly served for natural systems; essentially loses the Vic Valley (and slightly weakens the Penedes) as a scarce valuable large agricultural landscape; loses the green nets that provide nature in rapidly growing areas, and prevent coalescence of municipalities with loss of identity; and reduces the magnificence and quality of the Great Park for Barcelona and neighboring municipalities.

A 64-page portion of the report maps and discusses planning solutions for each of 16 Sections within the Greater Barcelona Region (Forman 2004a). Sections were selected for convenience and clarity of presentation: Calaf Valley; Manresa Valley; Cardona-Sallent Area; Serralada Transversal–Llucanes Area; Vic Valley; Igualada-Miralles Area; Montserrat-Sant Llorenc-Montseny; Foix-Penedes; Garraf-l'Ordal; Valles-Collserola; Grenollers-Sant Celoni Valley; Lower Tordera Valley; Maresma; Lower Llobregat Floodplain; Llobregat Delta; Barcelona. These do not represent additional plans for each section, but rather portray more clearly the details and combined solutions for the three regional plans.

Flexibility and adaptability for regional stability

Flexibility to "roll with the punches" and *adaptability* to gradually change over time remained as an underlying planning goal. Adaptability in this case is, to a certain extent, a function of people and their institutions. However providing flexibility on the land may increase the options for adapting to change.

At a general level, *systems*, as networks with components connected by flows, are a useful way to provide flexibility. Many systems, especially those dominated by living organisms, provide adaptability. Nature's system (the emerald network),

Table 10.1 *Major differences in the three plan options for the Greater Barcelona Region*

Most-promising Plan	Solid Plan	Minimal Plan	Attribute provided by a plan
11	10	8	Number of large natural-vegetation-patch emeralds
3+11	2+11	0+9	Connection types between emeralds: 1st number = reconnection zones (stronger); 2nd number = ribbons
6	6	5	Number of large food-producing areas
6+1	7+0	5+0	Number of agriculture-nature parks: 1st number = small; 2nd number = large
21	17	13	Number of small (& medium in 1st plan) PEIN-protected areas integrated into, or connected to, the emerald network
Much	Much	Some	Streams across the GBRegion: improvement in aquatic ecosystems and valuable refugia for biodiversity
Much	Much	Little	Ter River and reservoirs: more clean water and reduced nitrogen/phosphorus eutrophication problem
Much	Much	Little	Foix reservoir: more clean water
Much	Much	Some	Tordera River: more watershed area protected for nature, clean water, and restored aquatic ecosystems
Much	Some	Little	Llobregat R.: more clean water and restored aquatic ecosystems
Much	Much	Some	Wetlands with wildlife and biodiversity: restoration in Tordera and Llobregat floodplains/deltas and across the GBRegion
Some	Some	Little	Coastal vegetation and undeveloped coastline: restoration
Some	Little	Little	Marine littoral zone: restoration
Much	Much	Some	Landscape connectivity across highways for wildlife & people
Much	Much	None	Green-net protection in area of nearby growing municipalities. Added benefits in specific areas within the GBRegion
Yes	Yes	No	Garraf to Foix reservoir to Ancosa: green ribbon
Yes	Yes	No	West and northwest of Garraf: threatened area strengthened
Yes	Yes	No	Lower Anoia: some vineyard area protected with Penedes
Yes	Yes	No	Ancosa-Miralles-Saburella area: 5 PEIN areas reconnected
Yes	Yes	No	Serra de Rubio north of Igualada: new emerald established
Yes	Yes	No	Vic Valley: protected against urbanization and fragmentation
Yes	Yes	No	North or northeast of Vic Valley: two agriculture-nature parks
Yes	Yes	No	El Moianes area: four small PEIN areas connected
Yes	Yes	No	North of Montseny: reconnection zone protected
Yes	Yes	No	Upper Tordera River Valley: reconnection zone protected
Yes	Yes	No	East of Mataro; relocate rail line away from coastline
Yes	Yes	No	El Prat/L'Hospitalet: river corridor with pearl
Yes	Yes	No	Lower Llobregat: traffic rerouted for Great Park BarcaBarcos
Yes	Yes	No	Llobregat delta: coastal resources protected as a whole
Yes	No	No	El Foix River: ribbon protection of waterway above reservoir
Yes	No	No	Collserola: eliminate threat of bisection/fragmentation
Yes	No	No	Prats de Llucanes area: ribbon protection of valuable stream
Yes	No	No	Ter River watershed: extra protection for river and reservoirs
Yes	No	No	Eastern Valles: more area of agriculture and nature sustained
Yes	No	No	Santa Coloma area: ribbon protection of key Tordera stream
Yes	No	No	Maresma slopes: reduce steep-slope damage by greenhouses

agricultural system, transportation system, groundwater system, stream/river system, and economic system are all examples present in the region. Negative feedbacks in systems inherently provide stability. Also several of the systems contain a hierarchy that may also provide stability.

Redundancy or multiple components is another way to offer flexibility. Maintaining five major agricultural landscapes rather than fewer, four important rivers with clean water rather than one or none, ten major emeralds rather than eight, two major floodplain/delta aquifers rather than one or none, and so on, provides for stability. Optional routes for flows and movements also help accomplish the broad goal. Emeralds with three or four connections to other emeralds mean that with, e.g., climate change or new urbanization pressure, species may move in different directions to more suitable locations.

Still another key dimension of flexibility and adaptability is *diversity*. Maintaining five types of major agricultural landscape, six types of natural emerald, five types of satellite city, and so forth is better than one type of each. When a disturbance or disease degrades or eliminates one type, the other four still sustain resources.

Current trends are certainly threatening the future of some resources, such as water-supply reservoirs, floodplain/delta aquifers, agricultural resources, and coastal water quality in the Barcelona Region. The diverse types of flexibility and adaptability built into the three plans should enhance stability for the region's future.

Major results and recommendations

Nine major results and recommendations resulting from specific analyses are listed (Forman 2004a). Other results from finer and broader lenses could be highlighted. Two succinct broad conclusions are illustrative. The urban region as a distinct area matters in a big way. Natural systems and their uses in the region matter even more:

(A) *The "emerald network" as the backbone of natural systems* in the region is a system of connected large natural park areas, which provide aquifer protection, wildlife movement, trails for walking and recreation, nature conservation, and other benefits for society.

(B) *A set of diverse large agricultural areas plus smaller "agriculture–nature parks"* provides food production, open land, important successional habitats, and economic flexibility.

(C) *Protected highest-quality stream valleys and widespread restored small wetlands* significantly enhance aquatic ecosystems, biodiversity, and paths for walking.

(D) *Three river solutions*: (a) the Tordera watershed provides especially high quality nature and natural systems as a magnet for visitors and nature tourism away from the crowded coast; (b) the Ter River, its critical reservoirs, and the Foix reservoir have better protection, more clean water, and less nitrogen/phosphorus-caused pollution; (c) the Llobregat River Valley at the heart of the region has more clean water, improved aquatic ecosystems, less flood hazard, and an enhanced richness of nature.

(E) *A set of strategic places for growth and for limited growth* is highlighted to enhance communities and to maintain or strengthen the key natural systems for the region.

(F) *An array of solutions to enhance towns and small cities* repeated across the region involve industry, transportation, traffic, parks, streams, stormwater, sewage, wetlands, nature, recreation, and town identity and distinctiveness.

(G) *An impressive park along the Lower Llobregat floodplain* offering vast recreational opportunity and a visible symbol to all travelers and the world is established and appealingly linked to the people of Barcelona and many neighboring municipalities.

(H) *Fine-scale spatial solutions for widely repeated small locations*, i.e., gullies, streams, highways, and the edges of towns, have an especially important cumulative effect across the region.

(I) *Flexibility and stability for the region's future* are provided with several strategies, including large agricultural areas, a diverse economic base, the emerald network, dispersion of natural systems, more clean water, protection of the two most important water resources, diverse flood-reduction mechanisms, and relatively self-contained municipalities.

The land mosaic outlined in this Greater Barcelona Region project and plan provides for an enormous improvement in nature and natural systems, as well as human uses of them. Clearly establishing the underlying principles for the spatial solutions, and applying them to the distinctive patterns of this region, is important for one to understand, apply, explain, and defend the patterns and plans.

History may record that Barcelona embarked on a pioneering trajectory in the opening decade of this century, a route that will reverberate far beyond for others to follow. The land of the Greater Barcelona Region, and its intelligent arrangement, represent capital, heritage, nature's system, inspiration, and home. Together these provide a vibrant future. A visionary land mosaic, where nature and people both thrive long-term, lies within grasp just over the horizon.

Reflections two years later

A series of additional instructive and important messages, some quite intriguing, has emerged in the two years since the planning project (Forman 2004a) was completed. For international readers, the concepts are useful, while the specific places and projects are included to provide real examples. These are succinctly organized around three phases: (1) the planning project process; (2) developments after plan completion; and (3) thinking about implementation and the initial steps.

The planning project process

Throughout the process I worked closely with a team of three regional experts and an organizer/GIS/planning specialist. They took me throughout the region, taught me, arranged for consultations with knowledgeable leaders, provided piles of information, and openly answered my never-ending questions. All became good friends. I can only think of eight things they didn't tell me, and the reasons are useful for such a project.

(1) *A high-speed rail line, Madrid to Barcelona to France.* This was under construction, so near the beginning I asked for a map of the route and station locations. "Don't give that to Forman," was the chief's response. Partly they did not know and partly the decision was politically charged, but mainly, I think, he was telling me to produce the best plan possible without being influenced by other existing or proposed plans. I liked that.

(2) *A long-proposed east–west highway across the large Collserola park.* Probably they did not mention it because the idea was nearly dead and construction highly unlikely.

(3) *A proposed new north–south highway across Collserola.* I never learned why this was not mentioned; perhaps the highway was considered unlikely.

(4) *France-to-Madrid highway being built to pass through Vic.* Probably not mentioning it was simply an oversight, or it got lost in our three languages.

(5) *Important aquifers in the region.* My team and also outside specialists searched, but came up with nothing for me. It is possible, though surprising, that the information does not exist; more likely it is buried somewhere and I needed to consult with a hydrologist.

(6) *Groundwater in the Maresma area along the coast.* The same reason as in the preceding case applies.

(7) *Gracious wife of one of my team members is the Catalan President's niece.* Probably they did not wish to affect or complicate my work.

(8) *Cultural symbols across the region.* In this case the information was provided but, uncharacteristically, it took a very long time to appear. I later learned that in an ongoing election process the main political party opposed to the interests of many of my contacts was recommending the expansion of cultural symbols.

With the benefit of two years of reflection plus three subsequent visits to the region and conversations with a range of people, I would have added or increased emphasis on the following issues. These are details. The plan presented, both in concept and in specifics, is almost exactly as I would present it now in hindsight:

(A) If I had visited the rural tourism areas outside the Barcelona region, minor changes in the Tordera portion would have been made.

(B) I would have pointed out the value of wildlife reintroductions, as successfully done for fallow deer, stork, and otter in a large park northeast of the Barcelona region.

(C) Agriculture–nature parks were highlighted for protecting rare migratory birds, but the large natural areas are especially important to protect rare resident Catalan birds.

(D) Protected estuaries by river outlets and deltas, such as of the Llobregat and Tordera rivers, are normally important nursery sources for many shellfish and fish along the coast.

(E) The France–Vic–Madrid highway under construction will relieve some truck traffic from metropolitan-area highways, but will place a large industrialization and urbanization pressure tending to degrade agricultural, water, and natural resources of the Vic Valley.

(F) An infrequent major wildfire, such as the 1994 Castelltallat fire, on the steep slopes around the three Ter River reservoirs could cause large erosion, sedimentation, water quality, and water capacity problems.

(G) A high-speed train station at Vilafranca, rather than at El Vendrell, would place large industrialization and urbanization pressures tending to degrade the valuable Penedes wine-growing area.

(H) A proposed new highway with tunnels north–south across the Collserola Park, is likely to cause fragmentation and degradation of its multiple resources so valuable to the Barcelona population and its future.

(I) The rather large Montnegre and El Corredor natural areas are still connected by surviving, but unprotected, woodland, which represents a strategic protection priority.

(J) Time constraints prevented real consideration of greenspaces within the City of Barcelona, where, in a European context, greenspace usable by people is limited. Also, converting the existing scattered green spots into a functioning system of integrated greenspaces, plants, and animals for people and nature would be of considerable value.

(K)   The plastic glasshouses in the Maresma warrant fuller evaluation of locations, slopes, types, crops, benefits, environmental problems, and values for the public.

(L)   Groundwater and water tables across the Maresma should be elucidated for addressing the torrentes and other land-use issues there.

(M)   The proposed growth area west of the lower Llobregat floodplain would be focused on the area with less-steep slopes, from about Torrelles and Santa Coloma to Gava/Viladecans/Sant Boi. A major emphasis on pioneering environmentally sensitive communities is appropriate to seriously protect natural water, soil, vegetation and biodiversity, and to make walking rather than motor vehicles central.

(N)   A greater emphasis on removal of buildings (many apparently illegal) on steep slopes in and around the Collserola Park would provide a wide range of major benefits.

(O)   A stronger emphasis on the convergent interests in addressing the strategic area around the lower Riera de Rubi would include clean water supply for Barcelona, and connectivity for walkers and wildlife between Collserola and major conservation areas to the north and west.

(P)   More insight into planning opportunities for the important Vallees landscape would be valuable, including its current distinctiveness as an agricultural landscape, the green-net concept, the recreational opportunities for a system of small woods connected by trails, a model area or two of a protected stream valley (e.g., the Conyas River), and at least partial solutions for the degraded Besos River and its tributaries.

(Q)   Describing the goals and regulatory conditions of the Agricultural Park would be useful, both for possible extrapolation to the Tordera floodplain and the proposed agriculture–nature parks, but also for market-gardening next to other cities worldwide.

(R)   A brief section on aesthetics in the region based on regional/local data and surveys could provide ideas for improving aesthetics in plans and projects, and also encouraging the public to take a greater interest in and care for the land and their communities.

(S)   Alternatives to an expanding Barcelona airport on the best aquifer and market-gardening area in the region might include a major new location near Manresa/Igualada/Calaf or might include a few specialized or regional locations in or near the region.

(T)   In view of urbanization trends over time, the string-of-pearls connection type seems especially useful in the inner portion of the region.

(U)   In view of urbanization trends over time, large underpasses or overpasses where wildlife crossing of highways is a significant factor seem most useful in the outer portion of the region.

(V)   Consistent with the urbanization models and patterns of Chapter 8, a greater emphasis on channeling future development to satellite cities seems valuable for the GBRegion.

(W)   The seeds of destruction of the Penedes and Vic agricultural landscapes are conspicuously planted, yet strong planning can prevent these valuable Catalunyan resources from mimicking the former Central Valles productive landscape.

(X)   In recognition of newer climate-change data and sea-level rise expectations of 0.5 to 5$^+$ m, identifying low-elevation areas ripe for inundation and storm surges, and gradually replacing built structures there with multiple-value wetlands, appear to be an increasingly important priority.

Developments after plan completion

When nearing completion, the plan was presented to specialists and the educated public at an evening presentation, with newspaper and television coverage in the following days. Several weeks later COAC, the professional association of architects and planners, reported in its bulletin on the plan and presentation, including one image of the proposed Great Park on the Llobregat floodplain.

A few months later the final plan report was turned in on time, with reviews and a few detailed revisions made shortly afterward. The report was translated into Spanish and published by a top quality Barcelona publisher, Editorial Gustavo Gili, as a 150-page, attractively designed book. The book was presented to the President, Chief Architect, and Mayor, who gave a formal talk about it to other leaders and the public at a ceremony in the new city Forum. The book was sold at bookstores in Spain, and I gave it some international exposure. Some knowledgeable interested people read and absorbed the book's themes and solutions. Some leaders expressed support or reservations and some grassroots groups used it as support for their objectives. No big controversy erupted.

Broader government, political, and personnel issues also played a role in the plan's effect. The Mayor, with a considerable interest in the plan, had just been reelected for another term. He encouraged key people in his administration and mayors of surrounding municipalities to absorb the ideas and use the book, and displayed it at expositions. The Chief Architect, as the prime catalyst and supporter of the project, wore two or three hats which progressively changed – from Chief Architect and Barcelona Regional Director, to Transportation and Urbanism Director, Barcelona Regional CEO, and Dean of the Swiss Mendrisio Architecture School – and big intra-city issues also consumed his time. Nevertheless, several leading lights in the Barcelona Regional planning operation worked with him to accomplish significant initiatives and begin gathering momentum. The Catalan President, originally a planner, was also just elected, but quickly he became committed to compelling national issues, while constantly balancing the differing interests of a three-party governing coalition. Still, he appointed many excellent people, including a leading geographer/urbanist, who, along with key people in the agencies from a previous long-term administration, made good progress in some areas related to the plan.

Thinking about implementation and initial steps

The initial post-plan phase attempting to link a complex plan with possible implementation is inherently full of pitfalls. In this case three disparate perspectives seem useful: (1) types of responses and actions; (2) nature, food, water, and built areas; and (3) some broader messages.

Types of responses and actions

Most plans end up on the shelf. Yet so much here was important on the ground. Several types of responses and actions are instructive:

(a) The Mayor as a top leader touted the book and its themes, discussed proposals in it with key mayors of the region, and encouraged regional collaboration.

(b) Local groups used the plan to push agendas in certain areas, e.g., for protection and restoration of torrentes in the Maresma and against industrialization and urbanization in the Penedes.

(c) Apparently no or little movement occurred for some pieces, e.g., related to commuter rail lines, El Vendrell, and the Tordera area ecotourism opportunities.

(d) Interest was expressed in the agriculture-nature park, but the concept was not elaborated fully enough, and thus some lack of understanding was expressed. Mentioning possible models or examples, as in Switzerland and Germany, would help.

(e) Some items consistent with the plan happened without the plan playing any or much role, e.g., eliminating a proposal to pipe in water from the L'Ebro River Basin to the west, and approving a land-use plan (long in the works) for an area north of Manresa and Calaf.

(f)   Some items facilitated or catalyzed by the plan moved rapidly ahead to completion, e.g., a blue-ribbon panel report on the multi-sectoral effects of climate change.

(g)   Other items facilitated or catalyzed by the plan were initiated, such as a group appointed to evaluate future transportation capacity needs within the region which would then be fit to the spatial plan, work on a government policy on biological corridors and connectivity, and work on an urbanization policy for the public to consider.

(h)   Some items previously on the table were energized or accelerated by the plan, e.g., a land plan being discussed for the Barcelona metropolitan area, a Catalan law of urban and land planning, a coastal zone land-protection plan with specific spots protected, removal of illegal houses on steep slopes especially on the eastern side of Collserola, and discussion of creating a regional corporation or planning commission with greater geographical coverage and mandate than the existing Barcelona Regional planning organization.

(i)   Some actions have occurred that are counter to the plan, e.g., expansion of the Barcelona airport on the valuable delta.

(j)   Some items consistent with the plan were not even in it, e.g., a large proposed agricultural park in the Vallees.

Nature, food, water, and built areas
Consider examples of post-plan developments relative to these four key themes of the plan. Progress in the nature or emerald-network area includes: (1) acceptance of the basic emerald-network idea by several organizations/agencies; (2) plans or progress for corridor connections both northwestward and eastward from the Collserola Park; (3) protection of small lands by Sant Celoni to increase natural connectivity with Montseny northward and the Maresma southward; (4) new corridor connection plans developed for several locations from Collserola to Serralada Transversal; (5) building permits temporarily frozen on the south side of Collserola; (6) accelerated building removal from steep slopes in Collserola; and (7) proposals for wildlife and walkers' overpasses to cross highways in various locations.

The food portion of the plan apparently has mainly brought recognition of the threats to large agricultural landscapes, including: (1) Penedes wine area threatened by industrialization, urbanization, and consequences of a possible high-speed-train station; (2) lower Llobregat floodplain and delta under a two-pronged threat, airport expansion and urbanization from adjacent municipalities; (3) towns north of Vic coalescing and pushing to link up more explicitly with Vic; (4) an accelerated east–west France–Vic–Madrid highway project threatening the Vic Valley resources; and (5) people intrigued by the agriculture–nature park concept but not fully understanding it.

The water portion of the plan has mainly generated interest relative to water quantity and water quality: (1) a proposal to transfer water to the region from the L'Ebro River Basin to the west has apparently died; (2) a water expert has identified more than a dozen feasible cost-effective ways to increase clean water in the region; and (3) a few of the ways involve cleaning up some of the widespread dirty surface water in streams, rivers, and reservoirs.

The built area and systems portion of the plan also relates to several developments: (1) the high-speed Madrid–Barcelona–France train may have stations at Vilafranca and the Barcelona airport, both locations being of exceptionally high natural-systems value; (2) the value of focusing growth on a handful of satellite cities, rather than further expansion of Barcelona, has been recognized; (3) transportation, especially rail lines, among the proposed-growth satellite cities is important; (4) at least one key mayor expressed support for the green-net concept; (5) Vic as a proposed satellite city will grow, perhaps too much, with completion of an east–west highway; and (6) eventually moving part of the Maresma rail line away from the coast makes good sense.

Some broader messages

Finally, some broader messages from the planning project have emerged in the past two years:

(A) No movement toward implementing the plan as a whole, or establishing a trajectory for it, is presently detectable.

(B) A Barcelona colleague recently quoted a phrase, "La Caixa runs everything," to me, apparently meaning that money or economics mainly determines what happens. Another colleague mentioned a big decision made by five political leaders in a room together who supposedly ignored all plans, apparently suggesting that actions are mainly determined by politics.

(C) A recent economic and construction boom lends urgency to several parts of the plan, especially the resources and opportunities that are moving toward a threshold of being no longer possible.

(D) Separate planning for different sectors continues, both in partial isolation and partial competition, in contrast to land planning where pieces serve multiple uses and fit together into a lucid logical whole.

(E) The plan, I was told, surprised everyone because of its comprehensiveness, its potentially implementable pieces, its emphasis on a functioning Greater Barcelona Region, and its compelling portrayal of the region's natural systems as important and people's uses and dependence on them as important. It was designed to be implementable as a whole, or in pieces and over time, and with the pieces fitting logically and well together.

(F)   A leading water expert pointed out to the public and press that this was the first such plan where water was one of the major central themes.

(G)   An ecology and land expert made perhaps the most astute observation when he pointed out that, although the plan as a whole would be difficult to implement, the new ideas have changed the frame of reference both for thinking about the region and its natural resources, and for all future plans ahead.

(H)   Finally, what's the value of doing a plan for such a large complex area as an urban region? The initial post-plan phase here highlights the following benefits. A plan catalyzes new action, facilitates or accelerates ongoing action, encourages people with ideas that are consistent with the plan, discourages proposals not consistent with it, puts new ideas on the table, highlights different priorities, and changes the frame of reference for thinking and for future plans. Taken together, the cumulative value even at the initial phase is considerable. Shouldn't all urban regions have such plans?

11

Gathering the pieces

Fancy being the cook with an unlimited budget preparing for an evening extravaganza. For days ahead you have visions of your visual and gustatorial creations, and begin gathering the pieces. Some fruit is not yet ripe, the fish not fresh, and your special chocolate unavailable. Yet unexpected surprises also appear – durands, anonas, sea cucumbers, and a glorious French wine. So, continually dropping and adding and sorting, you accumulate the ingredients to combine into magical culinary masterpieces.

The time has arrived in this book to begin gathering the pieces for promising urban-region land mosaics. The countless and infinitely diverse patterns appear from all of the preceding chapters and elsewhere. This chapter only begins the gathering process, as the reader, like the cook, will accumulate many other useful components, before fitting them together into masterpieces.

The lead-off section (Settings and forms of urban regions) uses a big-picture lens to identify useful patterns and processes. Then the section (Ability to extrapolate the Barcelona solutions) evaluates which of the many patterns provided for Barcelona (Chapter 10) apply widely to urban regions. The third section (Local communities, ecology, and planning) highlights the importance of the finer-scale building blocks in understanding and creating an urban region. The final section begins to explicitly evaluate the diverse pieces of the puzzle, placing them in three piles (The good, the bad, and the interesting).

Settings and forms of urban regions

Identifying major urban-region types is useful, both to get past the infinite variations evident around thousands of cities, and to identify some common threads or themes of wide applicability. Four broad categories of urban-region types are introduced. The first (Riverside cities and coastal cities) highlights the physiographically determined setting, and its implications for process and pattern in an urban region. The second category (Metro area and urban-region ring) focuses on the primary internal form of an urban region, which affects flows and land use. The third grouping (Other diverse types of urban regions) is effectively a basketful of delicacies waiting to be examined for their interesting patterns and processes. Most of the urban-region types identified in this section are illustrated in Color Figures 2–39.

A final section (Effect of and effect on other regions) places an urban region in the context of other regions, near and far. This emphasizes interactions and the role of the boundary between regions.

Overwhelmingly urban regions manifest a rather rounded compact form, mainly a result of urbanization from a downtown nucleus plus strong influences spreading outward over the surroundings. In fact, the basic spatial model for an urban region is a donut. In the *donut model*, the hole represents the all-built metropolitan area and the delicious portion around the hole is the urban-region ring (Figure 11.1).

Riverside and coastal cities

Riverside cities

The most prominent modification of the basic model form is a slice or line through the donut's center, representing a major river. The consequences and insights from this *sliced-donut* form, with river bisecting city and metropolitan area, are considerable. Riverwater flows one way, entering the urban region at a point on its boundary, flowing across the ring, entering the metro area, leaving the metro area, and flowing out of the region. Thus a mass of clean water or polluted water from an adjacent region enters at a point. Land uses in the urban-region ring then commonly add agricultural runoff and perhaps stormwater pollutants to the river water. The metro area normally adds industrial pollutants, sewage outflow, and much more stormwater pollution to the river. Polluted water extends far downriver across the urban-region ring, and may continue into the adjoining region.

In addition, the two halves of the urban region tend to tilt toward one another, with streams flowing across the surfaces to the river. Streams that flow into the metro area usually disappear, as streamwater combines with polluted stormwater and flows through pipes into the river. Hence the arrangement of land uses across the region may affect each stream, and each river section, differently. Upriver of the metro area, riverwater is normally cleanest, so natural vegetation cover such as forest/woodland in this portion of the region has

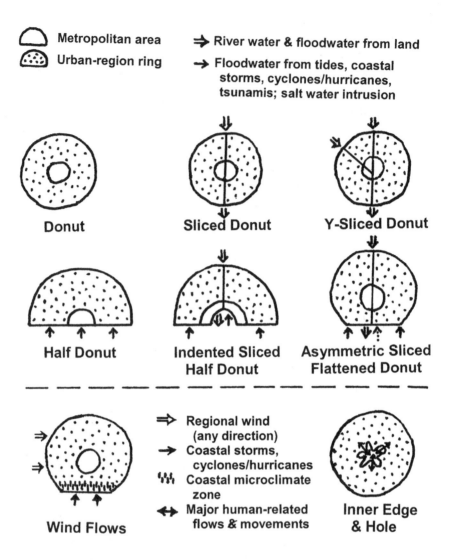

Figure 11.1 Donut model variations representing major structural and functional differences among urban regions. Straight line within urban region is a major river and straight line on edge of region is a seacoast or lakeshore.

particular value in maintaining a relatively clean river. That enhances nearby recreation/tourism opportunity and sometimes water supply. Natural vegetation across the downriver land can provide clean water that dilutes and helps clean the polluted riverwater.

Fish moving along the river are typically blocked by severe water pollution downriver of, or in, the metro area. Yet if riverwater is clean enough, fish populations moving from stream to stream may be sustained, a goal especially attainable in the upriver portion.

The sliced-donut form provides insight into key locations for numerous other characteristics of importance to natural systems and their uses by society – groundwater to provide clean water supply, flood hazard areas, riverside recreation sites, where riparian vegetation can have the greatest benefit, areas to protect aquatic ecosystems, and the best and worst areas for urbanization. The following sliced-donut types represent variations on the basic model, each with readily estimated implications for people and nature:

> *Large river bisecting city (and metropolitan area).* Bangkok, Brisbane (Australia), Paris, Cairo, Columbus (USA), Delhi, Edmonton, Novosibirsk (Russia), Bamako, Rome, Warsaw
>
> *City located at intersection of large rivers.* Lyon (France), Minneapolis/St. Paul (USA), St. Louis (USA), Portland, Seville (Spain)
>
> *City overwhelmingly on one side of a river.* Quebec, Vienna, Memphis (USA)
>
> *City in a major valley between high ridges.* Caracas, Bogota, Zaragoza (Spain), Guatemala City, Kayseri (Turkey), Grenoble (France)
>
> *City on a delta.* St. Petersburg, Karachi, New Orleans (USA), Rangoon (Myanmar/Burma), Rotterdam (Netherlands), Cairo, Cuttack, Lagos, Vancouver (Canada), Calcutta, Shanghai

Coastal cities

Flattening one side of a donut to represent a coastline makes the donut model work well for coastal cities. The *flattened-donut* form varies in the degree of flattening and thus length of coastline. An extreme flattening creates a semicircular metro area and region, such as Barcelona or Toronto. The city is located on the coastline subject to, e.g., hurricanes/cyclones, tsunamis, and underground saltwater intrusion, but also with the glories of a seaside or lakeshore location. More moderate donut-flattenings represent conditions where the city is set back from the coast, somewhat protected from the big problems just mentioned.

A second geometric component of the flattened donut is to add the central river slice or line so that it is perpendicular to the coastline. Most coastal cities originated where a river met the sea or a large lake. A third key spatial component, an indentation or nibbled-out area at the intersection of the lines, mimics the typical coastal bay or natural harbor present where a river meets the sea. In short, the *flattened-sliced-indented donut* model represents most coastal urban regions.

Just as for the riverside cities, these coastal donut geometries are good indicators of how the urban region is structured and how it works. The directional river flow, tilted land surfaces, and stream flows as described above are evident. In addition, along the coast the land commonly tilts and streams flow toward the sea. Water from the urban area pollutes a harbor or bay, especially one with limited circulation. Irrespective, pollution normally extends into the near-shore area beyond the coastline, where polluted water is then spread by currents or wind particularly in one direction along the coast.

A coastal microclimate, often extending inland 5 km (3 mi) or more, brings haze or fog, on- or off-shore breezes, and cooler or warmer conditions, which change seasonally according to the temperature difference between seawater and land. Coastlines in an urban region are usually lined with recreational resort development, and protected natural areas along the coast tend to be small and ever-threatened by human activities. Also coastal cities commonly have a "flattened semicircle" major highway around them for through travel and transport along the coast.

Onshore winds and storms and waves threaten and periodically damage coastal areas. Low-lying areas where river meets sea, especially around bays, formerly were largely wetlands, but today often much urbanized. Flooding from coastal storms, particularly at high tide, tends to cover these low areas. Ironically, the river brings floodwaters from inland storms to the same low areas.

In addition, the *direction of prevailing winds* indicates much about a region. Downwind of a metro area, considerable air pollution (and sometimes a bit more rain) is common, so recent development may be more on the upwind side. Wind direction is critical in locating airports and flight patterns, and reducing hazards such as a sudden large pollutant release from a nuclear power plant. Furthermore the upwind side of an urban region is strongly influenced by conditions in the adjoining region, from eroded-soil particulates to wildfire, industrial pollution, or cooling by a large lake or forested area. In short, the flattened side of the donut, as well as the central slice, indentation, and prevailing wind, provide considerable insight into pattern, process, ecology, and planning in an urban region. The following represent variations on the coastal city theme:

> *Coastal city without a prominent bay.* Barcelona, Toronto, Chicago, Rabat
> (Morocco), Algiers
> *Coastal city with a prominent bay.* Buenos Aires, San Diego, Marseilles,
> Lagos, Helsinki, Guayaquil (Ecuador), Tampa (USA), Lisbon
> *Coastal city with islands providing partial storm-protection.* Kuala Lumpur,
> Boston, Port Moresby (Papua New Guinea), Stockhom, Seoul

City on a peninsula. Manado (Indonesia), Freetown (Sierra Leone), Monrovia (Liberia), Beira (Mozambique)

Coastal city on a strip between water bodies or mountain ranges. Miami, Kobe (Japan)

City just upriver from coast, typically subject to floodwaters from both directions. Bangkok, London, Sapporo, Kuala Lumpur

Metro area and urban-region ring

The *inner-edge-and-hole of the donut model* also tells us much about an urban region. Thus a large expanding metropolitan area in the center tends to squeeze a narrow urban-region ring, which sometimes is also squeezed by urbanization in an adjoining region. Narrow rings are limited in resources, such as appropriate space for a water-supply source, sufficient recreation/tourism sites, a suitable heavy-industry-center location, market-gardening areas, biodiversity-rich large natural areas, and differing farmland landscapes. More generally, narrow rings around a large metro area lack flexibility and stability for a region's future.

Interestingly, the edge or border between metro area and urban-region ring is a useful indicator of key flows and movements in the region (Forman 1995). A compact metro area may largely result from a greenbelt or urban growth boundary or certain transportation planning. Compact metro areas may have much movement within them, but relatively little or modest radial movement out to, and in from, nearby unbuilt areas. As a consequence, the nearby surrounding urban-region ring seems to maintain its own functioning land-use integrity.

In contrast, a metro area with several or many prominent built lobes and greenspace wedges suggests considerable interaction in radial directions between metro area and urban-region ring. People living in the city have ready walking/biking access to greenspace that connects them to surrounding countryside for recreation. Conversely, wildlife and other species use the coves to continually move inward and enrich city parks. The cooling effect of greenspace wedges minimizes a city's heat-island effect, and wind channeled along certain coves helps keep the city's air clean. Also, built lobes usually contain significant radial transportation routes connecting city and surroundings.

The compact form of a metro area may result in part from the absence of a major ring highway. Ring highways seem to open up the surrounding inner portion of the urban-region ring to development. A diffuse metro area border often results. The diffuse border suggests the presence of an intensive network of local roads, perhaps a sprawl of homes on relatively large lots, extensive habitat loss, fragmentation of natural areas, proliferation of non-native species, degraded stream networks, and polluted groundwater for wells. Indeed over time the ring road frequently gets increasingly absorbed by a growing metropolitan area.

In summary, the metro area and urban-region ring category, like the categories for riverside cities and coastal cities, provides important insight into the structure, functioning, and potential planning of urban regions. The donut's two inner-edge-and-hole indicators are the form of the inner edge of the urban-region ring, plus the relative size of the metro area and urban-region ring. These are illustrated as follows:

> *Large metro area and narrow urban-region ring.* Philadelphia, Chicago
> *Small metro area and wide urban-region ring.* Abeche, East London, Erzurum
> *Compact metro area and wide urban-region ring.* Bucharest, Winnipeg (Canada), Edmonton
> *Metro area and urban-region ring interlinked by major greenspace coves and built lobes.* Moscow, Stockholm, Brasilia, Santiago, Kuala Lumpur
> *Diffuse border between metro area and urban-region ring.* Atlanta, Milan (Italy), San Antonio (USA), Washington, San Jose (Costa Rica)
> *Metro area delimited by protected greenspace of urban-region ring.* London, Portland

Other diverse types of urban regions

Using the same approach and logic as for the preceding three urban-region categories, the reader can readily learn much about the diverse urban-region types following. In fact, the patterns that emerge in this whole section on settings and forms of urban regions are easily gathered together to understand and creatively plan regions:

> *Prominent volcano providing aesthetics and recreation (and air pollution).* Naples, Kagoshima, Kayseri (Turkey), San Jose (Costa Rica), Sapporo, Mexico City
> *Metro area adjacent to mountain range.* Florence (Italy), Kobe (Japan), Denver (USA), Santiago, Sapporo, Valencia (Venezuela), Calgary (Canada)
> *City ringed by mountains.* Guatemala City, Grenoble (France), Ulaanbaatar
> *City on a prominent environmental gradient.* Madrid, Nairobi, Jabalpur (India), Kuala Lumpur
> *Metro area ringed by a natural vegetation matrix.* Iquitos, Tehran, Jodhpur (India), Abeche
> *Metro area ringed by a cropland matrix.* London, Nantes, Bangkok, Indianapolis (USA)

No major surface water body by metropolitan area. Madrid, Mexico City, Tehran, Bogota, Beijing, Athens, Quito, Johannesburg, La Paz, Santiago, Abeche

Large wetland by metro area. Buenos Aires, Miami, New Orleans (USA)

Metro area subdivided into major sections by water or greenspace. Brasilia, Birmingham (USA), Asahikawa (Japan), New York, Amsterdam

Urban region strongly affected by a nearby city. Glasgow, Baltimore (USA)

Urban region with megacity (>10 million population) in center. Delhi, Shanghai, New York, Jakarta, Tokyo

Urban region within a megalopolis of major cities. Osaka (Japan), Philadelphia, Utrecht (Netherlands)

Urban region as one of a ring of cities surrounding a large greenspace. Amsterdam, Rotterdam, Utrecht, The Hague (Netherlands)

Urban region in two or more nations, states, or provinces. San Diego/Tijuana, Strasbourg (France), Ottawa, Portland, Philadelphia, Cincinnati (USA)

Urban region in a national district/territory. Canberra, Mexico City, Washington

Form of metro area and urban-region ring strongly affected by regional planning. Canberra, Brasilia

Metro area strongly molded by regional planning but now pattern much obscured by urbanization. Curritiba (Brazil), Ankara, Washington, New Dehli, Chandigarth (India)

Urban region showing some evidence of regional planning. London, Berlin, Beijing, Sapporo, Singapore, Rome, Moscow

Effect of and effect on other regions

A city's region never exists in isolation, but is tightly linked to regions both adjoining and distant. The arrangement of outside regions largely determines the directionality of inputs and outputs, both positive and negative, for an urban region. Such directional flows in turn strongly influence patterns and processes within an urban region.

All sorts of things move or flow across boundaries, from groundwater and air pollutants to wildlife, walkers, cars, goods transported by trucks, trains, stream water, and fish. The source locations, directions of movement, rates of flow, and sink locations of these are so diverse that detailed observation and mapping are normally required for planning. Still, the flows are usefully grouped into three spatial categories (Forman 2004a): edge area of a region, adjacent region as a whole, and distant regions.

The *edge area just inside and outside the urban-region boundary* is a key zone because sources are so close to, and effects so likely to cross, the boundary. Boundary issues often require careful watching, as flows and movements can quickly enter or leave a region. Often clusters of sources and flows can be identified at a number of locations around a region's boundary that can be targeted for management and solution (Forman *et al.* 2004). Furthermore some urban-region boundaries gradually expand or shrink. That may add, or remove, certain boundary issues from a region, and therefore noticeably alter the region's concern or responsibility for an issue.

Flows and movements originating from a metropolitan area or major portions of an urban region are often more important than boundary issues. Upon crossing a regional boundary, these flows may affect most of the edge portion of a region or diffuse widely across the region. A particular urban region, of course, focuses primarily on the flows entering rather than leaving. Yet, in a broader multi-region context, entering and leaving effects are both important.

While interactions between adjoining regions are more obvious, distant changes can also significantly affect an urban region. An economic giant that sneezes may flood a distant region with certain goods, or open a big new market. A government policy change involving immigration or transportation type, for instance, may strongly reverberate in a distant region. Environmental change elsewhere may alter migratory bird patterns, livestock disease spread, public health problems, or nature-based tourism. In short, this section highlights the importance of three areas – boundary edge zone, adjoining regions, and distant regions – for understanding the ecology and planning the future of an urban region.

Ability to extrapolate the Barcelona solutions

Ecologists generally do not count the needles on a giant spruce tree or the eye-blinks of an eagle or the ants caught by an aardvark just to learn more about the species. To do such analyses for even a thousandth of the species on Earth would require eons of drinks from a "fountain of youth." Rather, the ecological measurements, along with the exhilaration of ecological discovery, are mainly made to develop principles and theories that apply widely to other species.

So it is with the detailed look at the Barcelona Region in the preceding chapter (Forman 2004a). How many of the solutions outlined in it apply only to that region? Do any of the solutions extrapolate effectively to cities around the world? Any solution with wide applicability could be quite useful for understanding and planning.

Each of the Barcelona Region pieces outlined in Chapter 10 is therefore placed in one of three categories: (1) solutions widely applicable to urban regions worldwide; (2) solutions widely applicable to certain sets of urban regions; and (3) solutions limited in applicability to the distinctive Barcelona Region.

Solutions widely applicable to urban regions worldwide

Twelve results from the Barcelona analysis (see Chapter 10 and Color Figures 2–39) fall readily into this category. These solutions seem applicable to most large and small cities irrespective of geography, land use, and culture:

> *The emerald network* as the backbone of natural systems applies well in virtually all urban regions. Where the matrix is natural vegetation (Iquitos, Tehran), the emerald network represents a spatial framework of priority areas to protect as urbanization and other land changes proceed over time. Where the matrix is cultivation (London, Chicago), the spatial framework indicates priority areas for restoration of natural systems. If emeralds are presently distributed across the urban-region ring, e.g., due especially to topography and geology (Barcelona), their enhancement, connection, and protection is the challenge.
>
> *Five types of connections* provide flexibility for connecting the large natural-area emeralds, a useful approach for the usual urban region with numerous built areas. Diverse connection types that provide connectivity across the land for both trail walkers and wildlife are especially valuable in the face of ongoing urbanization.
>
> *Agriculture–nature parks* are mainly established on aggregations of small farms, which exist in almost all urban regions (partially illustrated in parts of Switzerland and Germany [Pegel 2007]). A set of protected agriculture–nature parks would be particularly valuable where the urban-region ring is small, or the matrix has been relatively monotonized by extensive agriculture (Chicago, London, Bangkok).
>
> *Water-supply protection using vegetation cover* applies widely, e.g., where the water source is aquifer, shallow groundwater, lake, reservoir, river, or stream. To provide relatively clean riverwater, strategic areas and stretches are targeted for natural vegetation, since covering an entire river basin with vegetation is often difficult in an urban region. The vegetation-cover solution is least applicable to a large deep aquifer or a very large lake.

Protected highest-quality stream valleys are valuable in virtually all regions. In arid areas much of the water runs underground, but gullies or washes or wadis are normally useful imprints on the land surface. Targeting protection to high-priority streams is an alternative to attempting to protect all streams by regulation and enforcement, a difficult task in urban regions.

Restoration of small wetlands is probably valuable and feasible in all urban regions, which long ago lost most of their wetlands to draining and filling. Small wetlands, especially created on floodplains, at specific sites by the bases of hills or mountains, and at the end of certain stormwater pipes draining numerous impermeable surfaces, provide value to natural processes and many ecosystem services to society.

A set of strategic places for growth applies universally in urban regions. To help channel future urbanization into particularly appropriate areas, a few well-separated satellite cities in the outer portion of the urban-region ring, and one area close to the big city may often be a good balance (see Chapter 8).

A set of strategic places for limited growth, no growth, and building removal also applies universally. The solution addresses the problem of valuable sites or areas imminently threatened by urbanization. In addition, locations are pinpointed where especially damaging or inappropriate urbanization has taken place relatively recently.

An array of solutions for widely repeated small locations is useful in all regions. Gullies, streams, highways, villages, and towns, which are often too numerous to analyze and plan individually, are typical locations. A few generic solutions addressing diverse human and natural systems issues are developed, and then a solution is tailored to the distinctiveness of each location. An important cumulative effect across the region can be expected.

Edge parks for towns and small cities provide benefits for nature as well as for both today's residents and tomorrow's newcomers. Therefore such parks are valuable for towns and small cities that are growing, a common case in an urban-region ring.

An impressive park to protect a large nearby unappreciated area applies widely in or adjacent to metropolitan areas. Creating a monumental cultural flagship park that provides an array of both human and natural values has the added benefit of minimizing imminent threats to an existing valuable resource.

Flexibility and stability for a region's future is a synthesizing solution that incorporates many of the others listed. From economics to aesthetics and water to biodiversity, the specific land-use solutions are, in part, determined by how well they provide flexibility and stability for a region's future.

Solutions widely applicable to certain sets of urban regions

Ten Barcelona solutions (see Chapter 10 and Color Figures 2–39) fit this category. For each the applicable type of urban region is indicated as follows:

A set of diverse large agricultural areas applies well to regions with good agricultural soil in large patches and of different types. If a single productive soil type predominates, different landscape areas can be established by maintaining different predominant crops. The solution applies poorly in regions with only small patches of farmland soil, such as in many glaciated and arid areas.

A river-watershed magnet for visitors and nature tourism applies well where the matrix, or an extensive area in the outer urban-region ring, is natural vegetation containing a large river basin or catchment. If these natural conditions no longer exist, the river-watershed magnet solution might be a useful framework for landscape restoration.

River restoration, which includes both water and valley, especially applies in regions where a river valley is upriver from the metropolitan area. Non-point pollutant sources from sewage, storm water, industry, and agriculture are significantly reduced in part through changes in land-use pattern, including land protection.

Floodplain riparian vegetation providing an array of human and natural benefits applies widely in regions with streams and rivers. In hilly/mountainous terrain, woody vegetation and other techniques can be targeted separately to upper tributaries and to lower floodplains.

A package of flood-hazard-reduction techniques applies in urban regions with the threat of river or stream flooding. The package addresses issues related to land and slopes, stream and river valleys, storm-water pipe systems in metro areas, and wetlands.

Separate stormwater and sewage systems is a solution applicable in most urban regions where large areas have a single pipe system combining sewage wastewater and storm water runoff. Water bodies tend to be heavily contaminated with incompletely treated human wastewater, so identifying priority areas for separating the two systems is an important step.

Relocating heavy industry to efficient heavy-industry centers is particularly valuable in urban regions where heavy industry is located alongside a major water body or upwind of a population center. Gradual transfer to a heavy-industry center with efficient power, water, waste disposal, and transportation is an employment and economic investment. It also enhances the water bodies and downwind communities.

A truck-transport center mainly applies to urban regions without one. Centralizing the loading and unloading of goods and agricultural products by both local and long-distance trucks provides cascading benefits to people as well as the environment, in the city and across the region.

Large underpasses or overpasses for walkers and wildlife are applicable to regions where highways are significant disruptions to crossing the urban-region ring by trail walkers and local residents, as well as key wildlife species (Figure 11.2). In view of ongoing urbanization, wildlife movement connectivity tends to be mainly important in the outer portion of an urban-region ring.

A green-net solution applies widely to regions where many nearby growing towns and small cities are threatening to coalesce. The green net provides a multitude of people-and-nature benefits at different scales from local community to region.

Solutions limited in applicability to the distinctive Barcelona Region

None. All solutions provided in the land mosaic plan for the Barcelona Region, while tailored to the specifics of the region, apply widely to urban regions globally.

This summary, done more than two years after plan completion, was a surprise. The widespread applicability of the Barcelona solutions seems to highlight the commonalities of urban regions. The plan addresses today's big problems and trends, and also outlines a land mosaic with flexibility to sustain the region in the future. No detailed fine-scale design is involved, and no cookie-cutter homogenization is proposed. The solutions could be applied, each in a different way, to a thousand cities, and the cities would look as distinctive as ever. They might all have cleaner rivers, valuable emeralds, more efficient, cleaner heavy industry, and a glorious monumental park. Yet the culture and socioeconomics in each

Figure 11.2 Two vegetation-covered overpasses across a multilane highway. Only 12 km from center city Barcelona, these provide connectivity for local residents, hikers, and wildlife, and were built when the Ministry of Environment concluded that a new highway should not disrupt connectivity of the landscape. Carretera de los tuneles. R. Forman photo courtesy of Josep Acebillo and Marc Montlleo.

urban region will have created these in a unique manner, and in synergy with the distinctive features of that region.

Local communities, ecology, and planning

Bats in the bedroom are not much fun. They brush into our hair, carry diseases, and leave messes. How odd it would be if we didn't notice them, or just continued brushing our hair, dusting a spot here, and walking carefully there, as if nothing were amiss. Local communities, especially residential or bedroom-commuter places, have largely been ignored in this book on urban regions. Much of a region's population lives in and takes much interest in these communities, while giant forces of urbanization, traffic flow, water degradation, and much more swirl across the region, seemingly unseen.

Yet local communities are important building blocks of an urban region, and understanding the blocks is useful to grasp the big picture. Three types of local communities will be examined: (a) satellite city (illustrated by Boulder, Colorado); (b) planned town in the urban-region ring (illustrated by several communities including Celebration, Florida); and (c) suburban town in the metropolitan area (illustrated by Concord, Massachusetts). The examples are relatively well-known places in the USA, not typical or representative, but providing useful insights for understanding or planning of urban regions. In each case we consider how the local community fits into the patterns and processes of the broader region.

Satellite city

Boulder, Colorado (population 90 000) lies in the inner portion of Denver's (city population 555 000) urban-region ring, and at the base of the Rocky Mountains adjoining North America's Great Plains. Most of the nearby Rocky Mountain area is federally protected land, whereas Boulder lies in one of the areas with the highest rate of expected sprawl in the nation (Burchell et al. 2005). Quality of life and environmental quality are of particular concern to Boulder residents.

A flagship feature of the city is its 12 000 ha (30 000 acre) open space system (Benfield et al. 2001, Peter Pollock, personal communication). Highlights are a greenbelt (outlined in 1910 by Frederick Law Olmsted, Jr.), averaging about 2 km in width, and greenways slicing across the central built area to interconnect with the greenbelt. The greenways are mostly stream corridors with walking/biking trails. Small parks are present in the central area, and the greenbelt and greenways provide connectivity for species movement to and among the parks. The open space or greenspace system includes city land, state land, federal land, cemeteries, and more, whereas intensive-use spots such as ball fields are managed separately. The land is zoned with management regulations related to ownership, as well as uses such as walking, dog-walking, and bicycling (Miller and Gershman 1998). Overall the system is greatly appreciated and much used by Boulder residents, and also by residents of nearby communities.

Deer populations are relatively dense and cougar (mountain lion) sightings not infrequent, but Boulderites are generally appreciative or tolerant of wildlife. The relative abundance of top predators helps maintain a diverse food web and rich biodiversity (Chapter 4).

Three major goals effectively created, and are accomplished by, this greenspace system: (1) protection of natural systems, which includes: native grassland, wildlife, and biodiversity; soil erosion control; and reduction of water runoff and flooding; (2) recreational opportunity, so that all residents live near a park, greenway, or greenbelt, and have ready walking access to the open greenbelt area; and (3) shaping the development of the city, which includes limiting urbanization and sprawl and "disciplining" urban growth in the central portion.

The open-space protection process was jump-started in 1959 when a "blue line" was drawn at the 1757 m elevation contour, above which water and sewer services would not be provided (Benfield *et al.* 2001). Although some of the greenspace began as federal land, the bulk of the system was acquired by the city. In 1967 citizens decided to tax themselves, using a sales tax, to acquire land for protection. Thirty years later one could walk around the city on its protected greenbelt.

The central portion of Boulder dominated by built areas has also achieved results of ecological and planning interest. In 1976 citizens chose to slow growth to 2 % by limiting the number of building permits issued annually. In 1995 citizens dropped that to 1 % growth, which, though strikingly different from surrounding rapidly growing communities, equaled the national population growth rate. New development is mainly next to existing development, a good cost saving. Infill housing continues, though greenspaces are valued by citizens so opposition on a case-by-case basis limits the rate of infill. Affordable housing is limited, as in most surrounding communities, despite an innovative array of approaches to overcome the problem. Spring flooding problems next to the mountains have been controlled by using the diversity and arrangement of root and stem systems characteristic of native grassland, rather than by engineered structures.

Commercial growth has skyrocketed. In general, job availability exceeds housing availability, so commuters arrive from surrounding communities. Significant traffic congestion occurs, both due to commuting workers and shoppers. These increases in commercial activity and traffic have generated some case-by-case opposition to further commercial development. In addition to the much-used walking/biking trails, public transport across the city is relatively widespread and efficient.

The Boulder story offers several *useful lessons*. Residents can determine the kind of community, rather than vice versa. The community chooses slow growth, rather than no growth, rapid growth, or uncontrolled market growth. Open space is used to spatially define urban land, and to create a compact built area with sharp edges. A greenspace system with scattered parks, a greenbelt, and greenways interconnecting the greenbelt provides connectivity across the city for walkers, bikers, wildlife, and other species. The system also protects quite natural nearby ecosystems, which in turn provide valuable ecosystem services to the community.

Boulder's early focus on the open space of its edge is reminiscent of another magical place, Frederick Law Olmsted's emerald necklace. This was created on the edge of Boston and is now near downtown, where today it provides important habitat, aesthetics, recreation, and connectivity for species and numerous people. Edges of communities are where big solutions, especially greenspace benefits, can often be implemented.

A broader spatial view shows Boulder in a sea of coalescing communities characterized by rapidly expanding sprawl. A "green net" with greenspace corridors around town or municipality borders was proposed for an area where many expanding communities were threatening to coalesce (Chapter 10; Forman 2004a). The Boulder greenbelt would be an extreme example, because of its width, of a municipality ringed by a greenspace corridor. Ringing the local communities in Denver's region with greenspace corridors or thin greenbelts would create a broad-scale green net that provides important value – local recreation, nature protection, local walking trails, regional trails, connectivity for wildlife, and helping to maintain the identity and distinctiveness of each community. Regional initiatives could center around Denver, or say a county (e.g., planning begun in 1978), or even the surrounding local communities that are the most relevant to Boulder, an approach considered in the suburban town case below.

Planned town

With roots in nineteenth-century utopian communities and 1960s–1970s planned communities, the idea of planned towns grows as an alternative to new sprawl in the USA. Generally referred to as "new urbanism" (or neo-traditional town planning in Britain), the development is not without its critics, who sometimes harshly note that overall it is neither new nor urban, emphasizes real-estate development for profit, relies heavily on private cars, produces pseudo-"theme parks," is akin to a cult, and has confused ecology with green marketing.

Nevertheless, the key question is how a *planned town* compares with the five basic alternatives: (1) values of the preceding land; (2) values of the land enhanced without a new community; (3) a sprawl community of the same area; (4) a sprawl community holding the same number of people as the planned town; and (5) a community planned so that both people and nature thrive long term. A list of pros and cons for each of the six options should be a *sine qua non*. Although the planned-town idea has been applied to a neighborhood within a city, suburban community in the metro-area portion of a small city (e.g., Davis, California), and a master-planned housing-development community (Vernez Moudon 1989), here we consider the typical case of a new town in the urban-region ring.

Almost all the planned towns emphasize: reduced vehicle use; pedestrian-scale walkability; local recreation; water as an amenity, and often a part of stormwater management; small house lots with little lawn; houses close together, sometimes with many front porches; variable-unit townhouses and apartments; small or few outdoor private spaces; slightly narrow streets; unobtrusive garages often entered from a back alley; manicured, rather uniform planted areas; a distinct town center with shops and restaurants; sections or neighborhoods somewhat separated by water or greenspace (e.g., riparian strip or golf course); real-estate marketing; a relatively high density of people; social interactions; upper middle class values and aesthetics; and a regulated ordered conformity produced by planning. Of course these and other emphases vary by town. For new urbanist towns these patterns emanate from 27 listed principles largely driven by land development and social community (Congress for the New Urbanism 2000, Duany *et al.* 2000). The social, economic, and planning dimensions are grist for much evaluation and discussion (Katz 1994, Duany *et al.* 2000, Lund 2003, Garde 2004).

In contrast, natural systems, habitat, and environmental dimensions overall have been of tangential interest. These towns take up space, either habitats for species or places for us, so a rigorous ecological evaluation is needed. Some preliminary observations here emphasize the point. Habitat, species diversity, and rare species are largely ignored or minimized. Environmental monitoring, management, and improvement are often absent. Hydrologic groundwater protection and habitat restoration are typically unaddressed. Habitat connectivity for regional wildlife movement is normally missed. Ecological impacts of traffic noise and pollutants (considerable traffic results from few or distant jobs) remain unaccounted for. Adaptive management for water conservation, stormwater runoff control, energy use, and water and air pollution is generally overlooked. Consider briefly some examples:

> *Reston, Virginia* was built in a rural landscape in the 1960s, then engulfed by sprawl, and recently a people-oriented small-city center with a high density of people, walkways, shops, offices, and restaurants was inserted. Over time a planned town changes, in this case a metamorphosis catalyzed by the transformed land around it.
>
> *The Woodlands, Texas*, begun in the 1970s with a major goal of controlling floods in the community, in one sense was a smashing success by tailoring development to the capacity of different local soil types to absorb rainwater. Natural vegetation supporting rich biodiversity remains in front yards and back yards of most houses. Yet, in the pre-landscape-ecology era, the town plan effectively designed against an important icon species, the red-cockaded woodpecker, which requires a large natural area containing large pines and thrived next door in Texas' first state park.

Seaside, Florida was designed to extend a long distance along, and provide close-by human access to, a coastal beach. But the beach lies in a key area for rare nesting sea turtles. The town could have been planned to minimize damage to the coastline and turtle population.

Kentlands, Maryland extinguished a large meadow-grassland in an area where a regionally rare bird, the upland sandpiper, requires large grasslands to survive. Later, runoff from the new town's concentrated impermeable surface seems to have significantly altered existing downslope wetlands.

Lake Carolina, South Carolina was built along the shore of the only relatively large lake/reservoir in the area, a location that without study can be expected to cause significant ecological degradation locally and regionally. A token, presumably ineffective lakeshore buffer zone separates the lake and stream tributaries from the built area. Stream flooding and increased lake sedimentation doubtless occur. Some large natural areas are left, apparently for future development rather than in permanent protection.

Celebration, Florida is owned, built, and continuously subsidized by the Disney Corporation (best known for its cartoon film character, Mickey Mouse), seemingly as a marketing and image investment (Beardsley 1997, Ross 1999). In addition to the usual planned town attributes, electric vehicles and abundant oft-deserted sidewalks are in evidence. Numerous tourists give a resort flavor. The so-called town hall is a real-estate center, where the options of large houses, condominiums, and apartments are clearly geared to the moderately wealthy.

Just as for sprawl, lots of land was consumed for these few thousand residents. Few jobs and little or no public transport are available, so vehicles are well used. Water is much in evidence, as expected where considerable wetland was degraded. Permanent conservation restrictions were placed on some large undeveloped wetland tracts.

More interesting are the linkages with land elsewhere. To receive approval for the development, Disney agreed to purchase and restore habitats of an extensive distant pastureland under the eye of a major conservation organization. In developing the community, often mature trees were planted which came from somewhere. I well remember days of moaning flatbed trucks each carrying three full-length palms to Celebration, a process that converted a beautiful distant diverse savanna into monotonous pastureland.

Plantings of street trees, front yards, and backyards in the town are all determined by narrow lists of approved plants. Everything outside seems rigorously manicured to fit a prescribed appearance; randomness and surprises are kept out. The controlling, regulated, ordered, homogenized, predictable, and monotonous place helps select the residents, and then shape the kind of lives they live. In some ways a "theme park," extreme Celebration highlights many unwelcome elements of planned towns.

Planned-town experts can point to other places and other patterns, but would generally agree that so far ecology has not been a major priority. Yet it could be. Then planned towns, where nature and people both thrive long term, would be clearly better than the two sprawl options and the existing planned-town option presented at the outset. Until then, one must conclude that the present cases (despite some unplanned ecological benefits) illustrate the typical "Nature gets the leftovers" development approach. Ecological criteria are often little more than open-space amenity or aesthetics, sometimes combined with stormwater management. No rigorous scientific meaning or use of ecological principles or standards or accountability exists. Ecology, environment, conservation, and habitat are used in superficial generic ways, essentially as green marketing. Transform that, and planned towns could be a significant solution across the land.

Suburban town

Open space (greenspace) planning and protection in *Concord, Massachusetts* highlights three particularly useful insights for a community in its region (Ferguson *et al.* 1993, Forman *et al.* 2004): (1) broad town-wide or landscape-wide patterns for identifying land protection priorities; (2) a promising approach for regional thinking and collaboration; and (3) dealing with the presence of a major highway in the community. Concord is an outer Boston suburb with a rural feel. Located 55 km northwest of downtown Boston on a commuter rail line, the town has 17 000 residents in 67 km^2 (26 mi^2). A historic town where the American Revolution began, it later became America's nineteenth-century literary center where R. W. Emerson, L. M. Alcott, N. Hawthorne, and H. D. Thoreau (one of the roots of ecology) lived. In 1928 the town was a pioneer in passing a zoning bylaw to help guide its growth and development, and since the 1960s natural resource and land protection have been especially important.

In 1992 an "Open Space Framework" was delineated in a town plan to identify, compare, integrate, and rank the major town-wide features and the smaller special sites of open-space importance. The *town-wide pattern* consisted of large areas or patches of three types (built, natural, and agricultural) plus major corridors of three types (water-protection, wildlife, and human). Small special sites of greenspace importance were then mapped on the town-wide pattern. The results highlighted 25 priority places for protection, the highest priorities being to protect the essential core of large natural and agricultural areas, and secondarily major water-protection and wildlife corridors. The major patch-and-corridor network identified provides a rich array of resources and benefits to residents. Then from 1992 to 2004 eight percent of the town's total land surface was protected.

The exceptional value of a large natural area for biodiversity, and especially interior species, is illustrated by a study of 22 hectares within one of the town's large natural patches (600 ha Estabrook Woods). Highlights included: forest-interior species rare in the town's region (porcupines, fishers, barred owls, black-throated green warblers, hermit thrushes, northern waterthrushes); other species of special interest (great horned owls, blue-winged warblers, several vascular plant species; two state-listed rare invertebrate species); a paucity of invasive exotic species; and more state-listed and locally rare species on an adjacent piece of the forest. This large natural area is one of only two in the Boston Region with such a dependable array of forest-interior species.

In 2004 the next open-space planning process had a strong emphasis on protecting agriculture, water, biodiversity, and nature-based trail recreation. Intriguingly, it strengthened the Framework by considering three spatial scales, regional, town, and neighborhood. Since a large portion of the town's issues involve one or more other towns, a regional approach seemed important. Yet traditional regional approaches have many familiar shortcomings, such as threat to local control, inadequate budget, short half-life, and many issues extending into a different region.

Therefore a new regional approach was developed. First, a long list of ways Concord interacts with other towns showed that the bulk of the interactions over the last couple of decades were with 18 surrounding towns. Thus a functional town-centered or *locality-centered region* was recognized, composed of 18 surrounding localities commonly interacting with Concord at the center (Figure 11.3). In this case the local region covered an area five times the town's diameter. Databases were gathered and maps drawn for the 19 town region showing several

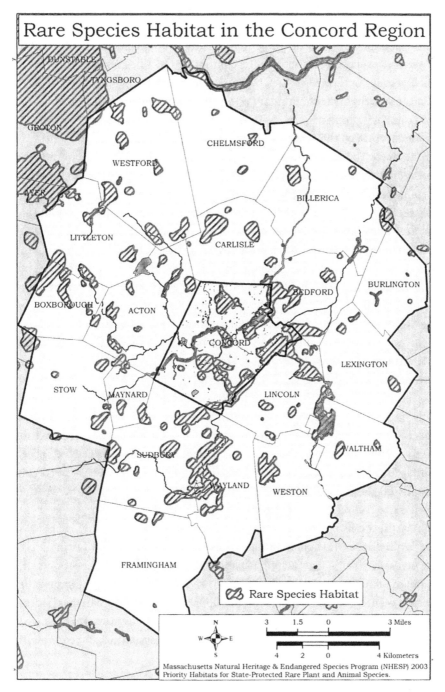

Figure 11.3 Locality-centered region and rare species habitat. Eighteen towns around Concord, Massachusetts (USA) form a functional region with frequent interactions. Priority habitats for state-protected rare plant and animal species are mapped to help understand regional patterns for greenspace planning in the central town. Mapped by Matthew Barrett using 2003 Massachusetts Natural Heritage & Endangered Species Program data. From Forman *et al.* (2004).

resources of importance to the town. These included protected open space, public trail systems, proposed regional trails, commuter rail lines, major roads, state-certified vernal pools, rare species habitat, and so forth.

Also the sources of cross-boundary effects were mapped to provide insight into where boundary problems are concentrated, that is, where things just outside the boundary affect the town, and things just inside the boundary affect an adjoining town. The regional maps and the cross-boundary-effects map thus put the preceding town-wide patterns in a broader spatial context, which clarified and strengthened the open-space planning priorities.

This regional approach, which did not threaten local control, also capitalized on a state requirement for developing open-space plans every five years. The 2004 Concord plan with maps and with databases identified was distributed to the surrounding towns, with the expectation that they will use many of the same regional databases in their planning. Within a few years most of the surrounding towns may well be thinking, and increasingly collaborating, regionally.

The third tough issue for local communities is the presence of a regional transportation route, in this town's case a bisecting *major highway* with considerable traffic. The 2004 planning effort highlighted the highway because its effect ramifies through so much of the land and the community – traffic noise in nearby residential areas, clogged adjoining roads, accidents, road salt, pedestrian crossing hazards, hazard to bicyclists, traffic noise degrading avian communities, and barrier to wildlife crossing. A range of partial solutions was proposed for each issue. Two years later the state had completed construction in Concord of its first four wildlife underpasses. Tracking studies of animals preceded construction, and diverse wildlife species quickly began using the underpasses that connected the two halves of the town.

Finally, the large patches-and-corridors framework for this suburban town closely parallels the multiple habitat conservation program for protecting biodiversity across the San Diego Region (Chapter 2), as well as the emerald network for a range of nature-and-people values in the Barcelona Region (Chapter 10). The locality-centered-region model should apply well in towns, parishes, shires, municipalities, local-government-areas, small cities, and counties. Perhaps with appropriate nudging, it could rapidly spread right across an urban region with widespread visible benefits.

Good, bad, and interesting patterns in urban regions

Imagine arriving at a party where the hostess has you reach into a paper bag with three types of squishy balls, feel around, and pull one out. You are delighted with the one you choose and will use it tomorrow. Alas, your friend picks a different type which promptly squirts you in the eye. Then the person behind you chooses a third unusual ball, and begins pondering it. The hostess knew all along which balls were good, bad, and interesting – and now you do too.

So it is with urban regions. The beneficial patterns, the detrimental ones, and those still worth evaluating for pros vs. cons are reasonably clear. There is no need for society to rediscover wheels around every city.

The selected patterns listed below are mainly grounded in the Color Figures of Chapter 5, the nature–food–water graphs of Chapter 6, the built-systems–built-areas graphs of Chapter 7, the urbanization attributes and analyses of Chapter 8, the principles laid out in Chapter 9, the case study results of Chapter 10, and many broader concepts of other chapters. Further patterns crowd the pages of this book. The astute reader will enjoy exploring these foundations and discovering more patterns of the good, the bad, and the interesting.

Good patterns

Benefits seem to clearly outweigh shortcomings for the following characteristics, which are grouped in several categories for convenience. Cities listed are illustrative, and more representative or more extreme examples can normally be found in the rich mixture of cities around the globe.

Metro area characteristics, city, and urban region

Urban-region area correlated rather closely with a province, state, or nation. Berlin, Singapore

City relatively close to several major landscape types that provide diverse resources. Abeche, Denver (USA), Kota (India), Kuala Lumpur, Madrid, Rahimyar Khan, San Diego/Tijuana, Santiago, Seville (Spain)

City on border of, rather than within, large valuable land-cover types. Beijing, Ottawa, Nairobi, Salt Lake City (USA)

City located by storm-protected coastal bay. Bandar Lampung (Indonesia), Boston, Buenos Aires, Edinburgh, Kagoshima, Lagos, Lisbon, San Diego/Tijuana, Seattle (USA), Sydney, Tunis

Relatively compact metropolitan area. Bucharest, Birmingham (UK), Christchurch (New Zealand), Edmonton, Manado (Indonesia), Nantes, Santiago, Winnipeg (Canada)

Planned city that has retained most key characteristics despite extensive growth. Canberra

Numerous small greenspaces in metro area. Atlanta, London, Portland, Philadelphia, Seoul

Greenway network tends to interconnect greenspaces for walkers/wildlife in metro area. Copenhagen, Minneapolis/St. Paul (USA), Ottawa, Toronto

Wide green corridor along major river bisecting metro area. Edmonton, Washington

Large central greenspace. Berlin, Christchurch (New Zealand), Mexico City, New York, Tokyo (mainly palace grounds)

Green roofs noticeably abundant in city. Basel (Switzerland), Zurich, certain German cities.

Relatively distinct border between metro area and surrounding countryside. Bucharest, Edmonton, Santiago, Shanghai, Sofia, Winnipeg (Canada)

Major natural-systems project (large edge park) achieved at one portion of metro-area border. Boston (Frederick Law Olmsted's Emerald Necklace), Barcelona (Antoni Gaudi's Parque Guell)

Urban-growth boundary that sharply reduces rate of outward urbanization. Portland

Scalloped metro area border providing benefits to nearby residents. Bucharest

Several green wedges projecting into metro area. Moscow, Brasilia, Buenos Aires, Cairo, Chicago, Jabalpur (India), Melbourne, San Diego/Tijuana, Stockholm

A single green wedge projecting into metro area. Birmingham (UK), Sofia

Especially wide and long green wedge, as a major source of wildlife for city. Portland

Greenbelt. London, Ottawa (around half of city)

Ring of large parks. Barcelona, Seoul (remnants of a former greenbelt)

Nearby mountain/hill slopes facing city mainly protected by forest/woodland. Erzurum, Portland, Sapporo, Ulaanbaatar

Towns mostly located near boundary of agricultural and natural areas. Ulaanbaatar, Tegucigalpa, Kuala Lumpur

Relatively low abundance or "density" of built-area borders in urban-region ring around large city. Cairo, Beijing

Regional land-use planning evident in the metropolitan area form. Canberra, Brasilia, Rome, Bucharest

Regional land-use planning evident in the urban-region ring. Brasilia, Canberra, London, Moscow, Beijing, Berlin

Nature, forest/woodland, and food production

Abundance of wooded landscapes in urban region. Berlin, Kagoshima

Large forest/woodland patches across region resulting from recent several-decade policy. Moscow, Berlin, Bucharest

Large semi-natural patch adjoining city or metro area. Barcelona, Brasilia (two, including much wetland), Edinburgh, Mexico City, Nairobi (grassland), Philadelphia, Paris

Metro area closely surrounded by forest/woodland and its resources. Anchorage (USA), Iquitos, Samarinda

Emerald network relatively well developed, though incomplete. Barcelona, San Diego

Abundance of protected sites of biodiversity importance. Chicago, Philadelphia

Abundance of sites for one-day recreation or tourism. Barcelona, Chicago, London, Seattle (USA), San Jose (Costa Rica)

Some Native Peoples' lands protected. Edmonton, San Diego/Tijuana, Cairns (Australia)

Different farmland-area types provide diversity of food products/socioeconomic values. Barcelona, Rome, Sacramento (USA), Seville (Spain)

Relatively large wooded patches within cropland landscapes. Bamako, Moscow, Bucharest

Market-gardening areas in proximity to city. Bangkok, Barcelona, Chicago, Kagoshima, London, Nairobi, Portland

Agriculture–nature park for market-gardening and aquifer protection next to city. Barcelona

Water

Drainage area around water supply ≥80% forest/woodland/natural cover. Boston, Canberra, Iquitos, Ottawa, Portland, Santiago, New York

Urban-region rivers and major streams mainly surrounded by natural vegetation cover. Berlin, Erzurum, Iquitos, Samarinda, Ulaanbaatar

Large reservoir or lake mainly embedded in metro area. Brasilia, Canberra, Madison (USA)

Natural vegetation abundant along the lakeshores present. Berlin, Sapporo, Stockholm

Extensive wetlands near metro area. Buenos Aires, Miami, Beira (Mozambique), Brasilia, Oslo

Transportation, development, industry, pollution

Public-transport commuter rail extending well beyond metro area. Berlin, Chicago, Kagoshima, London, Philadelphia, Rome, Sapporo, Stockholm

Bicycle parks and community (allotment) gardens near commuter rail stations to serve city residents. Amsterdam, Utrecht

Transit-oriented development along a commuter rail line. San Diego, Sydney

Reticulate rail network across urban region providing for non-radial passenger travel. London

Low level of annual vehicle-kilometers (miles)-traveled per person. Berlin, Shanghai, Singapore

Traffic-congestion programs in the city to limit air pollution and greenhouse gases. London, Singapore, Rome

Main highways often along border of, rather than within, major land-cover types. Sapporo

Vegetated overpasses that provide connectivity across highways for wildlife and walkers. Barcelona, Amsterdam (for wildlife)

New development in urban-region ring mainly occurring next to existing development. Berlin

Only, or last, nuclear-power plant in urban region decommissioned/de-fueled. London, Portland

Heavy industry concentrated around a satellite city rather than metro area. Sapporo

Heavy industry concentrated near shipping/ferry port outside metro area. Sapporo

Major diverse pollution sources mainly on downwind edge of metro area. Edmonton, Chicago, Melbourne

Bad patterns

Although some benefits, of course, are present, negatives seem to clearly outweigh positives for the following characteristics.

Metro-area characteristics, city, and urban region

Urban region split between two nations. San Diego/Tijuana, Strasbourg (France)

Urban region split between two states or provinces in a nation. Chicago, Ottawa, Philadelphia, Portland, Washington

Megacity with significant urban infrastructure and other problems. Delhi, Mexico City, Sao Paulo

City with competitive/combined regional effects due to another nearby major city. Asahikawa (Japan), Philadelphia, Kagoshima, Ottawa

Metro area within an agglomeration of separated cities, each with >250 000 people. Amsterdam, Seoul, Tokyo

Unusually large metro area relative to population and area of urban region. Atlanta, Chicago, Philadelphia

Small urban-region ring to support the city. Chicago, Philadelphia, Kagoshima

Low-population-density metro area associated with sprawl. Atlanta, Chicago, Philadelphia, Toronto

Metro area patterns strongly reflecting political division into two parts. Berlin (previous division), Jerusalem, Nicosia (Cyprus)

Linear metro area spread along coast. Kagoshima, Miami, Kobe (Japan)

Elongated metro area separating nearby wooded landscapes. Brasilia, Berlin, Oslo, Ljubljana

Rather limited greenspace in metro area relative to population. Barcelona, Bucharest, Erzurum, Mexico City, Santiago

Diffuse metro-area border area with development gradually decreasing outward. Baltimore (USA), Denver (USA), Milan, San Jose (Costa Rica), Valladolid (Spain), Yerevan (Armenia)

Creeping development threatening large forest/woodland patch adjacent to metro area. Barcelona, Mexico City

Abundance or "high density" of built-area borders in urban region. Chicago, Philadelphia

Nature, forest/woodland, and food production

Little forest/woodland remaining in urban region. Bangkok, Bucharest, Chicago, Copenhagen, London, Minneapolis/St. Paul (USA), Nantes

Primary old-growth forest in urban region being rapidly logged. Iquitos, Manaus (Brazil), Samarinda

Few natural beaches/dunes, vegetation areas, or wetlands remaining along coast. Barcelona, San Diego/Tijuana, Brisbane (Australia), Miami

Recreation and tourism sites mainly at some distance from city. Cuttack

Limited cropland area producing food in urban region. Abeche, Brasilia, Canberra, Erzurum, Phoenix (USA), Omsk (Russia)

Only one or two main farmland-area type, providing little diversity in food products. Atlanta, Chicago, Edmonton, Erzurum, London, Xi'an (China)

Cropland more concentrated near metro-area border than across the urban-region ring. Cairo, Rahimyar Khan, Edmonton, Erzurum, Santiago, Portland

Outward urbanization removing scarce cropland area. Cairo, Philadelphia

Water

Drainage basin for water supply with significant area outside urban region. Philadelphia, San Diego/Tijuana (much water piped from Northern California and Colorado River), Sydney

Drainage basin for water supply mainly covered by intensive cropland. Kagoshima, London, Mexico City

Water-supply drainage basin close to metro area. Abeche, Bamako, Kagoshima, Seoul, Tegucigalpa

Main reservoir for water supply, recreation, and aesthetics polluted. Beijing (one of two), Brasilia (replaced by small distant reservoirs)

Water supply mainly from streams subject to drought and human impacts. Tegucigalpa

Water supply from groundwater wells subject to pollution and drought. Abeche, Bamako

Best aquifer in relatively dry region threatened by development. Barcelona

Water table dropped and water bodies dried out in central portion of region. Mexico City

Wetlands (subject to drought) providing major water source for irrigation system. Bangkok

Few rivers/major streams still surrounded by natural vegetation cover. Atlanta, Bucharest, Cairo, Chicago, Cuttack, Edmonton, Kagoshima, Mexico City, Rome, Sapporo

Main rivers/streams usually reduced to low flow or trickle. Barcelona, El Paso (USA), Mexico City

Lakes mainly ringed by cropland, and exhibiting sedimentation/water quality problems. Rome, Minneapolis/St. Paul (USA), Chicago

No major surface water body near city. Abeche, Beijing, Houston (USA), Johannesburg, Madrid, Mexico City, Milan, Nairobi

Transportation, development, industry, pollution

Large number of major radial highways, all reaching urban region boundary. Atlanta, London, Moscow

Ring highway mainly outside metro area, with extensive development beyond. Bangkok, Beijing (concentric ring roads), Boston, Mexico City, Milan, Rome, San Antonio (USA)

Two-lane ring road likely to be widened and catalyze widespread development. Bucharest, Moscow

Main roads and bridges periodically impassable due to flooding. Abeche, Cuttack, Rahimyar Khan

Limited main-road access between metro area and much of region. Abeche, Cuttack, Erzurum

High level of annual vehicle-kilometers (miles) traveled per person. Atlanta

Shipping/ferry port located far from city center. Bangkok, Cairo, Santiago

Built areas surround many streams and rivers. San Diego/Tijuana, Philadelphia, Chicago

High edge density (border length) of built areas in urban-region ring. Barcelona, Chicago, London, Moscow, Philadelphia

Towns and small cities in large areas of region threatening to coalesce by urbanization. Barcelona, Raleigh/Durham (USA), Toronto

Dispersed-site development the predominant model of urbanization. Atlanta, Chicago, Philadelphia, San Diego/Tijuana

Slopes near metro area much covered by built area. Brasilia, Beijing

Considerable squatter housing/shantytown areas. Brasilia, Rio de Janeiro, Lagos, Johannesburg, Jakarta, Kolkata (Calcutta)

Heavy industry concentration close to city. Beijing, Edmonton, Kagoshima, Seoul

Severe air pollution over metro area. Beijing, Buenos Aires, Cairo, Caracas, Delhi, Jakarta, Mexico City, Shenyang (China), Los Angeles

Extensive coastal near-shore water pollution. Bangkok, Barcelona, Buenos Aires, Marseilles

Rivers/streams extensively polluted by sewage, industry, and agriculture. Barcelona, Rome, Seoul, Tegucigalpa, Guangzhou (China)

Large mine-waste areas in urban region. Berlin, Johannesburg, Miami, Prague

Hazards

Riverside city subject to serious flood hazard. Bangkok, Cincinnati (USA), Jakarta, Iquitos, Santiago, Ulaanbaatar, Winnipeg (Canada)

On multi-channel river delta with extensive flooding. Cuttack, Dhaka, Ho Chi Minh, Lagos, New Orleans (USA), Rangoon, Rotterdam, Shanghai, St. Petersburg, Vancouver

Significant hazard from hurricane/cyclone. Havana, Miami, Osaka, Taipei

Significant hazard from coastal flooding. Bangkok, Manila, Mumbai (Bombay)

Significant area subject to inundation by sea-level rise. Amsterdam, Bangkok, Buenos Aires, Kolkata (Calcutta), Dakha, London, New Orleans (USA), New York

Significant earthquake-hazard area. Managua (Nicaragua), Kobe (Japan), Mexico City, San Francisco

Hill/mountain slopes widely cultivated, with erosion/sedimentation/flooding problems. Rome, Port-au-Prince (Haiti), Tegucigalpa

Threat of landslides. Ankara, Caracas, Hong Kong, San Salvador (El Salvador)

Uncertainty near major migrating river and moving desert dunes. Rahimyar Khan

Built areas close to frequent woodland fires. Barcelona, Canberra, Los Angeles, Philadelphia, San Diego/Tijuana, Montpelier (France)

Active nuclear-power facility in urban region. Kiev

Metro area close to military concentrations and demilitarized zone. Seoul

Interesting patterns

Many of these are relatively uncommon land-use patterns, where the balance between pros and cons seems somewhat equal, or more study is warranted.

Metro-area characteristics, city, and urban region

Two major cities nearby but separate. Glasgow and Edinburgh, Baltimore and Washington

A single government for essentially the whole urban region. Beijing, Brisbane (Australia)

Metro-area population more than twice the city population. Lagos, Los Angeles, Lyon (France), New York, Cincinnati (USA), Seoul, Tokyo

Megacity with somewhat manageable urban problems. London, Moscow, New York, Tokyo

Metro area in part a product of recent wars and a several-decade partition. Berlin

Relatively distinct semi-circular metro area. Manado (Indonesia), Toronto

Metro area with unusually long and convoluted border. Chicago, Atlanta, Philadelphia

Several built-area lobes on perimeter of metro area. Moscow, Stockholm, Santiago, Kuala Lumpur

No major greenspace wedges on metro-area perimeter. London, Barcelona, Mexico City, Beijing, Samarinda

Several unique/distinctive natural-systems-related features dose to metro-area border. Brasilia, Barcelona, Portland, London

Planned city seemingly "overrun" by extensive urban growth. Brasilia, Chandigarth (India), Curritiba (Brazil), Washington

Major sections of city somewhat separated by elongated greenspaces. Asahikawa (Japan), Brasilia (sections with distinctive shapes), Canberra, Harare (Zimbabwe), Johannesburg

Abundance of small, linear greenspaces in metro area. San Diego/Tijuana, Brasilia, Bamako, Seoul

Long narrow metro area. Miami, Piura (Peru), Ulaanbaatar

Regularly distributed villages over much of region. Bahawalpur (Pakistan), Xi'an (China)

Total built-area border length in urban region largely due to towns. Barcelona, Abeche, Erzurum, Rahimyar Khan

Considerable built-area border length in urban region due to satellite cities. San Diego/Tijuana, Bucharest

Only one satellite city, or none, in urban region. Abeche, Edmonton, Erzurum, Iquitos

Satellite cities mainly in inner urban-region ring. Seoul, Barcelona, Bangkok, Berlin, Kuala Lumpur, London

Satellite cities mainly in outer urban-region ring. Bucharest, Mexico City, Rome

Parallel ridges and valleys with streams slicing through metro area. Birmingham (USA)

Most of metro area with a fairly regular grid of streets/roads. Beijing, Minneapolis/St. Paul (USA), San Antonio (USA), Xi'an (China), Toronto

Ring of regional shopping malls distributed near metro-area border. Boston

Nature, forest/woodland, food production, and water

Only large forest/woodland area in region largely protected for recreation/biodiversity. Edmonton

Many gaps and narrow connections present between natural landscapes. Berlin, Kagoshima, Sapporo

Large wildlife park next to city providing greenspace and recreation/tourism values. Nairobi

Very little forest/woodland present mainly due to grassland or desert climate. Abeche, Buenos Aires, Amarillo (USA), Erzurum, Jodhpur (India), Tehran

Wooded area present is concentrated near metro-area border. Barcelona

Region with prominent large agricultural fields resulting from recent several-decade policy. Berlin, Bucharest, Moscow

Nearest cropland landscapes far from city center. Nairobi, Abeche

Agricultural land overwhelmingly surrounds streams and rivers. Sapporo, Bucharest, Mexico City, Cairo, Edmonton

Distant water supply. Nairobi, Santiago

Water supply depending on a canal(s). Berlin, Moscow

Extensive canal irrigation system. Bangkok, Cairo, Fresno (USA)

Aquaculture areas in proximity, as source of food and water pollution. Bangkok

City located on coast without a major bay or natural harbor. Barcelona, Chicago, Toronto, Algiers

Transportation, industry, and pollution

Large number of major airports in urban region. London, Moscow, San Diego/Tijuana

Nearest major airport relatively far from city center. Stockholm, Kagoshima, Cuttack, Milan

Access to outside world limited to air or boat travel. Iquitos

Several prominent concentric ring highways. Beijing

Commuter rail system essentially restricted to metro area. Beijing, Cairo, Portland, Edmonton

Shipping/ferry port for nearby offshore oil field. Samarinda, New Orleans (USA), Maracaibo (Venezuela)

This array of good, bad, and interesting patterns for urban region ecology and planning is, of course, incomplete. No summary is perfect or complete, and each reader is likely to note controversial items. A perusal of the figures through this book underlies how many more patterns could be added. Nevertheless, what can be done with such highlighted patterns?

Put them right to work improving urban regions. Two points are important. First, the patterns do not point to a single big best solution for shaping cities or urban regions. No magic solution or cookie-cutter approach exists. Who would want to visit Europe or Asia and find cities all look similar or the same? Putting principles to work helps avoid pitfalls, but most importantly creates solid foundations. From these, distinctiveness and glory can evolve and persevere.

Second how do we proceed? In simplest terms for a particular city, sort through the good and bad patterns and select a good batch. Perhaps evaluate different combinations, gradually discarding the least useful items. The enjoyable weaving process begins when a few of the collected patterns mesh nicely with the existing mosaic in the urban region of interest. Try roughly outlining, or sketching, or model-building the evolving mosaic, as items are continually added and discarded. Keep the changing outlines to marvel at the progress. The resulting tapestry should certainly transcend today's urban region. The Barcelona Region plan (Chapter 10) emerged from a primitive version of this process.

12

Big pictures

If I lived in a romantic castle atop a mountain, periodically I would charge around flinging open windows to let in light and air, and to gain inspiration from the glorious views around. This book is the castle. It is time to open those windows [ten of them today] and see our subject in broader challenging, delight-fully diverse perspectives.

These big-picture frameworks for urban regions and natural systems are pre-sented in three heterogeneous groups, though each of the ten broad perspectives stands on its own.

(1) The first group of big pictures, Garden-to-gaia; Urban sustainability; Dis-asters, highlights our major spatial arrangements with nature, plus the periodic disruptions.

(2) The second group, Climate change; Species extinction; Water scarcity, represents the gathering giant environmental challenges.

(3) The third group, Big-ideas–regulations–treaties–policy–governance; Megacities; Sense of place, brings strong social and cultural connec-tions to the forefront.

The final section, Awakening to the urban tsunami, attempts to identify the giants lying just over the horizon, and discover the best route ahead for us and for the land.

Garden-to-gaia, urban sustainability, disasters

These three challenging perspectives highlight the roles of spatial scale and critical linkages across the land in developing solutions for urban regions.

The third topic, the dreaded overnight catastrophe of particular importance to urban regions, pinpoints disruptive forces that must be accounted for in societal solutions.

Garden to gaia

Most of us can relate to a tiny garden at home, digging, planting, weeding, watching, harvesting, and eating with special pleasure. Satellite images of progressively larger areas – a house (p)lot, meadow or woodlot, neighborhood, locality/town, broad landscape, region, continent, and globe – almost always have a relatively extensive green background. In effect these areas, widely differing in spatial scale, are all productive gardens, with soil, plants, animals, water, and usually people (Lovelock 2000).

Which of these scales do we care most about? Typically loyalty to family is central, and one's commitment progressively decreases from neighborhood to town/city, state, nation, and globe. Although worldwide newscasts, economic globalization, and climate change force us to think globally, hardly anyone has a major allegiance to the planet.

Suppose one wished to improve the world in some way that is both visible and perseveres (e.g., to avoid having lived unnoticed or unrecorded by history). What scale would be optimal? Certainly one could have a visible effect on a tiny home garden, but there is almost no chance that the spot would remain in similar form over, say, decades or human generations. (Alternatively, hardly anyone can affect the whole globe as did, for example, Genghis Khan and Christopher Columbus.) Yet, considering the long history of predicted armageddons that never occurred, the globe is likely to muddle along in somewhat similar form for eons.

The "paradox of management" oft-faced by industry reflects this quandary (Forman 1995). Small spaces are easily changed, but inherently unstable. Large spaces are hard to change, yet have considerable stability.

The best solution seems to focus on *mid-size spaces, such as landscapes and regions.* At these scales one's improvement efforts may address both sides of the paradox, achieving an effect that is clearly visible in the short term and perseveres for the long term. For example, an agricultural or large-wetland landscape, or even a region of landscapes, such as New England (USA), Southern Sweden, or Central America, is a promising target for such dual success. Landscapes and regions are simply big gardens to be invested in and cared for.

This logic suggests that in urban regions the central business district is too small for a sustainability focus. The city is a possibility because it has a single government that can address multiple-sector issues (White 2002). Yet today's city is normally buried in and inseparable from its metropolitan area. The metro area is a distinctive visual unit (see Color Figures 2–39), but as repeatedly seen in previous chapters, is expanding outward and is thoroughly affected by in-and-out interactions with its surroundings. The metro area fundamentally depends on its urban-region ring for water, recreation/tourism, mineral resources, and much more. Within an urban region, farmland and forest landscapes tend to be peppered with towns and small cities, most growing, which obliterate much of the previous distinctiveness and integrity that the landscapes may have had. Consequently, planning at the level of the urban region itself appears to be the optimum solution to the basic question or quandary of where best to focus efforts for an effective mesh of nature and people in and around cities.

This fundamental point fits nicely in my bumper sticker (Forman 1995): "Think Globally, Plan Regionally, and *Then* Act Locally." Keep the globe in mind when making daily decisions. But most importantly, create a plan for every land-scape and every region that provides sustainably for nature and people. Then with the broad plan in hand, make the important local changes and refinements that fit effectively into the big picture.

Urban sustainability

Many scholars have noted that the term urban sustainability is essen-tially an oxymoron. It is extremely difficult to envision a city with thousands or millions of people packed together that provides a thriving balance for both people and nature. One might consider urban sustainability an idealistic goal or endpoint which we seek, but never reach. Or, since the basic concept, like trying to nail applesauce to a tree, seems vague and hard to pin down, every-one tends to define it to suit a particular purpose. I essentially avoid the term urban sustainability, but recognize that useful ways to think about it might be developed (Braat and Steetskamp 1991, Forman 1999, Forman 2002a, Berkowitz *et al.* 2003, Blowers 2003, Pezzey 2004, Rogers 2006, Moore 2007, Wu 2007).

Here are three approaches, with the last one offering the most promise. The first builds on the idea that a *multitude of tiny fine-scale solutions*, when added together, make a difference for a whole city or metropolitan area. These tiny solu-tions could be of a single type multiplied together many-fold, or of an array of types with potential synergies. Energy-efficient building materials, use of public transport, recycling of wastes, water conservation techniques, and food-growing on balconies and window boxes in the city are commonly cited examples. One could add "biophilic design" of buildings with green roofs (Stuttgart, Germany; Basel and Zurich, Switzerland) and a profusion of plants inside (Kellert and Wilson 1993, Peck and Kuhn 2003, English Nature 2003, Dunnett and Kingsbury 2004, Brenneisen 2006). City streets could be rife with storm-water swales, porous pavement, rich biodiversity, and aesthetic design (Beatley 2000, France 2002, Brandt *et al.* 2003, Hough 2004). Even at a somewhat broader scale, a city could be densely peppered with parks and greenspace corridors (Jacobs 1961, Cityspace 1998), so non-urban species can move relatively unimpeded throughout the city (Chapter 4).

A massive implementation of one of these fine-scale solutions, or several examples of all the types, could create a city where the packed people daily encounter and are attuned to the environment. Still, it would be an anthropocentric result. Only shreds of nature could thrive long term. Perhaps only a massive implementation of all the fine-scale solutions, admittedly an idealistic goal, could fit the test of urban sustainability.

A second sustainability concept takes a *systems view of the city*, somewhat analogous to "urban metabolism" (Sukopp and Werner 1983, Haber 1993, Tjallingii 1995, Ravetz 2000, Grimm *et al.* 2003). Consider a city as a huge box bulging with people, and in which some products are manufactured and traces of food grown. If all the holes in the box are blocked up, the people die, because the internal production is totally inadequate to sustain the population. Also very little has been stored in the city to handle such a crisis as plugged-up holes. Normally the holes of the box are open so huge amounts of things, from food to water, building materials, vehicles, and people, enter. And immense amounts leave, including garbage, pollutants, vehicles, and people. The city, as part of a larger system, is a box with inputs and outputs.

The functioning of Hong Kong, in 1971 a coastal city with four million residents (Boyden *et al.* 1981, McNeill 2000), is a vivid example. Its daily atmospheric inflows and outflows (in thousands of tons) were: oxygen 27 and 0, respectively; carbon dioxide 0 and 26.5; carbon monoxide 0 and 0.16; sulfur oxides 0 and 0.31; nitrogen oxides 0 and 0.12; and dust 0 and 0.04. The city's in-and-out water flows were: freshwater 1068 and 819; sewage water 0 and 819; and solids in sewage 0 and 6.3. Other major inputs and outputs included: petroleum 11.7 and 0; food 6.3 and 0; food waste 0 and 0.8; miscellaneous cargo 18 and 8.1; and people 0.53 and 0.52. The city was a giant sponge, every day absorbing tons of freshwater, petroleum, food, and cargo goods. And every day it sent tons of pollutants, sewage, and diverse materials mainly to surrounding areas.

The larger the inputs and outputs relative to production and storage within the box, the less stable or sustainable a city is (Tjallingii 1995, Forman 1995). A truckers' or garbage-removal-workers' strike, or a major breakdown of the water-supply or sewage-treatment system, causes enormous disruption. All of these have happened in various cities to the dismay of residents and policymakers alike. Nevertheless, in addition to reducing population, each of the basic components of the city system can be improved. Stability or sustainability can be enhanced by: (1) decreasing consumption by people within the city; (2) increasing production in the city; (3) increasing storage in the city; (4) decreasing inputs to the city; and (5) decreasing outputs from the city.

A third approach for urban sustainability highlights a city's "prime footprints" or imprint areas, and may be the only case where sustainability has a reasonable chance of attainment. First, identify the primary landscapes or sites that provide most of the inputs to a city. These might include water supply from a forested drainage basin in the urban region, key mining and industrial-production sites, a grain-production landscape, a pastureland area, a region exporting tropical fruits, and a market-gardening area near the city. Similarly, identify the areas receiving most of the outputs, i.e., the major solid-waste site, air-pollutant deposit areas, nearby recreation/tourism sites, and so forth.

The *prime-footprints model* or concept, in other words, refers to the primary source-and-sink areas connected to a city or urban region by routes of inputs and outputs. Establishing and maintaining a balance, where nature and people thrive in the prime-footprints system as a whole, would achieve urban sustainability for the city, even though in isolation the city supports only shreds of nature. In addition, decreasing the number of input-and-output routes, shortening the routes, and reducing the input-and-output amounts would all be steps toward sustainability.

Ecological footprint analysis (Chapter 3) (Wackernagel and Rees 1996, Costanza 2000, Rees 2003) is an important preliminary step in this direction. It identifies the total equivalent area used to support the people in a city. The prime footprints approach takes the next big steps by highlighting the specific footprint locations, the amounts and routes of inputs and outputs linking key locations to the urban region, and the importance of planning each footprint and the urban region together as an integrated system. In short, plan each of the prime-footprint areas linked to the city, plus the city, so the set as a whole sustains a suitable nature-and-people balance.

Disasters

The "ten bad ones" – wildfire, volcanic eruption, earthquake, tsunami, flood, hurricane (cyclone/typhoon), industrial-pollutant release, nuclear-power-plant radiation release, bombing, and disease outbreak – are particularly serious in urban regions where people and human structures are so concentrated. Landslides/avalanches, economic depressions, radical strong-government transformations, wars/conflicts, and massive immigrant arrivals (e.g., war or environmental refugees) could be added to the list. Nevertheless, we begin with characteristics common to disasters in urban regions, then sequentially glimpse each of the ten disaster types, and end with some guidelines targeted to urban regions.

Disasters are sudden events causing great loss or damage. Humans are impacted, often for a prolonged period, through property damage, illness, and death (Kreimer *et al.* 2003). Disrupted infrastructure networks – gas pipelines, oil pipelines, powerlines, telephone lines, water pipelines, sewer lines, rail-roads, and roads/highways – particularly cause effects to ramify across a region. Clean water supply as a daily human need is especially significant after a disaster.

Sometimes effects on natural systems are more severe, such as radioactivity in soil and water around Chernobyl and Kiev which will degrade nature for eons. Ecosystem services (Chapter 4) to the region's people are reduced. The degree of loss or damage from a disaster also depends on preceding human activities and land uses. A massive 2005 New Orleans (USA) flood, which was triggered by a hurricane, mainly resulted from broken levees constructed to raise a huge adjoining lake level 5 m above major portions of the city (Costanza *et al.* 2006). So now let us turn to the ten bad disasters.

Wildfire

If you enter an area where almost every plant has fire adaptations, e.g., rapid multi-stem resprouting, the landscape has a long history of frequent fire. If you see a few buildings and it's warm or windy, watch out. Attempts to stop the history have created a large fuel buildup ripe for an abnormally big fire. So it is with the Pine Barrens by the Philadelphia metro area, eucalypt forest around Canberra, chaparral engulfing San Diego and Los Angeles, and oakland by Barcelona and the Mediterranean. A 2003 wildfire eliminating koala, emu, and other wildlife swept into Canberra's green wedge and compact neighbor-hoods. In both 2003 and 2007, wildfire wiped out >2000 homes where sprawl had encroached on the fire landscape around San Diego.

Volcanic eruption

Lava flows, massive mud or debris flows, landslides, dense particulate air pollution that coats the lungs, and longer-term deposits of volcanic ash are the culprits. The Italian city of Pompeii was covered by an eruption of Mount Vesuvius so quickly that archaeologists centuries later felt they were looking at a moment in daily life frozen in time. In 1985 Pereira (Colombia) was devastated by a mud flow from a volcanically melted alpine glacier.

Earthquake

Consider some urban disasters: 526 Antioch (Syria, now Turkey); 1703 Tokyo; 1755 Lisbon; 1812 Caracas; 1905 Kangra (India); 1906 San Francisco; 1908 Messina (Italy); 1923 Kanto (Japan); 1939 Erzincan (Turkey); 1948 Ashgabat (Turkmenistan); 1964 Anchorage (USA); 1976 Guatemala City; 1976 Tangshan (China); 1985 Mexico City; 1995 Kobe (Japan); 2003 Bam (Iran). Rolling waves or surface ruptures with lateral or upward/downward movements last for seconds or minutes. Utility lines rupture, houses fall, mid-rise buildings collapse, highways give way, buildings sink, fires break out, and landslides occur. Damage from San Francisco's earthquake was mainly because the water pipes broke and a huge fire swept across the city. The long narrow city of Kobe was Japan's busiest seaport and one of Asia's top ports. Its earthquake killed 6400 people outright, left 300 000 homeless, crumbled downtown skyscrapers, toppled elevated highways, and destroyed large parts of the port and city. A 2006 *National Geographic* world map suggests that a third of the 38 cities analyzed lies in an earthquake "high risk" zone. Overall the two primary strategies in earthquake hazard reduction are: (1) identifying areas of high seismic risk, and (2) designing structures to withstand shaking. Progress in both of these has occurred in a few countries.

Tsunami

Earthquakes, as well as underwater volcanic eruptions and landslides, often trigger seismic waves or tsunamis that race across the ocean at some 500 km/hr (300 mi/hr). Upon reaching shallow water by coastlines, the powerful waves swell to great heights and rush inland threatening coastal cities. In 365 AD an earthquake leveled one of the world's "wonders," the Pharos (Lighthouse) of Alexandria (Egypt), and sent a tsunami across the Eastern Mediterranean. Maintaining protective coastal wetlands (Farber 1987, Danielsen *et al.* 2005) and mainly building at higher elevations are key coastal guards against tsunami damage.

Flood

In addition to tsumanis, four types of flooding devastate cities. The New Orleans case introduced above had a perched lake/reservoir at a higher elevation, so the city was flooded when the levee/dam broke (Costanza *et al.* 2006). Second, cities alongside rivers or on river deltas are inundated by water from upriver (Figure 12.1). Deforestation or development on hill slopes and mountain slopes removes the absorptive capacity of natural vegetation, and accelerates waterflows over the land surface into streams and rivers (Chapter 4) (Jared 2004). More water arrives faster, causing a higher peak flow or flood. Third, a coastal city is subject to flooding, especially during high tides, when huge windstorms push seawater toward a coastline, effectively raising the water level and inundating onshore areas. The fourth case is a riverside city a short distance inland from the coast, where the severe onshore windstorm pushes water right up the river to inundate the city. Thus Bangkok (Color Figure 5) gets flooded by water from both directions, approaching from upriver and approaching from the coast downriver. Cities partly built on low areas (Amsterdam, Cairo, Dhaka) of course are especially subject to floods.

Figure 12.1 Flood caused by extensive loss of natural vegetation. Without nature's absorptive capacity, water from a 30 cm (1 ft) rainstorm on relatively flat clayey soil readily flowed to, and accumulated in, extensive low areas. Texas. Photo courtesy of USDA-Soil Conservation Service.

Hurricane (cyclone or typhoon)

This strong large-diameter rotating wind, commonly strengthening as it moves over warm seawater, is especially dangerous to coastal cities, though inland cities are also damaged. Devastating cyclones hit Hong Kong in 1937 and Dhaka (Bangladesh) in 1942, 1961, 1963, 1965, 1970, and 1991. High winds of course cause lots of property damage. If the hurricane is moving slowly, often heavy rain precedes and follows the eye of the storm. Thus floods are common, and saturated earth contributes to extensive tree blowdowns. A 1938 hurricane in Central Massachusetts (USA) caused the most tree damage on the southeast slopes and tops of hills, and on the northwest ends of large openings such as ponds and meadows (Foster and Boose 1992). These sites appear to have had the highest wind velocities, probably just after the hurricane's eye had passed. The New Orleans hurricane passed over an extensive area where wetland marshes had been eliminated (Farber 1987). Yet one estimate suggests that the storm surge could have been reduced 4.7 cm for each kilometer of marsh in front of the city, due to the presence of both vegetation and shallow water (Farber 1987, Danielsen *et al.* 2005, Costanza *et al.* 2006). A 6 km wide protective marsh would have dropped the approaching seawater by 30 cm (a foot).

Industrial air pollution

The sudden release of toxic substances into the city air by industries has been a chronic problem, though periodically the type and amount of pollutant is considered a disaster (e.g., in Mexico City, Bangkok, Kuala Lumpur). Industries tend to aggregate close to cities where workers live. The effects on human health and on the survival of wildlife populations in the urban region of course depend on the substance emitted. As perhaps the worst case, in 1984 a pesticide-producing industry in the heart of *Bhopal* (India), a city of 1.5 million, released 40 tons of a toxic gas (methyl isocyanate or MIC), which is heavier than air. Some 3000 people are reported to have died quickly, perhaps another 15 000 over the following two decades, and the total continues to climb. Reports suggest that a third of the city's population was exposed to the chemicals and a significant portion had related health problems. Contamination is apparently still present at the factory site and its dumping grounds, as well as in drinking water. Incentives for heavy industries to relocate away from water bodies and dense human populations, to heavy-industry centers with efficient energy, water, and waste disposal (Chapter 10), should pay manifold dividends.

Radiation releases from nuclear power plants

The 38 urban regions analyzed are in 32 nations worldwide, over half of which have operating nuclear-power reactors (John P. Holdren, personal communication). Most are in France, Japan, USA, United Kingdom, and Russia. Releases of radioactivity into the air and water apparently are not infrequent, and no safe level or threshold exists. Because the radioactivity often accumulates through the food chain, insect-eaters and meat-eaters are generally most affected. The radioactivity is incorporated into tissue and damages proteins and DNA, which leads to radiation disease and death. Altered DNA can be passed to future generations with consequent genetic and health problems. Occasionally large releases of radiation have occurred, such as at Windscale (United Kingdom) in 1957 and Three-Mile-Island well west of Philadelphia in 1979. A massive 1986 release at *Chernobyl* near the small city of Prypyat and 130 km north of Kiev (USSR, now Ukraine) resulted in elevated rates of many illnesses, especially cancer, that continue to appear two decades later. A 27 km radius (17 mile) "exclusion zone" exists (including the former city of Prypyat) where radioactivity levels are high and access is highly restricted; three other nuclear reactors still operate there.

Much of the debris from the exploded power plant was buried nearby, especially in a floodplain where groundwater is now contaminated. Strontium, cesium, and heavy plutonium fell on agricultural land. Radioactive soil later blew over extensive areas including Kiev, where radioactivity threatens children, adults, and wildlife (Henry A. Nix and David Hulse, personal communications). Chernobyl is humanity's most indelible big stain on Earth – a 25 000 year legacy.

Bombing

Bombs dropped from aircraft have been used to destroy major portions of cities, such as London in 1942, Cologne (Germany) in 1942, and Hanoi in the 1960s. Resulting fires caused considerable damage as well. Atomic bombs destroyed much of *Hiroshima and Nagasaki* (Japan) in 1945, and residual radiation caused illness and death for decades thereafter. The 2001 destruction of the World Trade Center by plane crashes killed about 3000 people at one spot in New York, yet the resulting clouds of pollutants spread widely causing illness and temporary physical damage across the city.

Disease outbreak

The fourteenth century Black Death or Plague spread rapidly through Medieval Europe, often cutting a city's population by a third or half. Rats and fleas and people were packed together in walled cities so transmission of the bacterial disease was rapid. The 1917–18 influenza virus pandemic killed millions in cities worldwide. With half the world's population urban today, and nearly 65% a generation from now, disease outbreaks that easily outrace medical research may be disasters just ahead.

What can be done in urban regions in the face of the "ten bad ones"? First, try drawing a large donut on a piece of paper, with one edge of the donut somewhat flattened, and a line perpendicular to the flattened edge that cuts the donut in half (Chapter 11). The donut's hole represents a metropolitan area, the yummy part is the urban-region ring, the flattened side a seacoast, and the line a major river bisecting the metro area. Now draw arrows for the *major directional flows*: prevailing entering wind; river water entering from higher land; tsunami from the sea; hurricane/cyclone; water pushed upriver by the hurricane; and any others. The goal is to identify and map the areas in the urban region that are particularly susceptible to disasters, and areas that are not. For example, radiation release on the upwind side of the donut's hole spreads over the city, whereas air pollution from an industry on the downwind side is less likely a problem. Perhaps place red checkers or marks on the areas most susceptible to disasters, and black checkers on the best places for locating key structures. In effect, directional flows across the urban region are a key to hazard reduction.

The *checkered donut* could get more interesting and useful by adding other sorts of sites or areas in different colors. What and where are the items, such as the perched lake next to New Orleans or perhaps a nuclear power plant, which normally should be avoided in an urban region under any circumstances? Where are all the low areas subject to flooding, where wetlands may be useful to provide valuable ecosystem services? Where are the strongest winds expected in a hurricane? Where could protective wetlands be established against coastal storm surges? Where could self-building sand dunes be maintained to protect against storm effects? Where are the dangerous specific earthquake-fault-lines, such as between hard and soft rock material? By now the donut looks complex, yet also some disaster-prone areas and other relatively unsusceptible areas should be evident.

Finally, for the disaster-reduction solution, add some *risk-avoidance techniques* from industry, game theory, or military strategy. For example, add redundancy or "don't put all eggs in one basket." The biggest gain in risk avoidance is going from one to two valuable objects, some benefit results from increasing to three, and very little is gained after five. Or, diversify so that instead of having three objects of the same type, have three different objects. Or, increasing the size of an operation decreases the per-unit product cost up to a point, but when the operation begins to get too big, decentralize. Or, increase the connections among objects so that if one object or one connection fails, the system is likely to keep going. Or, increase the number of loops or circuits in the system to provide more alternative pathways. The list of potential disaster-reduction techniques goes on. Some of these are useful in dealing with the sudden event, while others provide resilience so the damage afterward is minimized. That combination should be particularly useful for disasters such as earthquakes, bombing, and volcanic eruption (Kreimer *et al.* 2003).

The donut map must be pretty interesting if all these dimensions for disaster reduction have been incorporated. Apocalypse and Armaggedon do not seem inevitable; the planned urban-region donut provides areas for optimism.

Climate change, species extinction, water scarcity

These three big pictures highlight the gathering giant environmental challenges before us. Under the hand of these forces the land and water we depend on are changing, ever more rapidly, as in the upward-turning portion of an exponential curve.

Climate change

Global climate change and its effects, especially associated with human activities, are the subject of daily news headlines and numerous treatises. Here the effects on and effects of climate change are briefly mentioned, but mainly we explore some promising responses for urban regions and their natural systems to the varied climate-change effects expected.

Vehicle usage, industry, and heating/cooling of buildings, as prime sources of CO_2 and other greenhouse gases that lead to human-caused climate change, are concentrated in cities. London and Singapore have some controls on vehicle usage. Thus solutions must focus on cities to cut down the giant-chimney-like upward rushing CO_2 flows. Certain German and Swiss cities are seriously investing in green buildings (Spivey 2002, Dunnett and Kingsbury 2004, Brenneisen 2006). Urban agriculture is growing worldwide (Smit 2006). These are exceptions, but at the same time possible prototypes. Addressing the much-documented heat island effect (von Stulpnagel et al. 1990, Arnfield 2003), which worsens with both urbanization and global warming, is particularly relevant now and long overdue. Certainly many investments and actions by governments, organizations, and individuals that decrease urban greenhouse-gas sources are warranted to reduce the proliferating effects of climate change (Gore 2006).

Three broad effects of anthropogenic climate change on urban regions are (McCarthy et al. 2001, Houghton et al. 2001, Climate Change Impacts on the United States 2001, IPCC *draft* summary for Policymakers 2007): (1) temperature increase, (2) sea-level rise, and (3) extreme weather events. For simplicity, precipitation is included with the first and third effects. Typical effects will be outlined, though of course variations exist for cities in every corner of the world.

The *temperature increase* expected is, in general, a few degrees Celsius (several degrees Fahrenheit) over a few decades. What does that mean, for instance, to the now-familiar city and region of Barcelona (Chapter 10) (Forman 2004a)? Average annual temperature is likely to rise about 2 °C (4.5 °F) by the 2020s, and 3 °C by the 2050s. Temperature increases in summer months will be even higher. Average annual precipitation is likely to decrease about 10–15 % by the 2020s, and 20 % by the 2050s. An increased frequency and intensity of summer heat waves, increase in summer drought risk, and greater frequency of intense precipitation events are likely.

Does that really matter to people, or to the land? Yes. These climatic changes would significantly squeeze the already limited water-supply system, parch the soil of the economically and culturally important wine-growing area, scorch the already-hot summer tourism season causing manifold problems for hotels, restaurants, and local economies, and produce debilitatingly hot summer conditions for all of Barcelona's residents. Indeed these effects appear to be imminent.

Temperature increase typically increases evapo-transpiration which dries the soil and desiccates plants. In built areas, that leaves less shade and less evaporative cooling, i.e., a hotter city. A massive increase in green roofs and other biophilic design solutions using drought tolerant plants could reduce the problem,

Figure 12.2 Koala, a rare species feeding on eucalypt (gum tree) leaves that is threatened by urbanization and climate warming. The much beloved animals are subject to dog, fox, and other predators on the ground, and thus have difficulty moving between habitat patches in urban regions. Gary M. Stolz photo courtesy of US Fish and Wildlife Service.

at the same time providing flooding and biodiversity benefits (Kellert and Wilson 1993, Peck and Kuhn 2003, Hien *et al.* 2007). Analogously, an extensive transformation of streetscapes using stormwater swales, porous pavement, biodiverse plantings, and other solutions could reduce heat buildup and provide other benefits (Beatley 2000, France 2002, Brandt *et al.* 2003, Hough 2004). More tree and shrub cover in greenspaces would also help.

At a broader scale, a hotter, drier climate is likely to leave some species "imprisoned" in fragmented patches of nature, and subject to slow decline and disappearance (Figure 12.2). The *emerald network solution* (Chapter 4) is an important step against climate change. A batch of large natural-vegetation areas spread across the region is the most important component. The emeralds are distributed across gradients from high to low temperature and much to little moisture, providing suitable locations for many or most terrestrial species to live. As climate warms or cools, species move and relocate in emeralds with suitable conditions. Vegetated corridors that connect the emeralds facilitate the movement of many, though not all, species (Noss 2003). In effect, the emerald network provides flexibility, redundancy, and stability in a changing world.

An alternative oft-proposed biodiversity solution is a major north–south corridor. This would facilitate species movement in response to changing temperature. Unfortunately a corridor is simply too narrow as a habitat to effectively support numerous key species. Furthermore a narrow habitat or strip is difficult to protect and sustain long term in an urban region, where nearby people are numerous, human activities extensive, and ongoing outward urbanization expected.

Sea-level rise, resulting mainly from continued melting of the Arctic ice cap, Greenland, and Antarctic shelves, is the second major effect of climate change. Sea-level rises of 0.5 to 5+ m (based on ice volume subject to melting) are commonly predicted by specialists, though the rate of melt is difficult to estimate (new data and new processes discovered keep increasing the projected rate). In any event, low-lying coastal areas are likely to be inundated. Coastal cities comprise a major proportion of the world's cities. They are usually located where a river meets the sea, and almost always have considerable low area. Cities somewhat upriver from the sea, such as Cairo, Dhaka (Bangladesh), and Jacksonville (USA), also have serious inundation problems.

Thus sea-level rise seems likely to cause a *massive relocation* of people and development in many coastal urban regions. This suggests the need for a regional, perhaps 20–30 year plan for sequentially removing people, buildings, roads, and railroads from low areas, and relocating them elsewhere, such as around certain satellite cities (Chapter 8). If planned well, the relocation of people can produce large environmental benefits as well. These might include protective coastal wetlands and self-building dunes, reestablishing biodiversity and recreation around long-lost wetlands, and relocating heavy industry away from water-bodies to designated industry centers with efficient water, power, transport, and waste disposal.

Unplanned relocations of people in coastal cities may accelerate outward urbanization and sprawl in the urban region. Alternatively, relocating people and development to areas close to the metro-area border could eliminate any existing sprawl there, and be the stimulus for a regional transit-oriented-development solution to transportation (Chapter 2) (Calthorpe 1993, Cervero 1998, Dittmar and Ohland 2004). Indeed sea-level rise augurs poorly for reliance on underground rail or road transportation systems in these cities.

Extensive redistributions of urban populations also provide the opportunity for implementing different transportation modes, such as diverse flexible people-movers, small raised monorails, horizontal ski-transport technologies, dual-level shop-front sidewalks, and other methods for relatively short- and medium-distance movement in metro areas.

Climate-change specialists also point to an expected increase, both in size and frequency, of *extreme weather events* such as intense winds and heavy precipitation. Floods and hurricanes/cyclones tend to cause especially serious problems in urban regions, because of the density of both people and solid human structures like buildings, roads, and bridges. These structures carefully designed by the engineer or architect for society typically have less give, less resilience, than a healthy wetland or natural area, so when confronted with a strong enough force they break.

One solution against extreme-weather events is to build using both nature and engineering (including bioengineering), and nature and architecture (biophilic design, including green roofs). Another approach is a higher investment priority in identifying, and placing nature in every location likely to be heavily impacted by extreme-weather events. Steep slopes and narrow floodplains would become building-free, wetland vegetation would cover low areas, and shrubby strips would cover riparian zones. Minimizing abrupt boundaries and providing multiple lines of defense are additional strategies against extreme-weather events. For example between deep water and uplands, protected gentle slopes, shallow water bodies, wetlands, and natural ridges all help to absorb or dissipate disruptive energy pulses (Costanza *et al.* 2006).

Climate-change scenarios offer an opportunity for urban regions to address many large knotty issues. Reducing carbon-dioxide production calls for citizens and cities alike to change. Implementing solutions to temperature increase, sea-level rise, and extreme-weather events does too. Educating a region's residents of the problems and laying out possible solutions is a key step. Then, capitalizing on the resulting expectation of change, a broader, bigger set of urban-region issues can be addressed than simply climate change.

Species extinction

Species extinction is forever, one of those rare phenomena, like death, that is irreversible. Habitat loss is the major cause, though habitat degradation and perhaps fragmentation are significant overall (Wilson 1992, Wilcove *et al.* 1998, Groom *et al.* 2006, Primack 2004, Lindenmayer and Burgman 2005). Loss of isolated habitats, such as on islands and in lakes, and of tropical rainforest that holds about half the Earth's species, has been the core of species extinction.

Human activities in the past century have caused the extinction rate to sky-rocket. Although the effects of species extinction, so far, have been minor on society, the seeds of pervasive societal problems are likely being sown, and the ethics of knowingly wiping away species is appalling. Continuing the current rate of biodiversity loss promises a convulsion equivalent to the replacement of dinosaurs by mammals around the globe 65 million years ago. Habitat removal has enormously favored *Homo sapiens* and about 40 domesticated species, plus range expansions of numerous "non-native" species (McNeill 2000).

Urban regions are generally not the centers of species extinction, though urbanization in about 30 ecoregions, mostly coastal and large-island areas with concentrations of endemic vertebrates, may be quite significant (Robert McDonald, personal communication). Rather, urban regions are centers of people, their activities, and their impacts that spread widely outward across the land. Market forces and politics may degrade distant areas, yet public support, votes, and funding in cities may also protect nature and species in distant areas.

Rare species probably exist in all urban regions and their protection is valuable (Beatley 1994). However, this has to be placed in a priority context of time and space. In the face of concentrated people, activities, and impacts, plus ongoing outward urbanization, the chance of "permanent" or long-term protection of rare species in an urban region is low. Still, even short-term protection may suffice if a rare species then spreads to more suitable areas beyond the region. Endemic species which are known only in an urban region (such as Capetown, South Africa and Perth, Australia) warrant special protection effort. Nevertheless, with funding and protection efforts always limited, conservation priorities should normally be in areas outside urban regions, where long-term protection against species extinction is more likely to succeed.

The emerald network extending across an urban region, as well as across a state or province or geographic region, was highlighted in the preceding section (also see Chapters 2 and 10) as a way to counteract the effects of climate change. These *large natural areas*, that can sustain large-home-range vertebrates and minimal viable populations of interior species (Chapter 4), are the keystones. Emeralds in an urban region also have the largest interior-to-border ratios, which greatly facilitates management and protection against habitat degradation by concentrated people and their activities.

Yet even emeralds can be degraded. Costa Rica has a widely known set of seven large conservation-park areas. One day the Minister of Natural Resources and Energy commented that the quality of these national parks is only as good as the nation's economy (Rene Castro Salazar, personal communication). If the economy turns sour and families run out of food and jobs, people look to these areas as safety nets. "Parques nationales" (national parks, which belong to the people) are where one can harvest resources, even colonize, if hard times arrive. Emeralds can be stolen, right in an urban region.

The *corridors* of an emerald network have a low interior-to-border ratio and, as noted in the preceding section, are difficult to protect in an urban region with so many people and ongoing urbanization (Briffett *et al.* 2000, Jongman and Pungetti 2004, Hilty *et al.* 2006). Perhaps the "string of pearls" corridor type (Chapters 4 and 10; Color Figure 41) (Forman 2004a), where a tree-lined path is well used and widely known by people, is most likely to persevere over time in an urban region. A sequence of small patches (pearls) of semi-natural vegetation along the path may persist long term as treasured neighborhood parks.

Urban regions are the great centers for *non-native and invasive species* (the non-native species that colonize and spread in natural ecosystems) (Kareiva *et al.* 2007). Ships and planes and trains and vehicles bring most non-native (exotic) species to a city. Rail yards are hotspots of non-natives (Muehlenbach 1979, Kowarik and Langer 2005), which also are typically abundant along railroads and roads (Forman *et al.* 2003). Of particular importance are the extensive residential areas and house lots surrounding a city, where non-native species are purposely and widely planted for diversity and human delight. This massive area teeming with exotics, which are continuously being transported by people and vehicles, is probably the prime source of non-native and invasive species into natural areas across the urban region, and beyond.

Some ecologists consider invasive species to be a major current threat to biodiversity, and others consider this to be a minor tangential issue compared, e.g., with habitat loss and habitat degradation due to human activities. Two guidelines seem useful in reducing the chance of biodiversity problems associated with invasive species. First, eliminate known and suspected invasive species from the market, especially the nursery business that supplies wholesale and retail plantings. Second, since emeralds are about the only areas in an urban region where sustained management against invasives is worthwhile for long-term biodiversity protection, surrounding these large natural areas with moderately large adjacent properties containing a low density of non-native species generally makes sense. Also some "invasives" are really naturalized species.

Connectivity for species movement across a whole urban region should reduce the chance of biodiversity loss there. This generally requires greenspace corridors, normally with walking paths, which interconnect with large semi-natural areas around the region's unbuilt donut (or coastal half-donut) geometry. The corridors should also appropriately connect to adjoining regions, e.g., in the cardinal directions. Transportation corridors can be crossed with major wildlife underpasses and overpasses (Chapter 4) (Trocme *et al.* 2003, Iuell *et al.* 2003, Forman *et al.* 2003). However, strip development along a highway requires a wide greenspace interruption to facilitate crossing by wildlife. Particular effort for maintaining species connectivity may be appropriate between two nearby cities, which could become connected by future strip development along a highway.

The outer portion of the urban-region ring furthest from the city is especially important for biodiversity protection. Mapping and evaluation of biodiversity patterns both in this outer area and in nearby areas of adjoining regions is important. Thus some species only in the region and some species primarily just outside of it may be effectively protected long term. Urbanization around outer satellite cities (Chapter 8) poses a threat, so the particular areas urbanized require careful planning, and concentric-zones urbanization around the major city may be preferable.

In short, overall biodiversity loss on the planet will not be arrested in urban regions. But in the face of huge increases in urban people in the next generation, much can be done in these regions to protect their species and especially reduce urban impacts on rare species of other regions.

Water scarcity

Humans keep growing in number and spreading out from cities over the land. People want clean freshwater to drink and for domestic use. We also need relatively clean freshwater for agricultural food production and for running industry. The planet operates with essentially a fixed amount of freshwater, which effectively keeps recycling (in a hydrological cycle) through the atmosphere, land, and sea. There seems to be no new freshwater to discover. With a fixed amount and growing human usage, the cost of water is rapidly rising and global water scarcity has arrived.

Water scarcity hits us unevenly. People in dry climates suffer first and worst. Prolonged droughts, dried-out vegetation, fires, wind erosion of soil, and atmospheric dust are familiar refrains. Human-caused desertification, salinized soils, and dropping water tables in the ground due to irrigation make headlines. But water scarcity also impacts urban regions, e.g., recent prolonged droughts in Canberra, Atlanta, Nairobi, and Brasilia, where so many people are packed together.

Cities are primarily surrounded by farmland in urban regions (Color Figures 2–39, and Chapter 6). To produce food products for the nearby population, and because land prices are high, irrigation is widely used in urban-region agriculture. Market-gardening near the metropolitan area is a classic example, where a high production of vegetables and fruits continues often year-round.

Industry, another major water-user, characterizes and often dominates cities. For instance, megacities in developing nations are effectively industrial cities. Water is used for cooling, and in some cases for power and waste disposal. Rising prices for scarce water are a serious economic problem for industry around cities, with widely reverberating effects.

Water supply for drinking and other daily domestic uses is particularly important in urban regions, because it requires rather clean water in an area bursting with pollution sources. Clean water may be piped long distances into an urban region as an "inter-basin transfer" (Thayer 2003, Ghassemi 2006) (San Diego, Sydney), but that is expensive, and the process dries out and degrades the source area. Desalinization of salt water can provide a limited amount of water, but at considerable cost. More promising for a city is to rigorously protect natural vegetation around a water-supply reservoir or lake in the outer urban-region ring (New York). This, of course, would be easier if a political/administrative perimeter corresponded closely with that of a dramage basin or catchment, as in the rare case of Florida's Water Management Districts, which have taxing authority and regulatory power.

Using river water as a water-supply source has the advantage of usually having ample water, but the significant disadvantage of typically being polluted, which requires expensive water treatment for cleaning. Pollution sources tend to be diverse and widely distributed, and it is extremely difficult to cover a river's extensive drainage basin with vegetation. In consequence clean rivers in urban regions are a rarity (Paul and Meyer 2001). A water supply based on pumping from streams may serve smaller communities, but for major cities (e.g., Tegucigalpa) the rate of pumping and the diverse acute pollution sources pose limitations. Yet the presence of dirty streams and rivers also means that freshwater is all around the inhabitants of urban regions. Cleaning it up provides relatively clean water for most urban uses. Water conservation and recycling are also potentially major clean-water sources.

Groundwater as a water-supply source is a special problem, because of potential diverse pollution sources in an urban region (Bamako) and limited hydrologic water pressure. Yet a well-protected aquifer with considerable hydrologic pressure can be a valuable source of drinking water, as in parts of the European Community and Russia (Margat 1994). Groundwater is also a major source for industrial use in Northern and Central Europe, South Korea, and Japan. In contrast, it is extensively used for irrigated agriculture in Mediterranean areas (e.g., Spain, Greece), Australia, India, and parts of the USA.

Groundwater aquifers should be covered with essentially continuous vegetation, especially over the upper portion (Gibert *et al.* 1994). Otherwise polluting chemicals from development or agriculture percolate into the groundwater, which (except in limestone areas) move slowly and tend to accumulate. The widespread impermeable surface associated with urbanization, and the rarity of remnant wetlands and stormwater-drainage swales, sends rainwater rushing off the land, thus reducing water recharge into the groundwater. That commonly results in contaminated water and a lowered water table under and around metropolitan areas.

An extensive wooded area upwind of an urban region evapo-transpires water into the atmosphere. When cooled, the moist air may produce rain for the urban region and its groundwater. Air pollutants from the urban region in turn may include particles and aerosols around which the moist-air water molecules coalesce, producing rain both in the region and downwind of it. Thus deforestation, overgrazing, or desertification of the broad upwind area may contribute to drying out of an urban region.

These threads suggest three issues linking water scarcity and urban regions. The growing worldwide scarcity of freshwater causes increasingly severe limitations and costs on clean water supplies for most urban regions. The concentration of people and water uses in urban regions in turn is accentuating the global water scarcity. These reciprocal challenges seem headed for crisis as urban populations are expected to skyrocket in the years just ahead.

Big-ideas–regulations–treaties–policy–governance, megacities, sense of place

This umbrella of big pictures highlights strong social and cultural connections. The first topic, "Big-ideas–regulations–treaties–policy–governance," introduces the human forces that help determine how an urban region is maintained and changed. Then "Megacities" are examined as the largest human concentrations on Earth. Finally, "Sense of place" considers a compelling way to bring the giant urban-regions topic to the scale of a person.

Big ideas–regulations–treaties–policy–governance

A sequence of big ideas provides a useful framework for urban-region environmental issues, which in turn lead to the regulatory approaches and policy that help to govern human behavior. Consider an overlapping sequence of *big ideas* through history (McNeill 2000): religion (in diverse forms), rationalism/science, nationalism, hard-work-making-land-productive (or rural righteousness), and communism. After about 1940, economic growth, with roots in ancient China, Europe and elsewhere, became perhaps the most powerful idea of the twentieth century. Embraced by capitalists and communists alike, the economic growth idea transformed society's focus, relegating land and natural systems to the status of background source of resources for growth (chapter 3).

Environmental ideas, again deeply rooted in history, overall mattered little to society until the 1970s catalyst of Rachel Carson's *Silent Spring* (Buell 1995).

From the 1960s to 1990s a whole set of issues – wetlands, wolves, foaming rivers, and choking air – caught the public's attention, especially in developed nations. *Environmentalism* had arrived as an embryonic big idea.

A closer look is useful (McNeill 2000). From the 1960s, environmental political parties with momentum appeared, and many significant initiatives by government and citizens began to reduce water pollution and air pollution. In 1972 an International Conference on the Environment, which led to establishing the United Nations Environment Program, put Stockholm on the map for the burgeoning international environment community. From the 1980s, environmental protection agencies began appearing in developing countries, India was rife with environmental organizations, and Kenya's Green Belt movement was in full swing. International cooperation on the environment accelerated, the World Bank was forced into environmental awareness, the Brundtland Commission Report stimulated interest in sustainable development, and the Montreal Protocol highlighted the valued linkage of good science and diplomacy. The 1990s brought the UN Conference on Environment and Development in Rio de Janeiro, where everyone spoke for the environment in development, but few nations were willing to take the necessary steps, and a wide divide between rich and poor nations was revealed. The Kyoto Accord produced wide international agreement on limiting greenhouse gases, yet some key developed nations and various developing nations refused to sign. More recently the important reports of the International Panel on Climate Change (McCarthy *et al.* 2001, Houghton *et al.* 2001) have highlighted human-caused effects, especially of greenhouse gases, and their likely consequences. Twentieth-century economic growth models are progressively modified or replaced by models with a healthy focus on ecological and environmental economics (Chapter 3). The idea of environmentalism is rapidly maturing.

The preceding conferences, agreements, and reports are designed to govern behavior of political units and people at the international level. Similarly, regulations, codes, laws, and agreements are drafted to govern behavior at national, state/province, and local levels. Yet regulations have to be policed and enforced, and budgets sustained over time to support the policing of our behavior. Consequently the effectiveness of the regulatory approach is uneven and frequently changing. Around Melbourne (Australia), where cameras record traffic speeds plus turns at busy intersections, and a parking ticket costs nearly seventy times that in Boston, drivers tend to follow the rules. Another familiar problem with the regulatory approach is that regulations themselves can appear, be altered, or disappear "overnight," in the proverbial midnight session of politicians.

Therefore, rather than being a litany of proposed regulations for urban regions and residents' use of natural systems, this book focuses on *land-use pattern* as a generally more promising and lasting approach. If a particular land use or mosaic of land uses makes good sense and is understood by the public, that pattern is unlikely to be changed overnight. It has staying power likely to be sustained. Of course if the land use does not make good sense or is not known by many people, it is in jeopardy and in some cases should be changed.

Land-use planning and regulatory approaches depend primarily on government, and history has provided a complex array of *governmental arrangements* in urban regions. One might recognize a typical case of a region with a central large city, other municipalities in the metropolitan area, a number of satellite cities, numerous towns and villages, and some overlapping political/administrative units, such as counties at intermediate spatial scales. Yet, San Diego/Tijuana and Strasbourg (France) have urban regions that straddle two nations, so regional policies must evolve through two governmental systems and cultures. Ottawa's region straddles two provinces and Kansas City (USA) two states, so differing province/state governments and allegiances are involved in developing policy. The Cincinnati Region (USA) includes parts of three states. In contrast, Beijing and Brisbane (Australia) each essentially have a single municipal government that controls the entire urban region. Regional urban planning should be fairly easy for these cities. However, the overnight-change issue exists and, with few checks-and-balances, a plan may be good or bad. The Mexico City, Canberra, and Washington regions, in contrast, are each within a separate federal district (outside of adjoining states/provinces) so, in theory, they control their regions. But in practice, as national capitals, their national governments have much say in planning or regional thinking. As an interesting model, Hannover (Germany) governs its whole region, with local governments for sub-areas.

London is an instructive case, because plans have been repeatedly drawn up for the city, and over time the area included has progressively grown. Today's London plans seem to be approaching, and in some cases exceeding, the urban region of Color Figure 21. Today's broad vision partially reflects population growth, but more important and salutary is that it increasingly embraces natural systems, their uses, and the threats to them. Drainage basins for water supply, a passenger rail network beyond the city, more passenger airports, regional recreational opportunities, and so forth are in the current plans. So, in comparing the several preceding cities, the spatial arrangement of political/administrative units and land uses emerges as a key determinant for policy and governance.

A different approach, where *national policy* tended to create the same pattern around many cities, is informative. In the twentieth century during the Soviet era, the land around, for example, Bucharest, Berlin (Nelson 2006), and Moscow, was molded by strong central policy. Isolated buildings and clusters of buildings were mostly eliminated, and people basically lived either in industrial collectives (large towns/small cities), agricultural collectives (villages), or scattered historic villages. Analogously, hedgerows and small wooded areas were mainly eliminated, leaving essentially two types of land, extensive field areas and large forest patches (see Color Figures 8, 10, and 23).

During almost the same period in the USA, government policy produced a repeated land-use pattern around most American cities. US policy was sort of a non-overt-policy or laissez-faire market control, together with numerous indirect incentives and subsidies. This pattern produced around cities covering 50 degrees of longitude and 20 degrees of latitude is widely called American sprawl (Chapter 2). In essence, for several decades two huge nations, the USA and USSR with a deep policy divide, each repeatedly stamped itself all over with a single cookie-cutter imprint, but the contrasting imprints used produced two lone giants unrelated to anyone or to each other.

Finally, given this breadth of settings for urban regions, let us briefly explore the possibilities for *regional governance* using the USA as an example. Both local government and national government create and finance considerable policy focusing on land use (National Research Council 1999, Irazabal 2005, Babbitt 2005). However regional thinking and initiatives have been sporadic. Local government particularly has little incentive to think regionally. State government, at the intermediate level, could be a force for regionalism, but typically considers it to be a threat. States are often at odds with large-city governments, and only occasionally establish policy for regions within a state.

So what can be done for urban regions? Typically the basic approach is to establish a regional organization with legal responsibility and political authority, either in a narrow somewhat-technical domain (e.g., water, recreational trail systems, airports) or in a broader domain relative to land use. Six alternatives might be suggested (Burchell *et al.* 2005):

(1) *Voluntary confederations*: no real power, but can highlight issues and provide information for the public.

(2) *Public–private coalitions*: similar to the preceding, but can also draw up alternative plans for evaluation by policymakers and the public.

(3) *Federally created regional agencies*: specialized functions, e.g., air quality, transportation, or biodiversity, with power and money to implement policies.

(4) *Regional bodies with broad authority*: established by state legislatures, though difficult to sustain reauthorizations and funding (e.g., Portland).

(5) *Regional governments established by merging municipalities/counties*: local governments are replaced by a regional one (e.g., Indianapolis, Miami/Dade Co.).

(6)   *Contracts between separate governments*: addresses specialized issues/functions, provides economy of scale, and does not require state legislation.

In the evolution of big ideas, could *urban-region planning* burst forth as the next big one? Motivators are there in the public mind – urbanization, sprawl, traffic, shrinking water supply, rising sea level, squatter settlements, not-in-my-backyard, and much more. Solutions targeted to the city or metro area are welcome, but are generally band-aids – too small and temporary. What will catapult urban regions and environmental planning together as the big solution in the public eye?

Megacities

Cities with a population exceeding 10 million (or 8 million) seem to be in a class of their own, not just quantitatively, but qualitatively. Putting aside the question of what area is included in a population estimate, the number of megacities with >10 million people has clearly accelerated over the past half-century (World Urbanization Prospects 2001):

1950	1	New York
1975	5	Tokyo, New York, Shanghai, Mexico City, Sao Paulo
2000	17	Tokyo, Sao Paulo, Mexico City, New York, Mumbai (Bombay), Los Angeles, Calcutta, Dhaka, Delhi, Shanghai, Buenos Aires, Jakarta, Osaka, Beijing, Rio de Janeiro, Karachi, Metro Manila

If an agglomeration (city and adjacent, or nearly so, cities around it) with >10 million is considered, the number rises to 23, with the addition of Seoul, Cairo, Moscow, London, Tehran, and Istanbul. Another analysis highlights 22 megacities, adding Paris and Lagos, while dropping Dhaka, Beijing, and Istanbul (McNeill 2000). The United Nations Population Division lists 20 such urban agglomerations in 2005: the 17 above plus Cairo, Lagos, and Moscow.

No obvious groupings of megacities stand out. More than half are near the sea and have considerable low-lying area subject to flooding. Several are in basins or other locations that accentuate air pollution problems (Los Angeles, Beijing, Mexico City). However, the usual way to divide megacities is their presence in developing nations (Jakarta, Dhaka, Cairo) and developed nations (Tokyo, New York, London). Before briefly considering major environmental dimensions, it is well to note that megacities also illustrate characteristics of cities in

Figure 12.3 Beloved dragons and ball in Beijing's "Forbidden City", the actual, symbolic, and cultural center for the people of a megacity. An unusual case, Beijing municipal government governs essentially the entire 100 km (65 mi) radius urban region (Color Figure 7). Two of the nine dragons on a brightly colored ceramic courtyard wall. R. Forman photo.

general. Thus any city may have a strong cultural heritage (Figure 12.3), depend on radial highways, have wetlands largely eliminated, or be in a disaster-prone area (Manila, Dhaka, Mexico City) (Steedman 1995).

However some *attributes characterize megacities* and seem to be important in almost all of them (Ezcurra and Mazari-Hiriart 1996, McMichael 2000, El Araby 2002). An extensive impermeable surface across the metropolitan area produces several major effects. The heat-island effect results in high temperatures, especially in summer (Chapter 4) (von Stulpnagel *et al.* 1990, Ichinose *et al.* 1999, Arnfield 2003). Heat accentuates certain air pollutants and their effects. The hard-surface area increases flooding, and the so-called 100 year flood may return quite often. With reduced water recharge into the ground, the water table drops over a large area, and soil subsidence under buildings and streets commonly occurs, sometimes quickly. Groundwater is contaminated under and around the massive impermeable surface area. Surface water quality is degraded over a major portion of an urban region (Ren *et al.* 2003).

Megacity transportation for the huge population also poses numerous problems, from streets clogged with pedestrians, bicycles, and vehicles (Beijing, Calcutta) to underground rail systems periodically subject to flooding (New York, Osaka, Moscow). An extensive area of high-vehicle-density and traffic jams is characteristic (Los Angeles, Sao Paulo, Tokyo). Megacities produce huge amounts of CO_2 contributing to global climate change. These giant cities have giant ecological footprints (Chapter 3). Within their nations, megacities are all dynamic centers for economic growth, social groups and institutions, politics and government, transportation, culture, jobs, and opportunity. Finally, megacities serve as powerful magnets attracting more people.

Megacities in developing countries are somewhat distinctive. A relatively high proportion of urban poor and an abundance of informal housing (squatter settlements) are characteristic. Effective urban infrastructure, such as public transport, clean water supply, and gas/electricity, only serves portions of the metro area. For example, sewage treatment serves a quarter of the people in Calcutta and Dhaka. Because low-lying areas are largely covered with development, floods commonly cause major problems.

Also, megacities in developing countries are normally prime centers of industrial production. Atmospheric particulate pollution, resulting from coal burning, industrial production, and other sources (e.g., fecal particles, dust from upwind soil erosion) tends to be severe. Public health problems are rampant. Street trees are relatively sparse overall, and parks may be invaded by squatters. Effective zoning surrounding the metro area is limited, so land uses are intermixed at a relatively fine scale. This in turn degrades or eliminates water and biodiversity values that depend on large semi-natural patches (Chapter 4).

On the other hand, *megacities in developed nations* tend to expand outward in a low-density form associated with extensive auto usage and thus cover much larger and more distant areas. This process may produce aggregations of buildings around numerous villages, towns, and small cities scattered across the land (London), or result in massive sprawl of houses on relatively large lots (Los Angeles, New York). Both patterns also create an extensive fine-scale road network across the land, with an extensive use of cars and trucks thereon. Building sprawl, road network, and vehicle use, when combined, cause extensive habitat loss and habitat degradation.

Megacities in developed nations also are more likely to be financial than industrial centers. Urban liveability or quality of life is often considered to be a goal (Costanza et al. 2006). A relatively high priority is given to investment in clean water supply (sometimes transported from a relatively distant forested drainage basin), sewage treatment, a separate pipe system for stormwater drainage and human wastewater, solid waste disposal (transported to distant sites), controls on air pollutants, controls on water pollutants, and the establishment and maintenance of greenspaces. Zoning and controls on urbanization may be effective in some areas. Regional planning is periodically attempted and occasionally somewhat effective in megacities of developed countries.

Like virtually all discussions of megacities, this glimpse cannot avoid the large number and large size of problems facing society (Fuchs *et al.* 1994, Main and Williams 1994, Stubbs and Clarke 1996, McMichael 2000). Looking for *positive signs* is a useful strategy, when a giant lies ill before us. The rate of population growth in megacities is slowing, though growth continues, perhaps because people realize these cities are not working well. Specific ambitious projects, such as housing and sanitation, affecting many people have provided inspiration, even hope (Altshuler and Luberoff 2003). For example, in downtown Seoul, a polluting multilane highway in 2004 was rapidly removed to reestablish a river with powerful aesthetic and recreational dimensions. Beijing, in preparing to host a summer Olympics, established and planted (most people hope "permanently") numerous greenspaces. Regional planning, in part driven by environmental considerations such as water supply, air pollution, sprawl, and greenspaces, seems to be growing in megacities and anticipating population growth (Chapter 2).

To these few positive signs of life, one can add some ideas for the future. To shift priorities toward low-cost flexible transportation systems or simple healthy safe aesthetic housing types might have positive reverberations. Or target investment to certain existing nodal areas, distributed across the region, to help catalyze the spread of vibrant neighborhoods and communities around them. Or perhaps provide economic opportunities and other incentives for residents of certain targeted areas to move to satellite cities, and in the vacated areas, create places of community value, from treasured playgrounds, trash-removal locations, and park-like meeting places to protected flood-reduction wetlands. Or reduce the dependence on an uncaring global economy, by investing more in local-to-national markets, which spring from local and regional culture and result in economic diversity. Certainly, placing a priority on ongoing high-visibility urban-region planning is a cost-efficient investment that helps everyone understand today's megacity, and think about a better tomorrow.

If one plots important environmental and human variables against city population size, so far no point is known where thresholds coalesce. Plato advised people to start a new city when one reaches 50 000. Today, is one 10 million megacity better than two 5 million or twenty half-million-people cities? Faced with the multitude of megacity region challenges, powerful positive ideas for the future must be found. Because many more megacities lie just over the horizon.

Sense of place

When abroad and asked where I am from, I usually say Boston. If the questioner says, "You live in the city?," I respond, "No, the Boston Region." A century or so ago farmland with distinctive rural towns mostly surrounded the city, but now they have largely coalesced into the city's region. Today Boston is a place for Bostonians, and the Boston Region a place the rest of us relate to.

People have an affinity, often strong, for a place (Tuan 1977, Jackson 1994, Nassauer 1997, Buell 2001, 2005). To develop that affinity or place-connectedness, they have lived there for a period, seeing, hearing, smelling, feeling, and experiencing the space. They know some of the people, as well as arrangements of buildings, roads, and greenspaces. A strong loyalty and caring about the place often develops. Hardly anyone expresses a *sense of place* for the planet or even a metropolitan area. Ignoring nationalism, rather few people express a strong sense of place for a nation or a state or province. The smaller the space the more likely a strong affinity develops, so a town or neighborhood or backyard often engenders a strong sense of place (Tuan 1977, Thayer 2003).

Does someone from the Tokyo Region or San Francisco Region or Sao Paulo Region relate to and care about the region as a whole? Perhaps, because we are bathed with regional TV–radio–newspaper stories, entertainment events, sports teams, cultural resources, air pollutants, and more. But "home range movements" are the real way to gain a sense of place, seeing, feeling, and experiencing a region. The daily home range of a person, analogous to an animal's home range (Chapter 4), includes the routes and sites visited during most days. Where walking, bicycling, or horse transport prevails, it correlates somewhat with a town or neighborhood. But with a huge net of paved roads used by motorized vehicles, the daily home range may include a few adjacent towns or much of a city.

Still, a person's *annual home range*, i.e., the spider web of routes and sites visited during most years, is perhaps more relevant where needs and activities are largely done by driving in a vehicle. The annual home range ties the urban region together in one's mind, creating a sense of place for a region.

What does a person feel deeply about in a place? Two components at the human scale may be central, an arrangement of human-made objects and the nature or natural systems interwoven with it. Unlike animals that live in habitat space, rarely does a person live surrounded only by nature, though many of us treasure and are inspired by it. The human-made objects, especially buildings, roads, walkways, vehicles, and so forth, are probably less important individually for a sense of place than is their collective presence and arrangement. Putting aside the problem of "whose place?," that design of anthropogenic objects may be sufficient to engender a sense of place for some people.

But perhaps most of us would find the space incomplete or sterile (Kaplan *et al.* 1998). Adding trees and birds and water and changing weather, for instance, is needed to bring the place alive. Biophilic design (Chapter 2) speaks to the value, even need, of people for nature. Plants and vegetation are especially familiar manifestations of nature, but wildlife, flowing water, or sky and weather may be central for some of us (Jackson 1994, Karr 2002, Orr 2002, Yu and Padua 2006). In short, a strong *sense of place* perhaps especially develops from the arrangement of human objects and nature in the space where a person lives for a period.

Finally, consider change. Implicit in the preceding is that the space or place remains in a reasonably similar form over the period of a person's experience there. That continuity or stability is at odds with the widespread urbanization processes of change in an urban region. People treasure and care about and fight for their town or place. Meanwhile the onrushing wave of population growth and outward development laps around them, undermines and degrades their place, and seems to roll right over it. What does a person with a strong sense of place do? Stay, and suffer? Or leave, and roll outward ahead of the tidal wave?

Awakening to the urban tsunami

Today the Earth has six billion people, half living in urban areas (*State of the World's Cities* 2006). One out of two urban people lives in a city of <500 000. A billion people, one out of three urban residents, live (in a "slum") with inadequate housing and no or few basic services.

By 2030, a single generation ahead, United Nations Population Division data point to a global population of eight billion, with 60 % in urban areas. Do the math: 50 % × 6 = 3 billion urbanites today, and 60 % × 8 = almost 5 billion tomorrow. Two billion humans, a doubling, are expected to live in urban slums.

In 2030, big cities will be noticeably more numerous (megacities increasing from 20 to nearly 30, and cities of 1–10 million increasing from about 400 to 600; Robert McDonald, personal communication), and more widely distributed over the land. Meanwhile small and mid-size cities are growing at even faster rates. Equally conspicuous, outward urbanization around nearly all the cities will have created more extensive metropolitan areas, more built-up urban regions, and more-severe heat-island effects. Add it all up. Huge urban population growth, more cities, more big-population cities, cities more widely distributed over the land, cities with larger metropolitan areas, outward urbanization from much longer metro area borders, and dispersed urbanization much further out. Does that add up to an "urban tsunami" on land?

In essence, more cities, more big cities, more widely distributed, with growing populations and with expanding metropolitan areas are all rapidly covering the land – an *urban tsunami* easily identifiable today. Our powerful turbulent eddies coalesce in places, as the wave sweeps swiftly, almost inexorably across the land.

Three other huge human forces are also testing the resilience of nature and the land. Water scarcity worsens leaving cities in short supply, cropland parched, water tables lowered, wildfires burning, and answers drying up. Species extinction accelerates as tropical forests shrink, coastal areas are developed, natural vegetation is removed, and key habitats are fragmented and degraded, all worrisome trends for society's ecosystem services. Climate change due to human activity becomes more obvious, as air temperatures increase, ice sheets melt, sea level rises, and droughts become more severe.

These *four giants* – urban tsunami, water scarcity, species extinction, and climate change – have begun joining forces. Climate change and water scarcity combine to accentuate species loss. The urban tsunami together with climate change makes water scarcity worse and the urban heat-island effect more severe.

When will the giant forces reach us? That is, when could the extensive degradation of natural systems overwhelm society's response capability, producing masses of environmental refugees and widespread disruption of social order? Thresholds or tipping points (Gladwell 2000) resulting in clearly identifiable crises may not appear. Each force, or all four together, may continuously and perhaps exponentially gather, until, as Aldo Leopold (1944) expressed the idea, "One simply woke up one fine spring to find the range dominated by a new weed." Expected scenarios for all four forces suggest grim global conditions in some two to three decades. Will they arrive in tandem or together? Or could the urban tsunami catch and overwhelm us first?

If one has full faith in growth and market economics (Chapter 3), with infinite substitutability, or in serendipity – something will come along and solve the problem – then there appears to be no problem. Society can continue as is, and natural systems can keep on degrading. The analyses in this book, along with the growing importance of ecological economics (Chapter 3), find that to be a flawed approach. Society can do better.

The four giants are certainly after us, and wisdom suggests that we ponder what can be done at least in today's urban regions. First a reality check is useful. An urban region teems with people (hundreds of thousands, millions), buildings of varied heights and sizes, streets and roads with an extensive permeating surface area, and vehicles in huge numbers consuming fuel and liberating wastes. Moreover, the people's ecological footprints cover large areas of land. Such a massive system, perhaps paradoxically, has both enormous inertia and extreme instability. Some components, such as an extensive impermeable surface or a road network, can resist almost anything. Meanwhile others, including a water-supply system or tall buildings, can change catastrophically fast.

Most people considering significant improvement of an object such as an urban region conclude that it is simply too big and too difficult. Wait for a crisis event ... out of water, massive disease outbreak, bombing, earthquake. Then address problems, rebuild, or even move away. Predicting or waiting for a crisis event is an unlikely strategy for improvement. "Muddling along" is probably the most likely scenario. Unfortunately, for such a large system this normally means continuing in a downward or worsening trajectory. "Entropy increases." "The second law is after us." The region is maintained by regulations and laws at the wrong scale, by indifference or lack of understanding, and by absence of vision. No implementable big idea or vision appears to fight entropy. Like the four giants above, the trajectory points toward crisis. Shadows creep across the land.

Two alternatives offer promise. Many people in their life want to *improve the world a bit*. Incrementalism, resulting from lots of people over time doing this, can produce useful results. What are the likely results? The first is illustrated by the reported responses of two leading environmentalists, when separately interviewed (shortly before they died) about their major career accomplishments. After brief reflection, each gave essentially the same response: "I think I slowed the rate of environmental degradation." They did indeed. A more ambitious goal would be to level off the trajectory of degradation so the world gets no worse. Or better still, turn things around so the trajectory is positive. Most ambitious would be to see a positive world. Incrementalism, while salutary, seems unlikely to achieve these ambitious goals.

The second alternative for improving the world is a new *vision* or visioning approach, occasionally used in planning exercises. One outlines a tangible vision of a positive future (e.g., for an urban region). Later the varied trajectories to get from here to there can be considered. The vision is spatial, so planners, decision-makers, and the public can all envision it. But the vision is outlined in generic form without details. Sketching out a vision is hard, requiring clear thought, broad perspective, and a big-picture solution. After undertaking the vision approach once, I added scattered clouds superimposed over my outline sketches to emphasize the difficulty in perceiving and portraying an optimal future. The vision approach is strengthened by several or many people representing diverse fields and cultures, each outlining a vision and then comparing and evaluating the results. Society should welcome the opportunity to select among competing visions. Indeed the public often follows people with vision.

Periodically in this book we have met a rhino in the restaurant, kangaroo in the kitchen, and bats in the bedroom. These beasts were rampaging about, while we carefully adjusted a rug here, wiped clean a spot there, and rearranged our favorite trinkets. In our urban regions, a life-support system, the natural systems upon which we depend daily and for our future, is being ravaged. Meanwhile society concentrates on a housing development here, a new road there, or an economic development project somewhere.

How could we miss the big picture before us? A giant bulldozer on automatic rampaging over our most needed land. Scarcely a speck in the urban tsunami . . .

Appendix I

Mapping procedure for satellite images

Five steps comprise the mapping procedure for satellite images of the 38 urban regions analyzed (Chapter 5).

(1) Locate data. Using an Enhanced Thematic Mapper Plan sensor, click on Map Search, toggle on "ETM+", and enter the Place (e.g., Cairo, Egypt).

(2) Download color band. After locating the area of interest, click on "Preview and Download."

(3) Select images. In the options for downloading, focus on ETM+ data and individually select an image based on its unique identifier (quadrangle), XXX–XXX.

(4) Download bands. Using the download interface, which is similar to a file transfer protocol (FTP) interface, download band 1, band 2 and band 3.

(5) Activate ArcView import. ArcView moves the work forward by extracting the winzip (compressed file), importing the geotif (geographically referenced image), activating the Arc Tool "Composite Bands", and scaling the map to the spatial scale of interest.

Appendix II

Urbanization models evaluating 18 attributes in 38 regions

Four alternative spatial models (Figure 8.2) were superimposed on the 38 urban region maps (Color Figures) to evaluate the effect of urbanization on 18 attributes measuring natural systems and their human uses. Attributes covered by urbanization were counted or estimated at each of the three time stages in a model. Amount recorded is an index of effect or degradation due to urbanization. The four models were ranked accordingly from best to worst (1 to 4) for each attribute in each region. See Table 8.2 for a summary of Appendix II.

The models (see Chapter 8) are: C=Concentric-zones model; S=Satellite-cities model; T=Transportation-corridors model; D=Dispersed-sites model. For Chicago, Philadelphia, and Atlanta, the dispersed-sites model is ranked based on the first time stage only; see text. A dash indicates that the attribute is absent or not mapped in the urban-region ring of that city.

Asterisks indicate the following. *Forest/woodland*: overall estimate based primarily on forest/woodland area covered, and secondarily on proximity to or fragmenting of forest/woodland. *Grassland/pastureland*: same approach as for preceding. *Desert/desertified area*: same approach as for preceding. *Rivers/major streams*: includes canals. *Marine coast*: along sea or bay. *Reservoirs/lakes*: an estimate based equally on the percentage of reservoirs/lakes affected and the total length of shoreline affected. *Drainage area for water*: refers to the area around a water-supply source expected to be especially important for protection, rather than to a particular or complete drainage basin/watershed/catchment. *Average distance to major highway*: interpreted as the further development is from a major highway the more land is impacted by roads and vehicles. *Degree of subdividing region*: overall estimate of disruption of connectivity for wildlife movement across a region due to strip development and development in other locations. *Edge density*: average length of built-area edge or border per unit area in the urban-region ring. *City center to metro-area border*: average distance from a point within the metro area to its border (the greater the distance, the less accessible surrounding countryside is to urban residents and the less likely outside species will reach city greenspaces). *Other attributes combined*: letters refer to the attribute present in a region; (a) =greenbelt; (b) = urban growth boundary; (c) = Native Peoples' land; (d) = aquaculture area; (e) = fire hazard area.

Urban regions	London	Berlin	Rome	Bucharest	Stockholm	Barcelona	Nantes	Chicago	San Diego/Tij'a	Philadelphia	Ottawa	Edmonton	Portland	Atlanta	Mexico City	Santiago	Brasilia	Tegucigalpa	Iquitos	Cairo	Nairobi	Bamako	East London	Abeche	Beijing	Moscow	Seoul	Tehran	Sapporo	Ulaanbaatar	Erzurum	Kagoshima	Bangkok	Kuala Lumpur	Cuttack	Samarinda	Canberra	Rahimyar Khan	Average
Biodiversity sites (%)																																							
C	3	3	2	3	4	1	3	1	2.5	3	2	3.5	3.5	2.5	3.5	2	3.5	2	–	2.5	2	1.5	1.5	2.5	3.5	3	4	3	2.5	3	1	1	1.5	3	1.5	1.5	3	1.5	2.38
S	1	2	2	3	4	3	1	2	1	2	1	1	1	2.5	1	2	1.5	1	–	2.5	1	3.5	1.5	2.5	1.5	1	1	1	1	1	3	3	3.5	1	1.5	1.5	3	3	1.78
T	3	1	4	4	2.5	4	2	3.5	2.5	1	4	2	2	1	3.5	4	1.5	4	–	2.5	4	3.5	3.5	2.5	1.5	2	2	2	4	2	4	4	3.5	3	4	3.5	1	4	2.93
D	3	4	3	2	2.5	2	4	3.5	4	4	3	3.5	3.5	4	2	2	3.5	3	–	2.5	3	1.5	3	2.5	3.5	4	3	4	2.5	4	2	2	1.5	3	3	1.5	3	1.5	2.91
Recreation/tourism sites (%)																																							
C	3	3	1.5	1.5	3	1.5	1.5	2.5	3	2.5	3	2	2.5	4	3	1	4	2	2.5	4	2	2	2.5	2	3.5	1.5	3	2.5	2.5	2	1.5	1	1.5	2.5	1.5	2	3	1.5	2.38
S	1	1	1.5	1.5	–	3	1.5	1	1	2.5	1	2	1	1	1	4	1	1	1	1	1	2	1	2	1	3.5	1	2.5	1	2	1.5	3	1.5	1	3	2	1.5	3	1.69
T	3	2	4	1.5	–	4	4	4	4	1	3	2	4	2	4	2	2	4	4	2	4	4	2.5	4	2	3.5	2	2.5	3	2	4	4	4	4	4	4	4	4	3.16
D	3	4	3	4	–	3	3	2.5	2	4	3	4	2.5	3	2	3	3	3	2.5	3	3	2	4	2	3.5	1.5	4	2.5	3	4	3	2	3	2.5	1.5	2	1.5	1.5	2.77
Forest/woodland*																																							
C	3.5	1	1	2	2	3	2	2.5	2	1	1	1.5	1	1	3	3	4	1	1	–	1.5	1.5	1	–	2.5	–	3	–	2.5	2.5	2	1	1.5	1	1	1	1	2	1.81
S	1.5	4	4	4	2	1	4	2.5	2	4	2	3	3	2.5	1	1	1	4	4	–	3.5	2.5	3	–	1	3	1	–	1	1	2	2	4	3	3	4	3	2	2.53
T	1.5	2	4	1.5	2	2	2	2.5	2	3	4	4	2	2.5	2	2	2.5	3	3	–	3.5	2.5	4	–	2.5	3.5	2	–	4	2.5	4	4	3	4	4	3	4	4	2.94
D	3.5	3	2	2	4	2	2	2.5	4	2	3	1.5	4	4	4	4	2.5	2	2	–	1.5	2.5	2	–	4	2	4	–	2.5	4	2	3	1.5	2	2	3	2	2	2.71
Grassland/pastureland*																																							
C	–	–	–	–	–	–	–	–	–	–	–	1	–	–	1	–	1	–	–	–	3.5	2	1.5	2.5	–	–	–	–	–	1	3	–	–	–	–	–	3	–	1.95
S	–	–	–	–	–	–	–	–	–	–	–	4	–	–	4	–	4	–	–	–	3.5	1	3.5	2.5	–	–	–	–	–	4	4	–	–	–	–	–	4	–	3.45
T	–	–	–	–	–	–	–	–	–	–	–	3	–	–	3	–	3	–	–	–	1	4	3.5	2.5	–	–	–	–	–	3	1	–	–	–	–	–	2	–	2.6
D	–	–	–	–	–	–	–	–	–	–	–	2	–	–	2	–	2	–	–	–	2	3	1.5	2.5	–	–	–	–	–	2	2	–	–	–	–	–	1	–	2.00

(cont.)

Desert/desertified area*

C	–	–	–	–	–	–	–	–	–	–	–	–	1	–	–	–	–	–	–	–	2.5	2.00
S	–	–	–	–	–	–	–	–	–	–	–	3.5	3.5	1	–	–	1	3	–	–	2.5	2.63
T	–	–	–	–	–	–	–	–	–	–	–	3.5	3.5	1	–	–	4	–	–	–	2.5	2.88
D	–	–	–	–	–	–	–	–	–	–	–	2	2	2	–	–	2	–	–	–	2.5	2.50

Nearby slopes facing city (% cover)

C	3.5	3	–	4	3.5	–	3.5	2.5	4	4	3	3.5	3.5	3	3.5	3.5	–	4	–	–	–	3.39
S	1	1	–	1	1	–	1	1	1	1	1	1	1	1	1.5	1	–	1	–	–	–	1.08
T	2	2	–	2	2	–	2	2.5	2	2	3	2	2	2	1.5	2	–	2	–	–	–	2.16
D	3.5	4	–	3	3.5	–	3.5	3	3	3	4	3.5	3.5	4	3.5	3	–	3	–	–	–	3.37

Rivers/major streams (km length)*

C	3.5	3	2.5	1	3.5	1	3.5	1	4	1	2.5	1	3	2.5	3	3.5	1.5	2	3	2.5	2	2.34
S	1	2	1	3	1.5	3	3.5	2.5	1	3.5	1	1	1	1	1.5	1.5	3	1	1	1	1	1.88
T	2	1	2.5	4	1.5	4	1.5	3	2	1	3	2.5	4	4	1.5	1.5	3	4	2	4	4	2.80
D	3.5	4	2	2	3.5	2	3.5	4	3	4	4	4	3	2.5	4	3.5	1.5	3	4	2.5	3	2.99

Major wetlands (% cover)

C	–	3.5	1	–	3.5	–	3	–	4	–	2.5	2.5	–	2	2.5	–	1.5	3	3	2	1.5	2.41
S	–	1.5	4	–	1.5	–	1	–	1	–	2.5	2.5	–	1	2.5	–	1.5	1	1.5	4	3	2.12
T	–	1.5	2.5	–	1.5	–	3	–	2	–	2.5	2.5	3.5	3.5	2.5	–	4	4	1.5	2	3.5	2.74
D	–	3.5	2.5	–	3.5	–	3	–	3	–	2.5	2.5	3.5	3.5	2.5	–	3	3	4	2	1.5	2.73

Flood hazard sites (%)

C	–	2	–	–	3	–	3	–	3	3	3	–	–	3	–	–	2	–	3	3	1.5	2.63
S	–	2	–	–	1	–	1	–	1	1	1	–	–	1	–	–	2	–	1	1	3	1.54
T	–	4	–	–	3	–	4	–	3	3	2	–	–	3	–	–	4	–	4	3	4	3.21
D	–	2	–	–	3	–	2	–	3	3	4	–	–	3	–	–	2	–	2	3	1.5	2.63

(cont.)

		London	Berlin	Rome	Bucharest	Stockholm	Barcelona	Nantes	Chicago	San Diego/Tij'a	Philadelphia	Ottawa	Edmonton	Portland	Atlanta	Mexico City	Santiago	Brasilia	Tegucigalpa	Iquitos	Cairo	Nairobi	Bamako	East London	Abeche	Beijing	Moscow	Seoul	Tehran	Sapporo	Ulaanbaatar	Erzurum	Kagoshima	Bangkok	Kuala Lumpur	Cuttack	Samarinda	Canberra	Rahimyar Khan	Average
Marine coast (km length)*	C	1	1	1	-	3	1	2	-	1	-	-	-	-	-	-	-	-	-	-	-	-	-	4	-	-	-	3	-	1	-	-	1.5	3	3.5	2	1.5	1.5	-	2.04
	S	4	3	3	-	1	2	4	-	4	-	-	-	-	-	-	-	-	-	-	-	-	-	1	-	-	-	2	-	2.5	-	-	3	2	1.5	4	4	-	-	2.71
	T	3	-	4	-	2	4	2	-	3	-	-	-	-	-	-	-	-	-	-	-	-	-	2	-	-	-	1	-	4	-	-	4	1	1.5	2	3	-	-	2.61
	D	2	2	2	-	4	3	2	-	2	-	-	-	-	-	-	-	-	-	-	-	-	-	3	-	-	-	4	-	2.5	-	-	1.5	4	3.5	2	1.5	-	-	2.64
Reservoirs/lakes (% and shoreline km)*	C	1.5	3	2.5	1	3	1.5	3	1	3.5	2	2.5	2.5	2	1	2	2.5	4	3.5	-	2.5	2.5	-	1.5	-	1.5	4	3	2.5	2	-	2	-	-	-	-	-	-	-	2.33
	S	3	2	2.5	2	2	4	1.5	3	1	4	2.5	1	1	3	4	2.5	1	1	-	2.5	2.5	-	1.5	-	4	1	1	2.5	2	-	2	-	-	-	-	-	-	-	2.27
	T	4	1	2.5	4	1	3	1.5	4	2	1	2.5	4	3	4	1	2.5	3	3.5	-	2.5	2.5	-	4	-	3	2	2	2.5	4	-	4	-	-	-	-	-	-	-	2.68
	D	1.5	4	2.5	3	4	1.5	4	2	3.5	3	2.5	2.5	4	2	3	2.5	2	2	-	2.5	2.5	-	3	-	1.5	3	4	2.5	2	-	2	-	-	-	-	-	-	-	2.72
Drainage area for water (% cover)*	C	3.5	-	-	-	4	-	-	-	4	-	4	3	3	2	-	2.5	3	3	4	-	-	4	1.5	-	2	-	3	3	-	-	-	4	-	-	-	-	-	-	3.08
	S	1	-	-	-	1	-	-	-	1	-	1.5	1	1	1	-	2.5	1	1	1	-	-	1	1.5	-	4	-	1	2	-	-	-	1	-	-	-	-	-	-	1.45
	T	2	2.5	-	-	2	4	-	-	2	2	1.5	2	2	4	-	2.5	2	2	2	-	-	2	4	-	2	-	2	1	-	-	-	2	-	-	-	-	-	-	2.29
	D	3.5	2.5	-	-	3	2	-	-	3	-	3	4	4	3	-	2.5	4	4	3	-	-	3	3	-	2	-	4	4	-	-	-	3	-	-	-	-	-	-	3.18
Market-gardening area (% cover)	C	1.5	-	-	-	-	4	-	-	-	3.5	-	-	3.5	-	-	-	-	-	-	-	2	4	-	-	-	-	1.5	-	-	-	2	2.5	3	-	-	-	-	-	2.75
	S	4	-	-	-	-	1	-	-	-	1	-	-	1	-	-	-	-	-	-	-	1	2	-	-	-	-	3.5	-	-	-	2	2.5	1.5	-	-	-	-	-	1.95
	T	1.5	-	-	-	-	2	-	-	-	2	-	-	2	-	-	-	-	-	-	-	4	2	-	-	-	-	3.5	-	-	-	4	2.5	1.5	-	-	-	-	-	2.50
	D	3	-	-	-	-	3	-	-	-	3.5	-	-	3.5	-	-	-	-	-	-	-	3	2	-	-	-	-	1.5	-	-	-	2	2.5	4	-	-	-	-	-	2.80

Table — Attribute scoring matrix (rows C, S, T, D per attribute; final column is the summary score)

Average distance to major highway (km)*

C	1	–	–	–	–	–	–	2	2	2	–	2.5	2	2.5	2	–	3	–	2	2	–	–	3	2	2	–	2	2	–	–	–	–	**2.20**
S	4	–	–	–	–	–	–	4	4	4	–	4	4	4	4	–	4	–	4	4	3	–	4	3	4	–	2	–	–	–	–	–	**3.57**
T	3	–	–	–	–	–	–	1	1	1	–	1	1	1	1	–	1	–	1	1	1	–	1	1	1	–	1	–	–	–	–	–	**1.13**
D	2	–	–	–	–	–	–	3	3	3	–	2.5	3	2.5	3	–	2	–	3	4	4	–	4	3	4	–	1	–	–	–	–	–	**3.10**
	(a)	(e)						(c)	(e)	(a)	(c)	(a)		(c)	(b)																		

Degree of subdividing region (high to low)*

C	2	2	2	2	2	2	2	2	2	2	2	2	2	2	2	2	2	2	2	2	2	2	2	**2.00**
S	1	1	1	1	1	1	1	1	1	1	1	1	1	1	1	1	1	1	1	1	1	1	1	**1.00**
T	4	4	4	4	4	4	4	4	4	4	4	4	4	4	4	4	4	4	4	4	4	4	4	**4.00**
D	3	3	3	3	3	3	3	3	3	3	3	3	3	3	3	3	3	3	3	3	3	3	3	**3.00**

Edge density (km length per km²)*

C	1	1	1	1	1	1	1	1	1	1	1	1	1	1	1	1	1	1	1	1	1	1	1	**1.00**
S	2	2	2	2	2	2	2	2	2	2	2	2	2	2	2	2	2	2	2	2	2	2	2	**2.00**
T	3	3	3	3	3	3	3	3	3	3	3	3	3	3	3	3	3	3	3	3	3	3	3	**3.00**
D	4	4	4	4	4	4	4	4	4	4	4	4	4	4	4	4	4	4	4	4	4	4	4	**4.00**

City center to metro-area border (km)*

C	4	4	4	4	4	4	4	4	4	4	4	4	4	4	4	4	4	4	4	4	4	4	4	**4.00**
S	1	1	1	1	1	1	1	1	1	1	1	1	1	1	1	1	1	1	1	1	1	1	1	**1.00**
T	2	2	2	2	2	2	2	2	2	2	2	2	2	2	2	2	2	2	2	2	2	2	2	**2.00**
D	3	3	3	3	3	3	3	3	3	3	3	3	3	3	3	3	3	3	3	3	3	3	3	**3.00**

Other attributes combined*

C	3.5	2	3.5	3.5	3.5	2	4	3	**3.20**
S	1	1	1.5	2	1.5	2	1	2	**1.40**
T	2	3	1.5	2	1.5	2	1	1	**1.80**
D	3.5	4	3.5	3.5	3.5	4	3	4	**3.60**
	(e)	(a)	(c)	(b)	(a)		(d)	(e)	

References

AASHTO. (2001). *A Policy on Geometric Design of Highways and Streets: 2001*. Washington, DC: American Association of State Highway and Transportation Officials.

Acebillo, J. and Folch, R. (directors) (2000). *Atles Ambiental de l'Area de Barcelona: Balanc de recursos i problemes*. Barcelona: Ariel Ciencia and Barcelona Regional.

Aghion, P. and Howitt, P. (1998). *Endogenous Growth Theory*. Cambridge, MA: MIT Press.

Ahern, J. (2002). *Greenways as Strategic Landscape Planning: Theory and Application*. Wageningen, Netherlands: Wageningen University Press.

Ahrens, C. D. (1991). *Meteorology Today: An Introduction to Weather, Climate, and the Environment*. St. Paul, MN, USA: West Publishing.

Altshuler, A. and Luberoff, D. (2003). *Mega-Projects: The Changing Politics of Urban Public Investment*. Washington, DC: Brookings Institution Press.

Anderson, M. G. (2003). *Ecoregional Conservation: A Comprehensive Approach to Conserving Biodiversity*. Boston: The Nature Conservancy.

Antrop, M. (2000). Changing patterns in the urbanized countryside of Western Europe. *Landscape Ecology*, **15**, 257–70.

Arnfield, A. J. (2003). Two decades of urban climate research: a review of turbulence, exchanges of energy and water, and the urban heat island. *International Journal of Climatology*, **23**, 1–26.

Arnold, C. L., Jr. and Gibbons, C. J. (1996). Impervious surface coverage. *American Planning Association Journal*, **62**, 243–58.

Asare Afrane, Y., Klinkenberg, E., Drechsel, P., *et al.* (2004). Does irrigated urban agriculture influence the transmission of malaria in the city of Kumasi, Ghana? *Acta Tropica*, **89**, 125–34.

Asomani-Boateng, R. (2002). Urban cultivation in Accra: an examination of the nature, practices, problems, potentials, and urban planning implications. *Habitat International*, **26**, 591–607.

Atauri, J. A. and de Lucio, J. V. (2001). The role of landscape structure in species richness: distribution of birds, amphibians, reptiles and lepidopterans in Mediterranean landscapes. *Landscape Ecology*, **16**, 147–59.

Atkinson, A., Davila, J. D., Fernandess, E., and Mattingly, M. (eds.) (1999). *The Challenge of Environmental Management in Urban Areas*. Aldershot, UK: Ashgate.

Avin, U. and Bayer, M. (2003). Right-sizing urban growth boundaries. *Planning*, **69** (February), 22–7.

Babbitt, B. (2005). *Cities in the Wilderness*. Washington, DC: Island Press.

Bacaria, J., Folch, R., Paris, A., *et al.* (1999). *Atlas Ambiental del Mediterraneo*. Barcelona: Institut Catala de la Mediterrania, Institut Cartografic de Catalunya, and ERF Gestio i Communicacio Ambiental.

Bahlburg, C. H. (2003). *A planning system of open spaces: The Berlin-Brandenburg Common Regional Plan (Germany)*, in Actes de III Simposi internacional sobre espais naturals i rurals en arees metropolitans i periurbanes, Barcelona: Initiative Communautaire Interreg III B, 85–90.

Bailey, R. G. (1995). *Descriptions of the Ecoregions of the United States*. Washington, DC: Miscellaneous Publication no. 1391, US Forest Service.

Bailey, R. G. (1998). *Ecoregions: The Ecosystem Geography of Oceans and Continents*. New York: Springer-Verlag.

Baker, W. L. (1989). A review of models of landscape change. *Landscape Ecology*, **2**, 111–33.

Barbier, E. B. (1987). The concept of sustainable economic development. *Environmental Conservation*, **14**, 101–10.

Barbour, M. G., Burk, J. H., and Pitts, W. D. (1987). *Terrestrial Plant Ecology*. Menlo Park, CA: Benjamin/Cummings.

Barker, T. and Sutcliffe, A. (eds.) (1993). *Megalopolis: The Giant City in History*. New York: St. Martin's Press.

Barraco, H., Pares, M., Prat, A., and Terradas, J. (1999). *Barcelona 1885–1999, Ecologia d'una Ciutat*. Barcelona: Ajuntament de Barcelona.

Bartels, J. M. (ed.) (2000). *Managing Soils in an Urban Environment*. Madison, WI: American Society of Agronomy.

Bartuska, T. J. (1994). Cities today: the imprint of human needs in urban patterns and form. In T. J. Bartuska and G. L. Young (eds.), *The Built Environment: Creative Inquiry Into Design and Planning*. USA: Crisp Publications, 273–88.

Beardsley, J. (1997). A Mickey Mouse utopia. *Landscape Architecture* (February), 76–93.

Beatley, T. (1994). *Habitat Conservation Planning: Endangered Species and Urban Growth*. Austin: University of Texas Press.

Beatley, T. (2000). *Green Urbanism: Learning from European Cities*. Washington, DC: Island Press.

Beatley, T., Brower, D. J., and Schwab, A. K. (1994). *An Introduction to Coastal Zone Management*. Washington, DC: Island Press.

Beesley, K. B. and Cocklin, C. (1982). *Perspectives on the Rural–Urban Fringe*. Ontario, Canada: Occasional Papers in Geography No. 2, University of Guelph.

Beier, P. and Noss, R. F. (1998). Do habitat corridors provide connectivity? *Conservation Biology*, **12**, 1241–52.

Benfield, F. K., Raimi, M. D., and Chen, D. D. T. (1999). *Once There Were Greenfields: How Urban Sprawl Is Undermining America's Environment, Economy, and Social Fabric*. Washington, DC: Natural Resources Defense Council and Surface Transportation Policy Project.

Benfield, F. K., Terris, J., and Vorsangeret, N. (2001). *Solving Sprawl: Models of Smart Growth in Communities Across America*. New York: Natural Resources Defense Council.

Bengston, D. N. and Youn, Y. C. (2006). Urban containment policies and the protection of natural areas: the case of Seoul's greenbelt. *Ecology and Society*, **11**, 3.

Bennett, A. F. (2003). *Linkages in the Landscape: The Role of Corridors and Connectivity in Wildlife Conservation*. Gland, Switzerland and Cambridge, UK: IUCN-The World Conservation Union.

Berg, P. and Dasmann, R. (1977). Reinhabiting California, *Ecologist*, **7** (December), 377–401.

Berger, A. (2006). *Drosscape: Wasting Land in Urban America*. New York: Princeton Architectural Press.

Berke, P. R., Godschalk, D. R., Kaiser, E. J., and Rodriguez, D. A. (2006). *Urban Land Use Planning*. Urbana: University of Illinois Press.

Berkowitz, A. R., Nilon, C. H., and Hollweg, K. S. (eds.) (2003). *Understanding Urban Ecosystems: A New Frontier for Science and Education*. New York: Springer.

Berling-Wolff, S. and Wu, J. (2004). Modeling urban landscape dynamics: a review. *Ecological Research*, **19**, 119–29.

Billington, C. and Tozer, E. W. (1997). *Ecological Inventory of NCC Urban Corridors*. Ottawa: National Capital Commission.

Binford, M. W. and Karty, R. (2006). Riparian greenways and water resources. In P. C. Hellmund and D. A. Smith (eds.), *Designing Greenways: Sustainable Landscapes for Nature and People*. Washington D.C.: Island Press, 108–57.

Bird, D., Varland, D., and Negro, J. (eds.) (1996). *Raptors in Human Landscapes: Adaptations to Built and Cultivated Environments*. San Diego, CA: Academic Press.

Blondel, J. and Aronson, J. (1999). *Biology and Wildlife of the Mediterranean Region*. New York: Oxford University Press.

Bloom, P. H. and McCrary, M. D. (1996). The urban buteo: red-shouldered hawks in Southern California. In D. M. Bird, D. E. Varland and J. J. Negro (eds.), *Raptors in Human Landscapes*. San Diego: Academic Press, 31–9.

Blowers, A. (ed.) (2003). *Planning the Sustainable City Region*. London: Earthscan.

Boada, M. and Capdevila, L. (2000). *Barcelona: Biodiversitat urbana*. Barcelona: Ajuntament de Barcelona.

Bolund, P. and Hunhammar, S. (1999). Ecosystem services in urban areas. *Ecological Economics*, **29**, 293–302.

Botterton, C. A. (2001). India's "Project Tiger" reserves: the interplay between ecological knowledge and the human dimensions of policymaking for protected habitats. In V. H. Dale and R. A. Haeuber (eds.), *Applying Ecological Principles to Land Management*. New York: Springer, 136–62.

Boulding, K. E. (1964). *The Meaning of the Twentieth Century*. New York: Harper & Row.

Boyden, S., Miller, S., Newcombe, K., and O'Neill, B. (1981). *The Ecology of a City and Its People: The Case of Hong Kong*. Canberra: Australian National University Press.

Braat, L. C. and Steetskamp, I. (1991). Ecological-economic analysis for regional sustainable development. In R. Costanza (ed.), *Ecological Economics: The Science and Management of Sustainability*. New York: Columbia University Press, 269–88.

Brandle, J. R., Hintz, D. L., and Sturrock, J. W. (eds.) (1988). *Windbreak Technology*. Amsterdam: Elsevier.

Brandt, J., Vejre, H., Mander, A., and Antrop, M. (2003). *Multifunctional Landscapes*. Southampton, UK: WIT.

Breen, A. and Rigby, D. (1996). *The New Waterfront: A Worldwide Urban Success Story*. New York: McGraw-Hill.

Brenneisen, S. (2006). Space for urban wildlife: designing green roofs as habitats in Switzerland. *Urban Habitats*, **14**(1).

Breuste, J., Feldmann, H., and Uhlmann, O. (eds.) (1998). *Urban Ecology*. Berlin: Springer.

Briffitt, C. (2001). Is managed recreational use compatible with effective habitat and wildlife occurrence in urban open space corridor systems? *Landscape Research*, **26**, 137–63.

Briffett, C., Sodhi, N. S., Kong, L., and Yuen, B. (2000). The planning and ecology of green corridor networks in tropical urban settlements: a case study. In J. R. Craig, N. Mitchell, and D. A. Saunders (eds.), *Nature Conservation 5: Nature Conservation in Production Environments: Managing the Matrix*. Chipping Norton, Australia: Surrey Beatty, pp. 411–26.

Browder, J. O. and Godfrey, B. J. (1997). *Rainforest Cities: Urbanization, Development, and Globalization of the Brazilian Amazon*. New York: Columbia University Press.

Buell, L. (1995). *The Environmental Imagination: Thoreau, Nature Writing, and the Formation of American Culture*. Cambridge, MA: Belknap Press of Harvard University Press.

Buell, L. (2001). *Writing for an Endangered World: Literature, Culture, and Environment in the U.S. and Beyond*. Cambridge, MA: Belknap Press of Harvard University Press.

Buell, L. (2005). *The Future of Environmental Criticism: Environmental Crisis and Literacy Imagination*. Oxford: Blackwell Publishing.

Bullard, R. D., Johnson, G. S., and Torres, A. O. (2000). *Sprawl City: Race, Politics, and Planning in Atlanta*. Washington, DC: Island Press.

Burchell, R. W., Downs, A., McCann, B., and Mukherji, S. (2005). *Sprawl Costs: Economic Impacts of Unchecked Development*. Washington, DC: Island Press.

Burel, F. and Baudry, J. (1999). *Ecologie du paysage: Concepts, methodes et applications*. Paris: Editions TEC & DOC.

Burgi, M., Hersperger, A. M., and Schneeberger, N. (2004). Driving forces of landscape change – current and new directions. *Landscape Ecology*, **19**, 857–68.

Busquets, J. (2005). *La Ciutat Vella de Barcelona: Un Passat Amb Future*. Barcelona: Ajuntament de Barcelona.

Calthorpe, P. (1993). *The Next American Metropolis: Ecology, Community, and the American Dream*. New York: Princeton Architectural Press.

Calthorpe, P. and Fulton, W. (2001). *The Regional City: Planning for the End of Sprawl*. Washington, DC: Island Press.

Campbell, S. (1996). Green cities, growing cities, just cities?: Urban planning and the contradictions of sustainable development. *Journal of the American Planning Association*, **62**, 296–312.

Campbell, S. and Fainstein, S. S. (eds.) (2003). *Readings in Planning Theory*. Cambridge, MA: Blackwell.

Carbonell, A. and Yaro, R. D. (2005). American spatial development and the new megalopolis. *Land Lines*, **17**(2), 1–4.

Cervero, R. (1993). *Ridership Impacts of Transit-Focused Development in California*. Berkeley: Monograph 45, Institute of Urban and Regional Development, University of California.

Cervero, R. (1998). *The Transit Metropolis: A Global Inquiry*. Washington, DC: Island Press.

Chen, A., Liu, G. C., and Zhang, K. H. (2004). *Urban Transformation in China*. Aldershot, UK: Ashgate Publishing.

Cheshire, P. C. (1988). *Urban Problems in Western Europe: An Economic Analysis*. London: Unwin Hyman.

Christaller, W. (1933). *Die zentraler Orte in Suddeutschland*. Jena: Fischer.

Cityspace: An Open Space Plan for Chicago. (1998). Chicago: City of Chicago, Chicago Park District, and Forest Preserve District of Cook County.

Clark, P. (ed.) (2006). *The European City and Green Space: London, Stockholm, Helsinki and St. Petersburg, 1850–2000*. Aldershot, UK: Ashgate Publishing.

Clevenger, A. P. and Waltho, N. (2005). Performance indices to identify attributes of highway crossing structures facilitating movement of large mammals. *Biological Conservation*, **121**, 453–64.

Climate Change Impacts on the United States (2001). *The Potential Consequences of Climatic Variability and Change. (2001)*. Cambridge: Cambridge University Press, Report of National Assessment Synthesis Team, US Global Change Research Program.

Colburn, E. A. (2004). *Vernal Pools: Natural History and Conservation*. Blacksburg, VA: McDonald and Woodward Publishing.

Collinge, S. K. and Forman, R. T. T. (1998). A conceptual model of land conversion processes: predictions and evidence from a microlandscape experiment with grassland insects. *Oikos*, **82**, 66–84.

Congress for the New Urbanism. (2000). *Charter of the New Urbanism*. New York: McGraw-Hill.

Corn, W. M. (1983). *Grant Wood: The Regionalist Vision*. New Haven, CT: Yale University Press.

Costa, J. E. and Baker, V. R. (1981). *Surficial Geology: Building with the Earth*. New York: John Wiley & Sons, Inc.

Costanza, R. (ed.) (1991). *Ecological Economics: The Science and Management of Sustainability*. New York: Columbia University Press.

Costanza, R. (2000). The dynamics of the ecological footprint concept. *Ecological Economics*, **32**, 341–5.

Costanza, R., Cumberland, J. C., Daly, H. E., *et al.* (1997a). *An Introduction to Ecological Economics*. Boca Raton. FL: St. Lucie Press.

Costanza, R., d'Arge, R., de Groot, R., *et al.* (1997b). The value of the world's ecosystem services and natural capital. *Nature*, **387**, 253.

Costanza, R., Mitsch, W. J., and Day, J. W., Jr. (2006). New vision for New Orleans and the Mississippi delta: applying ecological economics and ecological engineering. *Frontiers in Ecology and the Environment*, **4**, 465–72.

Costa-Pierce, B., Desbonnet, A., Edwards, P., and Baker, D. (eds.) (2005). *Urban Aquaculture*. Wallingford, UK: CABI Publishing.

Cothrel, S. R., Vimmerstedt, J. P., and Kost, D. A. (1997). *In situ* recycling of urban deciduous litter. *Soil Biology and Biochemistry*, **29**, 295–8.

Covich, A. P. (1976). Analyzing shapes of foraging areas: some ecological and economic theories. *Annual Review of Ecology and Systematics*, **7**, 235–57.

Craul, P. J. (1999). *Urban Soils: Applications and Practices*. New York: John Wiley.

Cronon, W. (1991). *Nature's Metropolis: Chicago and the Great West*. New York: W. W. Norton.

Cuperus, R., Backermans, M. G. G. J., Udo de Haes, H. A., and Canters, K. J. (2001). Ecological compensation in Dutch highway planning. *Environmental Management*, **27**, 75–89.

Daily, G. and Ellison, K. (2002), *The New Ecology of Nature: A Quest to Make Conservation Profitable*. Washington, DC: Island Press.

Daily, G. C. (1997). *Nature's Services: Societal Dependence on Natural Ecosystems*. Washington, DC: Island Press.

Dale, V. H. and Haeuber, R. A. (eds.) (2001). *Applying Ecological Principles to Land Management*. New York: Springer.

Daley, R. and City of Chicago. (2002). *A Guide to Rooftop Gardening*. Chicago: Chicago Department of Environment.

Daly, H. E. (1990). Toward some operational principles of sustainable development. *Ecological Economics*, **2**, 1–6.

Daly, H. E. and Cobb, J. B. (1989). *For the Common Good: Redirecting the Economy Toward Community, the Environment and a Sustainable Future*. Boston: Beacon.

Daniels, T. (1999). *When City and Country Collide: Managing Growth in the Metropolitan Fringe*. Washington, DC: Island Press.

Danielsen, F., Sorensen, M. K., Mette, F., *et al.* (2005). The Asian tsunami: a protective role for coastal vegetation. *Science*, **310**, 643.

Dasgupta, P. S. and Heal, G. M. (1974). The optimal depletion of exhaustible resources. *Review of Economic Studies, Symposium on the Economics of Exhaustible Resources*. pp. 3–28.

Davis, B. N. K. (1976). Wildlife, urbanization and industry. *Biological Conservation*, **10**, 249–91.

Davis, J., Ossowski, R., Daniel, J., and Barnett, S. (2001). Stabilization and savings funds for non-renewable resources: experience and fiscal policy implications. Washington, DC: Occasional Paper 205. International Monetary Fund.

De Blij, H. J. (1977). *Human Geography*. New York: John Wiley.

Decamps, H. and Decamps, O. (2001). *Mediterranean Riparian Woodlands*. Arles, France: Tour du Valat.

Decamps, H. and Decamps, O. (2004). *Au Printemps des Paysages*. Paris: Buchet/Chastel Ecologie.

Diamond, H. L. and Noonan, P. F. (1996). *Land Use in America*. Washington, DC: Island Press.

DiGregoria, J., Luciani, E., and Wynn, S. (2006). Integrating transportation and resource conservation planning: conservation banking. In C. L. Irwin, P. Garrett, and K. P. McDermott (eds.), *Proceedings of the 2005 International Conference on Ecology and Transportation*. Raleigh, USA: CTE, North Carolina State University, 101–10.

Dittmar, H. and Ohland, G. (2004). *The New Transit Town, Best Practices in Transit Oriented Development*. Washington, DC: Island Press.

Donahue, B. (1999). *Reclaiming the Commons: Community Farms and Forests in a New England Town*. New Haven, CT: Yale University Press.

Dramstad, W., Olson, J. D., and Forman, R. T. T. (1996). *Landscape Ecology Principles for Landscape Architecture and Land-use Planning*. Washington, DC: Island Press and Harvard University Graduate School of Design.

Dreier, P., Mollenkopf, J., and Swanston, T. (2004). *Place Matters: Metropolitics for the Twenty-First Century*. Lawrence, KS: University Press of Kansas.

Duany, A., Plater-Zyberk, E., and Speck, J. (2000). *Suburban Nation: The Rise of Sprawl and the Decline of the American Dream*. New York: North Point Press.

Dunnett, N. and Kingsbury, N. (2004). *Planting Green Roofs and Living Walls*. Portland, OR: Timber Press.

Dwyer, J. F. and Chavez, D. J. (2005). The challenges of managing public lands in the wildland-urban interface. In S. W. Vince et al. (eds.), *Forests at the Wildland-Urban Interface: Conservation and Management*. Boca Raton, FL: CRC Press, 269–83.

Easterlin, R. A. (2003). Explaining happiness. *Proceedings of the National Academy of Science*, **100**, 11176.

Eaton, M. M. (1997). The beauty that requires health. In J. I. Nassauer (ed.), *Placing Nature: Culture and Landscape Ecology*. Washington, DC: Island Press, 85–106.

El Araby, M. (2002). Urban growth and environmental degradation: the case of Cairo, Egypt. *Cities*, **19**, 389–400.

El Serafy, S. (1991). The environment as capital. In R. Costanza (ed.), *Ecological Economics: The Science and Management of Sustainability*. New York: Columbia University Press, 168–75.

Elmqvist, T., Colding, J., Barthel, S., et al. (2004). The dynamics of social-ecological systems in urban landscapes: Stockholm and the National Urban Park, Sweden. *Annals of the New York Academy of Sciences*, **1023**, 308–22.

English Nature. (2003). *Green Roofs: Their Existing Status and Potential for Conserving Biodiversity in Urban Areas*. Peterborough, UK: English Nature Research.

Epstein, D. (1973). *Brasilia, Plan and Reality*. Berkeley: University of California Press.

Erickson, D. (2006). *MetroGreen: Connecting Open Space in North American Cities*. Washington, DC: Island Press.

Ezcurra, E. and Mazari-Hiriart, M. (1996). Are megacities viable? A cautionary tale from Mexico City. *Environment*, **38**, 6–35.

Fahrig, L. (2003). Effects of habitat fragmentation on biodiversity. *Annual Review of Ecology and Systematics*, **34**, 487–515.

Fainstein, S. S. and Campbell, S. (eds.) (1996). *Readings in Urban Theory*. Cambridge, MA: Blackwell.

Farber, S. (1987). The value of coastal wetlands for protection of property against hurricane damage. *Journal of Environmental and Economic Management*, **14**, 143–51.

Farina, A. (2005). *Principles and Methods in Landscape Ecology*. Berlin/New York: Springer.

Ferguson, J., Connelly, M., Forman, R., *et al.* (1993). *1992 Open Space Plan*. Concord, Massachusetts, USA: Natural Resources Commission.

Fernandez-Juricic, E. (2000). Local and regional effects of pedestrians on forest birds in a fragmented landscape. *Condor*, **102**, 247–55.

Fischer, J., Lindenmayer, D. B., and Manning, A. D. (2006). Biodiversity, ecosystem function, and resilience: ten guiding principles for commodity production landscapes. *Frontiers in Ecology and the Environment*, **4**, 80–6.

Folch, R. (2000). *Socio-Economic Considerations of Territorial Zoning in the Barcelona Metropolitan Area: With General Objectives, by Josep Acebillo*. Barcelona: Estudi Ramon Folch.

Forman, R. T. T. (1964). Growth under controlled conditions to explain the hierarchical distribution of a moss, *Tetraphis pellucida*. *Ecological Monographs*, **34**, 1–25.

Forman, R. T. T. (ed.) (1979a). *Pine Barrens: Ecosystem and Landscape*. New York: Academic Press.

Forman, R. T. T. (1979b). The Pine Barrens of New Jersey: an ecological mosaic. In R. T. T. Forman (ed.), *Pine Barrens: Ecosystem and Landscape*. New York: Academic Press, 569–85.

Forman, R. T. T. (1987). The ethics of isolation, the spread of disturbance, and landscape ecology, in M. G. Turner (ed.), *Landscape Heterogeneity and Disturbance*. New York: Springer-Verlag, 213–29.

Forman, R. T. T. (1995). *Land Mosaics: The Ecology of Landscapes and Regions*. New York: Cambridge University Press.

Forman, R. T. T. (1999). Horizontal processes, roads, suburbs, societal objectives, and landscape ecology. In J. M. Klopatek and R. H. Gardner (eds.), *Landscape Ecological Analysis: Issues and Applications*. New York: Springer-Verlag, 35–53.

Forman, R. T. T. (2002a). Envisioning a land mosaic where both nature and people thrive. In *Jardines insurgentes: Arquitectura del paisaje en Europa 1996–2000*. Barcelona: Edicion Fundacion Caja de Arquitectos, 34–8 and 48–57.

Forman, R. T. T. (2002b). The missing catalyst: design and planning with ecology roots. In B. R. Johnson and K. Hill (eds.), *Ecology and Design: Frameworks for Learning*. Washington, DC: Island Press, 85–109.

Forman, R. T. T. (2004a). *Mosaico territorial para la region metropolitana de Barcelona*. Barcelona: Editorial Gustavo Gili.

Forman, R. T. T. (2004b). Road ecology's promise: what's around the bend? *Environment*, **46**, 8–21.

Forman, R. T. T. (2005). Roadside redesigns – woody and variegated – to help sustain nature and people, *Harvard Design Magazine*, Fall 2005/Winter 2006, 35–41.

Forman, R. T. T. (2006). Good and bad places for roads: effects of varying road and natural patterns on habitat loss, degradation, and fragmentation. In *Proceedings of the 2005 International Conference on Ecology and Transportation.* Raleigh, USA: Center for Transportation and the Environment, North Carolina State University, 164–74.

Forman, R. T. T. and Alexander, L. E. (1998). Roads and their major ecological effects. *Annual Review of Ecology and Systematics*, **29**, 207–31.

Forman, R. T. T. and Collinge, S. K. (1996). The "spatial solution" to conserving biodiversity in landscapes and regions. In R. M. DeGraaf and R. I. Miller (eds.), *Conservation of Faunal Diversity in Forested Landscapes.* London: Chapman & Hall, 537–68.

Forman, R. T. T. and Deblinger, R. D. (2000). The ecological road-effect zone of a Massachusetts (USA) suburban highway. *Conservation Biology*, **14**, 36–46.

Forman, R.T.T. and Godron, M. (1981). Patches and structural components for a landscape ecology. *BioScience*, **31**, 733–40.

Forman, R. T. T. and Godron, M. (1986). *Landscape Ecology.* New York: John Wiley.

Forman, R. T. T. and Hersperger, A. M. (1997). Ecologia del paessaggio e pianificazione: una potente combinazione. (Landscape ecology and planning: a powerful combination). *Urbanistica*, **108**, 61–6.

Forman, R. T. T. and Mellinger, A. D. (2000). Road networks and forest spatial patterns: comparing cutting-sequence models for forestry and conservation. In J. L. Craig, N. Mitchell, and D. A. Saunders (eds.), *Nature Conservation 5: Conservation in Production Environments: Managing the Matrix.* Chipping Norton, Australia: Surrey Beatty, 71–80.

Forman, R. T. T., Reeve, P., Beyer, H., *et al.* (2004). *Open Space and Recreation Plan 2004: Concord, Massachusetts.* Concord, MA: Natural Resources Commission.

Forman, R. T. T., Reineking, B., and Hersperger, A. M. (2002). Road traffic and nearby grassland bird patterns in a suburbanizing landscape. *Environmental Management*, **29**, 782–800.

Forman, R. T. T., Sperling, D., Bissonette, J. A., *et al.* (2003). *Road Ecology: Science and Solutions.* Washington, DC: Island Press.

Fortin, M.-J. (1999). Spatial statistics in landscape ecology. In J. M. Klopatek and R. H. Gardner (eds.), *Landscape Ecological Analysis: Issues and Applications.* New York: Springer, 253–79.

Foster, D. R. and Aber, J. D. (2004). *Forests in Time: The Environmental Consequences of 1000 Years of Change in New England.* New Haven, CT: Yale University Press.

Foster, D. R. and Boose, E. (1992). Patterns of forest damage resulting from catastrophic wind in Central New England, USA. *Journal of Ecology*, **80**, 79–99.

France, R. L. (ed.) (2002). *Handbook of Water-Sensitive Planning and Design.* Boca Raton FL: Lewis Publishers.

France, R. L. (2003). *Wetland Design: Principles and Practices for Landscape Architects and Land-use Planners.* New York: Norton.

Franklin, J. F. and Forman, R. T. T. (1987). Creating landscape patterns by forest cutting: ecological consequences and principles. *Landscape Ecology*, **1**, 5–18.

Frazer, L. (2005). Paving paradise: the peril of impervious surfaces. *Environmental Health Perspectives*, **113** (7).

Freemark, K., Bert, D., and Villard, M.-A. (2002). Patch-, landscape- and regional-scale effects on biota. In K. J. Gutzwiller (ed.), *Applying Landscape Ecology in Biological Conservation*. New York: Springer, 58–83.

Friedmann, J. (1973). *Retracking America*. New York: Anchor Press/Doubleday.

Frumkin, H., Frank, L., and Jackson, R. (2004). *Urban Sprawl and Public Health: Designing, Planning, and Building for Healthy Communities*. Washington, DC: Island Press.

Fuchs, R. J., Brennan, E., Chamie, J., Lo, F. C., and Uitto, J. I. (eds.) (1994). *Mega-City Growth and the Future*. Tokyo: The United Nations University.

Galatas, R. (2004). *The Woodlands: The Inside Story of Creating a Better Hometown*. Washington, DC: Urban Land Institute.

Garde, A. (2004). New urbanism as sustainable growth? A supply side story and its implications for public policy. *Journal of Planning Education and Research*, **24**, 154–70.

Garreau, J. (1991). *Edge City: Life on the New Frontier*. New York: Doubleday.

Geddes, P. (1915). *Cities in Evolution*. London: Williams and Norgate.

Germaine, S. S., Rosenstock, S. S., Schweinsburg, R. E., and Richardson, W. S. (1998). Relationships among breeding birds, habitat and residential development in greater Tucson, Arizona. *Ecological Applications*, **8**, 680–91.

Getting to Smart Growth. (2002). Washington, DC: Smart Growth Network.

Getting to Smart Growth II. (2003). Washington, DC: Smart Growth Network.

Ghassemi, F. (2006). *Inter-Basin Water Transfer: Case Studies from Australia, United States, Canada, China and India*. Cambridge: Cambridge University Press.

Gibert, J., Danielopol, D. L., and Stanford, J. A. (eds.) (1994). *Groundwater Ecology*. San Diego: Academic Press.

Gilbert, O. (1991). *The Ecology of Urban Habitats*. London: Chapman & Hall.

Gingrich, S. E. and Diamond, M. L. (2001). Atmospherically derived organic surface films along an urban-rural gradient. *Environmental Science and Technology*, **35**, 4031–7.

Girot, C. (2004). Eulogy of the void: the lost power of Berlin landscapes after the wall. *DISP (ETH Zurich)*, **156**, 35–9.

Gladwell, M. (2000). *The Tipping Point: How Little Things Can Make a Big Difference*, London: Littlebrown.

Godde, M., Richarz, N., and Walter, B. (1995). Habitat conservation and development in the city of Dusseldorf (Germany). In H. Sukopp, M. Numata, and A. Huber (eds.), *Urban Ecology as the Basis of Urban Planning*. Amsterdam: SPB Academic Publishing, 163–71.

Godron, M. and Forman, R. T. T. (1983). Landscape modification and changing ecological characteristics. In H. A. Mooney and M. Godron (eds.), *Disturbance and Ecosystems: Components of Response*. New York: Springer-Verlag, 12–28.

Goldstein, E. L., Gross, M., and DeGraaf, R. M. (1981). Explorations in bird-land geometry. *Urban Ecology*, **5**, 113–24.

Gomez-Ibanez, J. A. (1999). *Essays in Transportation Economics and Policy: A Handbook in Honor of John R. Meyer*. Washington, DC: Brookings Institution.

Goodman, S. W. (1996). Ecosystem management at the Department of Defense. *Ecological Applications*, **6**, 706–7.

Gordon, P. and Richardson, H. (1997). Are compact cities a desirable planning goal? *Journal of the American Planning Association*, **63**, 95–106.

Gore, A. (2006). *An Inconvenient Truth*. New York: Rodale.

Green, B. and Vos, W. (eds.) (2001). *Threatened Landscapes: Conserving Cultural Environments*. London: Spon Press.

Greenberg, J. (2002). *A Natural History of the Chicago Region*. Chicago: University of Chicago Press.

Grimm, N. B., Baker, L. J., and Hope, D. (2003). An ecosystem approach to understanding cities: familiar foundations and uncharted frontiers. In A. R. Berkowitz and K. S. Hollweg (eds.), *Understanding Urban Ecosystems: A New Frontier for Science and Education*. New York: Springer, 95–114.

Groffman, P. M., Bain, D. J., Band, L. E., *et al.* (2003). Down by the riverside: urban riparian ecology. *Frontiers in Ecology and the Environment*, **1**, 315–21.

Groom, M., Meffe, G. K., Carroll, R., and Contributors. (2006). *Principles of Conservation Biology*. Sunderland, MA: Sinauer Associates.

Gross, G. (2002). The exploration of boundary layer phenomena using a nonhydrostatic mesoscale model. *Meteorologische Zeitschrift*, **11**, 701–10.

Grove, A. T. and Rackham, O. (2001). *The Nature of Mediterranean Europe: An Ecological History*. New Haven, CT: Yale University Press.

Groves, C. R., Jensen, D. B., *et al.* (2002). Planning for biodiversity conservation: putting conservation science into practice, *BioScience*, **52**, 499–512.

Gu, C. and Kesteloot, C. (1998). Beijing's socio-spatial structure in transition. In J. Breuste, H. Feldmann, and O. Ulhmann (eds.), *Urban Ecology*. Berlin: Springer-Verlag, 288–93.

Gutzwiller, K. J. (ed.) (2002). *Applying Landscape Ecology in Biological Conservation*. New York: Springer.

Haber, W. (1993). *Okologische Grundlagen des Umweltschutres*. Bonn, Germany: Economica Verlag.

Haggett, P., Cliff, A. D., and Frey, A. (1977). *Locational Analysis in Human Geography*. New York: John Wiley.

Hahs, A. K. and McDonnell, M. J. (2007). Selecting independent measures to quantify Melbourne's urban-rural gradient. *Landscape and Urban Planning* (in press).

Hall, P. G. (2002). *Cities of Tomorrow: An Intellectual History of Urban Planning and Design in the Twentieth Century*. Oxford: Blackwell.

Handy, S. (1992). Regional versus local accessibility: neo-traditional development and its implications for non-work travel. *Built Environment*, **18**, 253–67.

Handy, S. (2005). *Critical Assessment of the Literature on the Relationships Among Transportation, Land Use, and Physical Activity*. Washington, DC: Transportation Research Board and Institute of Medicine, TRB Special Report 282.

Hansen, A. J. (2002). Ecological causes and consequences of demographic change in the new West. *BioScience*, **52**, 151–62.

Hansen, A. J. and Rotella, J. J. (2001). Nature reserves and land use: implications of the "place" principle. In V. H. Dale and R. A. Haeuber (eds.), *Applying Ecological Principles to Land Management*. New York: Springer, 54–72.

Hanski, I. A. and Gilpin, M. E. (eds.) (1997). *Metapopulation Biology: Ecology, Genetics, and Evolution*. San Diego: Academic Press.

Hara, Y., Takeuchi, K., and Okubo, S. (2005). Urbanization linked with past agricultural landuse patterns in the urban fringe of a deltaic Asian mega-city: a case study in Bangkok. *Landscape and Urban Planning*, **73**, 16–28.

Hardin, G. and Baden, J. (1977). *Managing the Commons*. San Francisco: W. H. Freeman.

Harris, L. (1984). *The Fragmented Forest: Island Biography and the Preservation of Biotic Diversity*. Chicago: University of Chicago Press.

Harris, L. D., Hoctor, T. S., and Gergel, S. E. (1996). Landscape processes and their significance to biodiversity conservation. In O. Rhodes, Jr., R. Chesser, and M. Smith (eds.), *Population Dynamics in Ecological Space and Time*. Chicago: University of Chicago Press, 319–47.

Heisler, G. M., et al. (1994). *Investigation of the influence of Chicago's urban forests on wind and air temperature within residential neighborhoods. In Chicago's Urban Forest Ecosystem: Results of the Chicago Urban Forest Climate Project*. Radnor, PA: General Technical Report NE-186, USDA Forest Service, 19–40.

Hersperger, A. M. (2006). Spatial adjacencies and interactions: neighborhood mosaics for landscape ecology planning. *Landscape and Urban Planning*, **77**, 227–39.

Hersperger, A. M. and Forman, R. T. T. (2003). Adjacency arrangement effects on plant diversity and composition in woodland patches. *Oikos*, **101**, 279–90.

Hien, W. N., Yok, T. P., and Yu, C. (2007). Study of the thermal performance of extensive rooftop greenery systems in the tropical climate. *Building and Environment*, **42**, 25–54.

Hilty, J. A., Lidicker, W. Z., Jr., and Merenlender, A. M. (2006). *Corridor Ecology: The Science and Practice of Linking Landscapes for Biodiversity Conservation*. Washington, DC: Island Press.

Hobbs, N. T. and Theobold, D. M. (2001). Effects of land-use change on wildlife habitat: applying ecological principles and guidelines in the Western United States. In V. H. Dale and R. A. Haeuber (eds.), *Applying Ecological Principles to Land Management*. New York: Springer, 37–53.

Hobbs, R. J. (1995). Landscape ecology. *Encyclopedia of Environmental Biology*, **2**, 417–28.

Hobbs, R. J. and Miller, J. R. (2002). Conservation where people live and work. *Conservation Biology*, **16**, 330–7.

Hodge, G. (1998). *Planning Canadian Communities – An Introduction to the Principles, Practice, and Participants*. Toronto: International Thomson Publishing.

Hong, S.-K., Song, I-J., Kim, H. O., and Lee, E. K. (2003). Landscape pattern and its effect on ecosystem functions in Seoul Metropolitan area: urban ecology on distribution of the naturalized plant species. *Journal of Environmental Science*, **15**, 199–204.

Houck, M. C. and Cody, M. J. (eds.) (2000). *Wild in the City: A Guide to Portland's Natural Areas*. Portland, OR: Oregon Historical Society Press.

Hough, M. (2004). *Cities and Natural Process: A Basis for Sustainability*. New York: Routledge.

Houghton, J. T., Ding, Y., Griggs, D. J., *et al.* (eds.) (2001). *Climate Change 2001: The Scientific Basis*. Cambridge: Cambridge University Press, Contribution of Working Group I to the Third Assessment Report of the International Panel on Climate Change.

Houston, P. (2005). Re-valuing the fringe: some findings on the value of agricultural production in Australia's peri-urban regions. *Geographical Research*, **43**, 209–23.

Howarth, R. B. and Norgaard, R. B. (1992). Environmental valuation under sustainable development. *American Economic Review*, **82**, 473–77.

Howe, J. (2002). Planning for urban food: the experience of two UK cities. *Planning Practice and Research*, **17**, 125–44.

Hulse, D., Gregory, S., and Baker, J., (eds.) (2002). *Willamette River Basin Planning Atlas: Trajectories of Environmental and Ecological Change*. Corvallis, OR: Oregon State University Press.

Ichinose, T., Shimodozono, K., and Hanaki, K. (1999). Impact of anthropogenic heat on urban climate in Tokyo. *Atmospheric Environment*, **33**, 3897–909.

Im, S.-B. (1992). Skyline conservation and management in rapidly growing cities and regions: successes and failures in Korea. *International Conference on Landscape Planning and Environmental Conservation Proceedings*. Tokyo: University of Tokyo.

Ingegnoli, V. (2002). *Landscape Ecology: A Widening Foundation*. New York: Springer.

Irazabal, C. (2005). *City Making and Urban Governance in the Americas: Curitiba and Portland*. Aldershot, UK: Ashgate.

Ishikawa, M. (2001). *City and Green Space*. (In Japanese), Tokyo: Iwanami Syoten Ltd.

Iuell, B. *et al.* (2003). *Habitat Fragmentation Due to Transportation Infrastructure: Wildlife and Traffic: A European Handbook for Identifying Conflicts and Designing Solutions*. Brussels: KNNV Publishers, COST 341.

Jackson, J. B. (1994). *A Sense of Place, A Sense of Time*. New Haven, CT: Yale University Press.

Jacobi, P., Drescher, A. W., and Amend, J. (2000). *Urban Agriculture: Justification and Planning Guidelines*. Canada: City Farmer.

Jacobs, J. (1992). *The Death and Life of Great American Cities*. New York: Vintage Books.

Jared, O. (2004). *Hazardous Metropolis: Flooding and Urban Ecology in Los Angeles*. Berkeley: University of California Press.

Jenerette, D. G. and Wu, J. (2001). Analysis and simulation of land-use change in the Central Arizona-Phoenix Region. *Landscape Ecology*, **16**, 611–26.

Jenks, M., Burton, E., and Williams, K. (1996). *The Compact City: A Sustainable Urban Form?* London: E&FN Spon.

Jim, C. Y. and Chen, S. S. (2003). Comprehensive greenspace planning based on landscape ecology principles in compact Nanjing City, China. *Landscape and Urban Planning*, **65**, 95–116.

Johnson, B. R. and Hill, K. (eds.) (2002). *Ecology and Design: Frameworks for Learning.* Washington, DC: Island Press.

Jones, C. I. (2002). *Introduction to Economic Growth.* New York: Norton.

Jongman, R. H. G. and Pungetti, G. (eds.) (2004). *Ecological Networks and Greenways: Concept, Design, Implementation.* Cambridge: Cambridge University Press.

Kahn, M. E. (2006). *Green Cities: Urban Growth and the Environment.* Washington, DC: Brookings Institution Press.

Kalff, J. (2002). *Limnology: Inland Water Ecosystems.* Upper Saddle River, NJ: Prentice-Hall.

Kaplan, R., Kaplan, S., and Ryan, R. L. (1998). *With People in Mind: Design and Management of Everyday Nature.* Washington, DC: Island Press.

Kareiva, P., Watts, S., McDonald, R. I., and Boucher, T. (2007). Domesticated nature: shaping landscapes and ecosystems for human welfare. *Science*, **316**, 1866–9.

Karl, T. L. (1997). *The Paradox of Plenty: Oil Booms and Petro-States.* Berkeley: University of California Press.

Karr, J. R. (2002). What from ecology is relevant to design and planning? In B. R. Johnson and K. Hill (eds.), *Ecology and Design: Frameworks for Learning.* Washington, D.C.: Island Press, 133–72.

Kasser, T. (2003). *The High Price of Materialism.* Cambridge, MA: MIT Press.

Katz, P. (1994). *The New Urbanism: Toward an Architecture of Community.* New York: McGraw-Hill.

Keddy, P. A. (2000). *Wetland Ecology.* Cambridge: Cambridge University Press.

Keiter, R. B. and Boyce, M. S. (eds.) (1991). *The Greater Yellowstone Ecosystem: Redefining America's Wilderness Heritage.* New Haven, CT: Yale University Press.

Kellert, S. R. (2005). *Building for Life: Designing and Understanding the Human–Nature Connection.* Washington, DC: Island Press.

Kellert, S., Heerwagen, J., and Mador, M. (eds.). (2007). *Biophilic Design: Theory, Science, Practice.* New York: John Wiley & Sons, Inc.

Kellert, S. R. and Wilson, E. O. (eds.) (1993). *The Biophilia Hypothesis.* Washington, DC: Island Press.

Kirkwood, N. (ed.) (2001). *Manufactured Sites: Rethinking the Post-Industrial Landscape.* New York: Spon Press.

Klopatek, J. M. and Gardner, R. H. (eds.) (1999). *Landscape Ecological Analysis: Issues and Applications.* New York: Springer.

Klotz, S. (1990). Species/area and species/inhabitants relations in European cities. In H. Sukopp and S. Hejny (eds.), *Urban Ecology: Plants and Plant Communities in Urban Environments.* The Hague, Netherlands: SPB Academic Publishing, 99–103.

Knaapen, J. P., Scheffer, M., and Harms, B. (1992). Estimating habitat isolation in landscape planning. *Landscape and Urban Planning*, **23**, 1–16.

Knight, R. L. and Gutzwiller, K. J. (1995). *Wildlife and Recreationists: Coexistence Through Management and Research.* Washington, DC: Island Press.

Konijnendijk C. C., Nilsson, K., Randrup, T. B., and Schipperijn, J. (eds.) (2005). *Urban Forests and Trees: A Reference Book.* Berlin: Springer.

Kowarik, I. and Korner, S. (eds.) (2005). *Wild Urban Woodlands: New Perspectives for Urban Forestry*. New York: Springer.

Kowarik, I. and Langer, A. (2005). Natur-Park Sudgelande: linking conservation and recreation in an abandoned railyard in Berlin. In I. Kowarik and S. Korner (eds.), *Wild Urban Woodlands: New Perspectives for Urban Forestry*. New York: Springer, 287–99.

Kraenzel, C. F. (1947). Principles of regional planning: as applied to the Northwest. *Social Forces*, **25**, 373–84.

Krebs, C. J. (1994). *Ecology: The Experimental Analysis of Distribution and Abundance*. New York: HarperCollins.

Kreimer, A., Arnold, M., and Carlin, A. (2003). *Building Safer Cities: The Future of Disaster Risk*. Washington, DC: World Bank.

Kremen, C. and Ostfeld, R. S. (2005). A call to ecologists: measuring, analyzing, and managing ecosystem services. *Frontiers in Ecology and Environment*, **3**, 540–8.

Kuan, S. and Rowe, P. (2004). *Shanghai: Architecture and Urbanism for Modern China*. New York: Prestel.

Kuhbler, A. (2000). *Grosser Tiergarten*. Berlin: L&H Verlag.

Lagro, J. A. (2001). *Site Analysis: Linking Program and Concept in Land Planning and Design*. New York: John Wiley.

Lal, R. (ed.) (1994). *Soil Erosion: Research Methods*. Ankeny, IA: Soil and Water Conservation Society.

Landsberg, H. E. (1981). *The Urban Climate*. New York: Academic Press.

Laurance, W. F., Vasconcelos, H. L., and Lovejoy, T. L. (2000). Forest loss and fragmentation in the Amazon: implications for wildlife conservation. *Oryx*, **34**, 31–45.

Lawrence, D. (2004). Erosion of tree diversity during 200 years of shifting cultivation in Bornean rainforest. *Ecological Applications*, **14**, 1855–69.

Layard, R. (2005). *Happiness: Lessons from a New Science*. New York: Penguin.

Layton, R. (1989). *Uluru: An Aboriginal History of Ayers Rock*. Canberra, Australia: Aboriginal Studies Press.

Lee, D. G., Kim, E. Y., and Oh, K. S. (2005). Conservation value assessment by considering patch size, connectivity, and edge. *Journal of Korean Environmental Research and Revegetation Technology*, **8**(5), 56–68.

LeGates, R. T. and Stout, F. (eds.) (2003). *The City Reader*. London: Routledge.

Leitao, A. B., Miller, J., Ahern, J., and McGarigal, K. (2006). *Measuring Landscapes: A Planner's Handbook*. Washington, DC: Island Press.

Lenzen, M. and Murray, S. A. (2001). A modified ecological footprint method and its application to Australia. *Ecological Economics*, **37**, 229–56.

Leopold, A. (1933). The conservation ethic. *Journal of Forestry*, **31**(6), 634–43.

Leopold, A. (1944). Cheat takes over. *The Land*, **1**, 310–13.

Leslie, M., Meffe, G. K., Hardesty, J. I., and Adams, D. L. (1996). *Conserving Biodiversity on Military Lands: A Handbook for Natural Resources Managers*. Arlington, VA: The Nature Conservancy.

Li, H., Franklin, J. F., Swanson, F. J., and Spies, T. A. (1993). Developing alternative forest cutting patterns: a simulation approach. *Landscape Ecology*, **8**, 63–75.

Liddle, M. (1997). *Recreation Ecology: The Ecological Impact of Outdoor Recreation and Ecotourism*. London: Chapman & Hall.

Lindenmayer, D. and Burgman, M. (2005). *Practical Conservation Biology*. Collingwood, Australia: CSIRO Publishing.

Lindenmayer, D. B. and Fischer, J. (2006). *Habitat Fragmentation and Landscape Change: An Ecological and Conservation Synthesis*. Washington, DC: Island Press.

Lindenmayer, D. and Franklin, J. F. (2002). *Conserving Forest Biodiversity: A Comprehensive Multiscaled Approach*. Washington, DC: Island Press.

Listhaug, O. (2005). Oil wealth dissatisfaction and political trust in Norway: a resource curse? *West European Politics*, **28**, 834–51.

Liu, J. and Taylor, W. W. (eds). (2002). *Integrating Landscape Ecology into Natural Resource Management*. New York: Cambridge University Press.

Lopez, H. (2003). Sprawl in the 1990s: measurement, distribution and trends, *Urban Affairs Review*, **38**, 325–55.

Losada, H., Martinez, H., Vieyra, J., *et al.* (1998). Urban agriculture in the metropolitan zone of Mexico City: changes over time in urban, suburban, and peri-urban areas. *Environment and Urbanization*, **10**, 37–54.

Losch, A. (1954). *The Economics of Location*. (Translated version). New Haven, CT: Yale University Press.

Lovelock, J. (2000). *Gaia: The Practical Science of Planetary Medicine*. Oxford: Oxford University Press.

Luck, G. W., Ricketts, T. H., Daily, G. C., and Imhoff, M. (2004). Alleviating spatial conflict between people and biodiversity. *Proceedings of the National Academy of Sciences (USA)*, **101**, 182–6.

Luck, M., Jenerette, G. D., Wu, J., and Grimm, N. B. (2001). The urban funnel model and spatially heterogeneous ecological footprint. *Ecosystems*, **4**, 782–96.

Luck, M. and Wu, J. (2002). A gradient analysis of urban landscape pattern: a case study from the Phoenix metropolitan region, Arizona, USA. *Landscape Ecology*, **17**, 327.

Ludwig, J., Tongway, D., Freudenberger, D., *et al.* (1997). *Landscape Ecology: Function and Management: Principles from Australia's Rangelands*. Collingwood, Victoria, Australia: CSIRO Publishing.

Lund, H. (2003). Testing the claims of new urbanism: local access, pedestrian travel and neighboring. *Journal of the American Planning Association*, **69**, 414–29.

Lynch, K. (1981). *A Theory of Good City Form*. Cambridge, MA: MIT Press.

Lynch, K. and Hack, G. (1996). *Site Planning*. Cambridge, MA: MIT Press.

Macionis, J. J. and Parillo, V. N. (2001). *Cities and Urban Life*. Upper Saddle River, NJ: Prentice Hall.

MacKaye, B. (1940). Regional planning and ecology. *Ecological Monographs* **10**, 349–53..

Magnusson, W. E. (2004). Ecoregion as a pragmatic tool. *Conservation Biology*, **18**, 4–5.

Main, H. and Williams, S. W. (1994). *Environment and Housing in Third World Cities*. New York: John Wiley.

Maki, S., Kalliola, R., and Vuorinen, K. (2001). Road construction in the Peruvian Amazon: process, causes and consequences. *Environmental Conservation*, **28**, 199–214.

Margat, J. (1994). Groundwater operations and management. In J. Gibert, D. L. Danielopol, and J. A. Stanford (eds.), *Groundwater Ecology*. San Diego: Academic Press, 505–22.

Marsh, W. M. (2005). *Landscape Planning: Environmental Applications*. New York: John Wiley.

Masud-Ul-Hasan, T. K. (*c.* 1965). *Hand Book of Important Places in West Pakistan*. Published by Pakistan Social Service Foundation, Karachi.

Mata, R. and Tarroja, A. (Coordinadores) (2006). *El paisaje y la gestion del territorio: Criterios paisajisticos en la ordenacion del territorio y el urbanismo*. Barcelona: Diputacio de Barcelona.

Matlack, G. R. (1993). Sociological edge effects: spatial distribution of human impact in suburban forest fragments. *Environmental Management*, **17**, 829–35.

Mauerer, U., Peschel, T., and Schmitz, S. (2000). The flora of selected urban land-use types in Berlin and Potsdam with regard to nature conservation in cities. *Landscape and Urban Planning*, **46**, 209–15.

Mayor Farguell, X., Quintana Gozalo, V., and Belmonte Zamora, R. (2005). *Aproximacio a la petjada ecologica de Catalunya (An Approximation to the Ecological Footprint of Catalonia)*. Barcelona: Consell Assessor per al Desenvolupament Sostenible, Generalitat de Catalunya.

McCarthy, J. J., Canziani, O. F., Leary, N. A., Dokken, D. J., and White, K. S. (eds.) (2001). *Climate Change 2001: Impacts, Adaptation, and Vulnerability*. Cambridge: Cambridge University Press, Contribution of Working Group II to the Third Assessment Report of the Intergovernmental Panel on Climate Change.

McCullough, D. R. (ed.) (1996). *Metapopulations and Wildlife Conservation*. Washington, DC: Island Press.

McDonald, R. I. and Urban, D. L. (2006). Forest edges and forest composition in the North Carolina Piedmont. *Biological Invasions*, **8**, 1049–60.

McDonnell, M. J., Pickett, S. T. A., *et al.* (1997). Ecosystem processes along an urban-to-rural gradient. *Urban Ecosystems*, **1**, 21–36.

McGarigal, K. and Cushman, S. A. (2005). The gradient concept of landscape structure. In J. A. Wiens and M. R. Moss (eds.), *Issues and Perspectives in Landscape Ecology*. Cambridge: Cambridge University Press, 112–19.

McGrath, B. and Thaitakoo, D. (2005). Tasting the periphery: Bangkok's agri- and aquacultural fringe. *Architectural Design*, **75** (June), 43–51.

McHarg, I. L. (1969). *Design with Nature, Garden City*. New York: Doubleday Natural History Press.

McIntyre. S. and Hobbs, R. (1999). A framework for conceptualizing human effects on landscapes and its relevance to management and research models. *Conservation Biology*, **13**, 1282–92.

McMichael, A. (2000). The urban environment and health in a world of increasing globalization: issues for developing countries. *Bulletin of the World Health Organization*, **78**, 1117–26.

McNeill, J. R. (2000). *Something New Under the Sun: An Environmental History of the Twentieth-Century World*. New York: Norton.

Meyer, W. B. and Turner, B. L. (1994). *Changes in Land Use and Land Cover: A Global Perspective*. Cambridge: Cambridge University Press.

Millennium Ecosystem Assessment: Current State and Trends. (2005). Washington, DC: Island Press.

Miller, C. K. and Gershman, M. D. (1998). Outdoor recreation and boundaries: opportunities and challenges. In R. L. Knight and P. B. Landres (eds.), *Stewardship Across Boundaries*, Washington, DC: Island Press, 141–58.

Miller, N. P. (2005). Addressing the noise from US transportation systems: measures and countermeasures, *TR News*, **240**, 4–16.

Milne, B. T. (1991a). Lessons from applying fractal models to landscape patterns. In M. G. Turner and R. H. Gardner (eds.), *Quantitative Methods in Landscape Ecology*. New York: Springer, 199–238.

Milne, B. T. (1991b). The utility of fractal geometry in landscape design. *Landscape and Urban Planning*, **21**, 81–90.

Mitsch, W. J. and Gosselink, J. G. (2000). *Wetlands*. New York: John Wiley.

Mladenoff, D. J. (2005). The promise of landscape modeling: successes, failures, and evolution. In J. A. Wiens and M. R. Moss (eds.), *Issues and Perspectives in Landscape Ecology*, Cambridge: Cambridge University Press, 90–100.

Moore, S. A. (2007). *Alternative Routes to the Sustainable City: Austin, Curitiba, and Frankfurt*. Lanham, MA: Rowman and Littlefield.

Moran, J. M. and Morgan, M. D. (1994). *Meteorology: The Atmosphere and the Science of Weather*. New York: Macmillan.

Moreira, F., Rego, F. C., and Ferreira, P. C. (2001). Temporal (1958–1995) patterns of change in a cultural landscape of northwestern Portugal: implications for fire occurrence. *Landscape Ecology*, **16**, 557–67.

Morgan, A. E. (1971). *Dams and Other Disasters*. Boston, MA: Porter Sargent.

Morgan, G. T. and King, J. O. (1987). *The Woodlands: New Community Development, 1964–1983*. College Station, TX: Texas A&M University Press.

Morin, P. J. (1999). *Community Ecology*. Malden, MA: Blackwell Science.

Mortberg, U. M. (2001). Resident bird species in urban forest remnants: landscape and habitat perspectives. *Landscape Ecology*, **16**, 193–203.

Mougeot, L. J. (ed.) (2005). *Agropolis: The Social, Political and Environmental Dimensions of Urban Agriculture*. London: Earthscan-IDRC.

Muehlenbach, V. (1979). Contributions to the synanthropic (adventive) flora of the railroads in St. Louis, Missouri, USA. *Annals of the Missouri Botanical Garden*, **66**, 1–108.

Munton, R. (1983). *London's Greenbelt: Containment in Practice*. London: Allen and Unwin.

Musacchio, L. R. and Wu, J. (2004). Collaborative landscape-scale ecological research: emerging trends in urban and regional ecology. *Urban Ecosystems*, **7**, 175–8.

Nash, R. (1982). *Wilderness and the American Mind*. New Haven, CT: Yale University Press.

Nassauer, J. I. (ed.) (1997). *Placing Nature: Culture and Landscape Ecology*. Washington, DC: Island Press.

Nassauer, J. I. (2005). Using cultural knowledge to make new landscape patterns. In J. A. Wiens and M. R. Moss (eds.), *Issues and Perspectives in Landscape Ecology*. Cambridge: Cambridge University Press, 274–80.

National Research Council. (1997). *Toward a Sustainable Future: Addressing the Long-Term Effects of Motor Vehicle Transportation on Climate and Ecology*. Washington, DC: National Academy Press.

National Research Council. (1999). *Governance and Opportunity in Metropolitan America*. Washington, D.C.: National Academy Press.

National Research Council. (2005). *Valuing Ecosystem Services: Toward Better Environmental Decision-Making*. Washington, DC: National Academies Press.

Nel.lo, O. (ed.) (2003). *Aqui, no!: Els conflictes territorials a Catalunya*. Barcelona: Editorial Empuries.

Nelson, A. (2006). *Cold War Ecology: Forests, Farms, and People in the East German Landscape, 1945–1989*. New Haven, Connecticut: Yale University Press.

Nilon, C. H. and Pais, R. C. (1997). Terrestrial vertebrates in urban ecosystems: developing hypothesis for Gwynns Falls watershed in Baltimore, Maryland. *Urban Ecosystems*, **1**, 247–57.

Norberg-Schulz, C. (1980). *Genius Loci: Towards a Phenomenology of Architecture*. London: Academy Editions.

Nordhaus, W. D. (1992). Lethal Model 2: The limits to growth revisited. *Brookings Papers on Economic Activity*, **2**, 1–59.

Noss, R. F. (2003). A checklist for wildlands network designs. *Conservation Biology*, **17**, 1270–75.

Noss, R. F. and Cooperider, A. Y. (1994). *Saving Nature's Legacy: Protecting and Restoring Biodiversity*. Washington, DC: Island Press.

Nowak, D. (1994). *Air pollution removal by Chicago's urban forest. In Chicago's Urban Forest Ecosystem: Results of the Chicago Urban Forest Climate Project*, Radnor. Pennsylvania: USDA Forest Service, General Technical Report NE-186, p. 63.

Odum, E. P. and Barrett, G. W. (2005). *Fundamentals of Ecology*. Belmont, CA: Thomson Brooks/Cole.

Odum, E. P. and Turner, M. G. (1990). The Georgia landscape: a changing resource. In I. S. Zonneveld and R. T. T. Forman (eds.), *Changing Landscapes: An Ecological Perspective*. New York: Springer-Verlag, 137–64.

Odum, H. T. (1973). Energy, ecology, and economics. *Ambio*, **2**(6), 220–7.

Odum, H. T. and Odum, E. C. (1981). *Energy Basis for Man and Nature*. New York: McGraw-Hill.

Odum, W. E. (1982). Environmental degradation and the tyranny of small decisions. *BioScience*, **32**, 728–9.

Office of Management and Budget. (2000). *Standards for Defining Metropolitan and Micropolitan Statistical Areas; Notice. Part IX. National Archives and Records Administration, USA*, Federal Register. **65**, 82228–38.

Oke, T. R. (1987). *Boundary Layer Climates*. London: Methuen and Co.

Oliveira, P. S. and Marquis, R. J. (eds.) (2002). *The Cerrados of Brazil: Ecology and Natural History of a Neotropical Savanna*. New York: Columbia University Press.

Olson, D. M., *et al.* (2001). Terrestrial ecoregions of the world: a new map of life on earth. *BioScience*, **51**, 933–38.

Omernik, J. M. (1987). Ecoregions of the conterminous United States. *Annals of the Association of American Geographers*, **77**, 118–25.

O'Neill, R. V., De Angelis, D. L., Waide, J. B., and Allen, T. F. H. (1986). *A Hierarchical Concept of Ecosystems*. Princeton, NJ: Princeton University Press.

Opdam, P., and Schotman, A. (1987). Small woods in rural landscape as habitat islands for woodland birds. *Acta Oecologia/Oecologia Generalis*, **8**, 269–74.

Opdam, P., Foppen, R. and Vos, C. (2002). Bridging the gap between ecology and spatial planning in landscape ecology. *Landscape Ecology*, **16**, 767–79.

Opdam, P., van Apeldoorn, R., Schotman, A., and Kalkhoven, J. (1992). Population responses to landscape fragmentation. In C. C. Vos and P. Opdam (eds.), *Landscape Ecology of a Stressed Environment*. London: Chapman & Hall, 147–71.

Orr, D. W. (2002). *The Nature of Design: Ecology, Culture, and Human Intention*. New York: Oxford University Press.

Orwin, C. S. and Orwin, C. S. (1967). *The Open Fields*. Oxford: Clarendon Press.

Osband, G. J. (1984). *Managing Urban Forests*. Cambridge, MA: Harvard University Graduate School of Design, Penny White Student Project.

Owen, J. (1991). *The Ecology of a Garden: The First Fifteen Years*. Cambridge: Cambridge University Press.

Ozawa, C. P. (2004). *The Portland Edge: Challenges and Successes in Growing Communities*. Washington, DC: Island Press.

Pandell, R., Martin, J., and Fulton, W. (2002). *Holding the Line: Urban Containment in the United States*. Washington, DC: Brookings Institution, Center on Urban and Metropolitan Policy.

Parc de Collserola: Llibre Guia. (1997). Barcelona: Patronat Metropolita del Parc de Collserola.

Park, C.-R. and Lee, W.-S. (2000). Relationship between species composition and area in breeding birds of urban woods in Seoul, Korea. *Landscape and Urban Planning*, **51**, 29–36.

Parsons, K. C. and Schuyler, D. (2000). *From Garden City to Greencity: The Legacy of Ebenezer Howard*. Baltimore, MD: Johns Hopkins University Press.

Parsons, K. C., Brown, S. C., Erwin, R. M., *et al.* (2002). *Waterbirds: Managing Wetlands for Waterbirds: Integrated Approaches*. Lawrence, KS, USA: Publication of the Waterbird Society.

Patterson, M. (2002). Ecological production based pricing of biosphere processes. *Ecological Economics*, **41**, 457–78.

Paul, M. J. and Meyer, J. L. (2001). Streams in an urban landscape. *Annual Review of Ecology and Systematics*, **32**, 333–65.

Pauleit, S., Jones, N., Nyhuus, S., Pirnat, J., and Salbitano, F. (2005). Urban forest resources in European cities. In C. C. Konijnendijk, K. Nilsson, T. B. Randrup, and J. Schipperijn (eds.), *Urban Forests and Trees: A Reference Book*. Springer, Berlin, 51–80.

Pearce, D. and Ulph, D. (1999). A social discount rate for the United Kingdom. In D. Pearce (ed.), *Economics and Environment: Essays on Ecological Economics and Sustainable Development*. Cheltenham, UK: Edward Elgar

Pearce, D. W. and Atkinson, G. D. (1993). Capital theory and the measurement of sustainable development: an indicator of "weak" sustainability. *Ecological Economics*, **8**, 103–8.

Peck, S. and Kuhn, M. (2003). *Design Guidelines for Green Roofs*. Ottawa: Ontario Association of Architects.

Pegel, H. (2007). Farming for nature in the Fehntjer Tief: a contribution to the sustainable development of an East Frisian cultural landscape. In B. Pedroli, A. von Doom, G. de Blust, *et al.* (eds.), *Europe's Living Landscapes: Essays Exploring Our Identity in the Countryside*. Wageningen, the Netherlands: KNNV Publishing, 225–38.

Peiser, R. (2001). Decomposing urban sprawl. *Town Planning Review*, **72**, 275–98.

Perlman, D. L. and Milder, J. C. (2005). *Practical Ecology for Planners, Developers, and Citizens*. Washington, DC: Island Press.

Perlman, J. E. (1976). *The Myth of Marginality: Urban Poverty and Politics in Rio de Janeiro*. Berkeley: University of California Press.

Perman, R., Ma, Y., McGilvray, J., and Common, M. (2003). *Natural Resource and Environmental Economics*. Harlow, UK: Pearson Addison-Wesley.

Perrings, C. (1991). Reserved rationality and the precautionary principle: technological change, time and uncertainty in environmental decision making. In R. Costanza (ed.), *Ecological Economics: The Science and Management of Sustainability*. New York: Columbia University Press, 153–66.

Peterken, G. F. (1996). *Natural Woodland: Ecology and Conservation in Northern Temperate Regions*. Cambridge: Cambridge University Press.

Peterken, G. F., Ausherman, D., Buchenau, M., and Forman, R. T. T. (1992). Old-growth conservation within British upland conifer plantations. *Forestry*, **65**, 127–44.

Pezzey, J. C. V. (2004). Sustainability policy and environmental policy. *Scandinavian Journal of Economics*, **106**, 339–59.

Pezzey, J. C. V., Hanley, N., Turner, K., and Tinch, D. (2006). Comparing augmented sustainability measures for Scotland: Is there a mismatch? *Ecological Economics*, **57**, 60–74.

Pezzoli, K. (1998). *Human Settlements and Planning for Ecological Sustainability*. Cambridge, MA: MIT Press.

Pickett, S. T. A. (2006). Advancing urban ecology studies: frameworks, concepts, and results from the Baltimore Ecosystem Study. *Austral Ecology*, **31**, 114–25.

Pickett, S. T. A., Cadenasso, M. L., Grove, J. M., *et al.* (2001). Urban ecological systems: linking terrestrial ecological, physical, and socioeconomic components of metropolitan areas. *Annual Review of Ecology and Systematics*, **32**, 127–57.

Pickett, S. T. A. and White, P. S. (eds.) (1985). *The Ecology of Natural Disturbance and Patch Dynamics*. New York: Academic Press.

Pilkey, O. H. and Dixon, K. L. (1996). *The Corps and the Shore*. Washington, DC: Island Press.

Pinchot, G. (1967). *The Fight for Conservation*. Seattle. WA: University of Washington Press.

Pinelands Commission. (1980a). *Comprehensive Management Plan for the Pinelands National Reserve (National Parks and Recreation Act, 1978) and Pinelands Area (New Jersey Pinelands Protection Act, 1979)*. New Lisbon, NJ: State of New Jersey.

Pinelands Commission. (1980b). *New Jersey Pineland Comprehensive Management Plan*. New Lisbon, NJ: Pinelands Commission.

Pino, J., Roda, F., Ribas, J., and Pons, X. (2000). Landscape structure and bird species richness: implications for conservation in rural areas between natural parks. *Landscape and Urban Planning*, **49**, 35–48.

Platt, R. H. (2004). *Land Use and Society: Geography, Law, and Public Policy*. Washington, DC: Island Press.

Platt, R. H., Rowntree, R. A., and Muick, P. C. (1994). *The Ecological City: Preserving and Restoring Urban Biodiversity*. Amherst, MA: University of Massachusetts Press.

Ponting, C. (1991). *A Green History of the World*. New York: Penguin Books.

Pouyat, R. V., McDonnell, M. J., and Pickett, S. T. A. (1997). Litter decomposition and nitrogen mineralization in oak stands along an urban-rural land-use gradient. *Urban Ecosystems*, **1**, 117–31.

Prat, N., Munne, A., Sola, C., *et al.* (2002). *La Qualitat Ecologica del Llobregat, El Besos, El Foix i la Tordera: Informe 2000*. Barcelona: Diputacio de Barcelona, Estudis de la Qualitat Ecologica dels ruis, Number 10.

Primack, R. B. (2004). *A Primer of Conservation Biology*. Sunderland, MA: Sinauer Associates.

Rackham, O. (1980). *Ancient Woodland*. London: Edward Arnold.

Radford, R. (coordinator) (1996). *Tom Roberts*. Adelaide: Art Gallery of South Australia.

Rapoport, E. H. (1993). The process of plant colonization in small settlements and large cities. In M. J. McDonnell and S. T. A. Pickett (eds.), *Humans as Components of Ecosystems: The Ecology of Subtle Human Effects and Populated Areas*. New York: Springer-Verlag, 190–207.

Ravetz, J. (2000). *City-Region 2020: Integrated Planning for a Sustainable Environment*. London: Earthscan.

Ray, D. (1998). *Development Economics*. Princeton, NJ: Princeton University Press.

Rees, W. E. (2003). Understanding urban ecosystems: an ecological economics perspective. In A. R. Berkowitz, C. H. Nilon, and K. S. Hollweg, (eds.), *Understanding Urban Ecosystems: A New Frontier for Science and Education*. New York: Springer, 115–36.

Register, R. (2006). *Ecocities: Rebuilding Cities in Balance with Nature*. Gabriola Island, British Columbia, Canada: New Society Publishers.

Reijnen, R., Foppen, R., and Meeuwsen, H. (1996). The effects of car traffic on the density of breeding birds in Dutch agricultural grasslands. *Biological Conservation*, **75**, 255–60.

Reijnen, R., Foppen, R., ter Braak, C., and Thissen, J. (1995). The effects of car traffic on breeding bird populations in woodland. III. Reduction of density in relation to the proximity of main roads. *Journal of Applied Ecology*, **32**, 187–202.

Ren, *et al.* (2003). Urbanization, land use and water quality in Shanghai 1947–1966. *Environment International*, **29**, 649–59.

Rich, C. and Longcore, T. (2006). *Ecological Consequences of Artificial Night Lighting*. Washington, DC: Island Press.

Ricklefs, R. E. and Miller, G. L. (2000). *Ecology*. New York: W. H. Freeman.

Rieley, J. O. and Page, S. E. (1995). Survey, mapping and evaluation of green space in the Federal Territory of Kuala Lumpur, Malaysia. In H. Sukopp, M. Numata, and A. Huber (eds.), *Urban Ecology as the Basis of Urban Planning*. Amsterdam: SPB Academic Publishing, 173–83.

Rieradevall, M. and Cambra, J. (1994). Urban freshwater ecosystems in Barcelona. *Verhandlungen Internationale Vereinigung Limnologie*, **25**, 1369–72.

Rigby, K. (2006). (Not) by design: utopian moments in the creation of Canberra. *Arena Journal Series*, **25/26**, 155–77.

Rink, D. (2005). Surrogate nature or wilderness? Social perceptions and notions of nature in an urban context. In I. Kowarik and S. Korner (eds.), *Wild Urban Woodlands*. Springer-Verlag, Berlin, 67–80.

Robertson, M. M. (2006). Emerging ecosystem service markets: trends in a decade of entrepreneurial wetland banking. *Frontiers in Ecology and Environment*, **4**, 297–302.

Robin, L. (1998). *Defending the Little Desert: The Rise of Ecological Consciousness in Australia*. Melbourne: Melbourne University Press.

Robinson, W. H. (1996). *Urban Entomology: Insects and Mite Pests in the Human Environment*. London: Chapman & Hall.

Roda, F., Retana, J., Gracia, C. A., and Bellot J. (1999). *Ecology of Mediterranean Evergreen Oak Forests*. Berlin: Springer-Verlag.

Rogers, P. P., Jalal, K. F., and Boyd, J. A. (2006). *An Introduction to Sustainable Development*. Cambridge, MA: Published by the Continuing Education Division of Harvard University and the Glen Educational Foundation.

Romer, P. M. (1990). Endogenous technical change. *Journal of Political Economy*, **98**, S71–S102.

Rosell Pages, C. and Velasco Rivas, J. M. (1999). *Manual de prevencio i correccio dels impactes de les infraestructures viaries sobre la fauna*. Barcelona: Generalitat de Catalunya, Numero 4, Departament de Medi Ambient.

Ross, A. (1999). *The Celebration Chronicles: Life, Liberty, and the Pursuit of Property Value in Disney's New Town*. New York: Ballantine Books.

Rowe, P. G. (1991). *Making a Middle Landscape*. Cambridge, Massachusetts: MIT Press.

Rowe, P. G. (2006). *Building Barcelona: A Second Renaixenca*. Barcelona: Barcelona Regional and ACTAR.

Saley, H., Meredith, D. H., Stelfox, H., and Ealey, D. (2003). *Nature Walks and Sunday Drives 'Round Edmonton*. Edmonton, Alberta, Canada: Edmonton Natural History Club.

Salvesen, D. (1994). *Wetlands: Mitigating and Regulating Development Impacts*. Washington, DC: Urban Land Institute.

Santamouris, M. (ed.) (2001). *Energy and Climate in the Urban Built Environment*. London: James and James.

Saunders, D. A. and Hobbs, R. J. (eds.) (1991). *Nature Conservation 2: The Role of Corridors*. Chipping Norton, NSW. Australia: Surrey Beatty.

Schmandt, J. and Clarkson, J. (eds.) (1992). *The Regions and Global Warming: Impacts and Response Strategies*. New York: Oxford University Press.

Schmid, J. A. (1975). *Urban Vegetation: A Review and Chicago Case Study*. University of Chicago, Department of Geography, Research Paper 161.

Schneider, A., Fiedl, M. A., McIver, D. W., and Woodcock, C. E. (2003). Mapping urban areas by fusing multiple sources of coarse resolution remotely sensed data. *Photogrammetric Engineering and Remote Sensing*, **69**, 1377–86.

Schonewald-Cox, C. M. and Bayless, J. W. (1986). The boundary model: a geographic analysis of design and conservation of nature reserves. *Biological Conservation*, **38**, 305–22.

Schrepfer, S. R. (1983). *The Fight to Save the Redwoods*. Madison, WI: University of Wisconsin Press.

Schwartz, H. (2004). *Urban Renewal, Municipal Revitalization: The Case of Curitiba, Brazil*. Alexandria, VA: Published by the Author.

Seddon, G. (1997). *Landprints: Reflections on Place and Landscape*. Cambridge: Cambridge University Press.

Sieghardt, M., Mursch-Rasdlgruber, E., Paoletti, E., *et al.* (2005). The abiotic urban environment: impact of urban growing conditions on urban vegetation. In C.C. Konijnendijk *et al.* (eds.), *Urban Forests and Trees: A Reference Book*. Berlin: Springer, 281–323.

Simmonds, R. and Hack, G. (eds.) (2000). *Global City Regions: Their Emerging Forms*. London: Spon Press.

Sit, V. F. S. (1995). *Beijing: The Nature and Planning of a Chinese Capital City*. New York: John Wiley.

Sklar, F. and Costanza, R. (1990). The development of dynamic spatial models for landscape ecology: a review and synthesis. In M. G. Turner and R. H. Gardner (eds.), *Quantitative Methods in Landscape Ecology*. New York: Springer-Verlag, 239–88.

Smit, J. (2006). Farming in the city and climate change: the potential and urgency of applying urban agriculture to reduce the negative impacts of climate change. *Urban Agriculture Magazine*, **15**.

Smit, J. and Nasr, J. (1992). Urban agriculture for sustainable cities: using wastes and idle land and water bodies as resources. *Environment and Urbanization*, **4**, 441.

Smith, R. L. (1996). *Ecology and Field Biology*. New York: HarperCollins.

Smith, W. H. (1981). *Air, Pollution and Forests: Interactions Between Air Contaminants and Forest Ecosystems*. New York: Springer-Verlag.

Snyder, G. (1990). *The Practice of the Wild*. San Francisco: North Point.

Soja, E. A. (2000). *Postmetropolis: Critical Studies of Cities and Regions*. Oxford: Basil Blackwell.

Song, I.-J. and Jin, Y.-R. (2002). A model of enhancing biodiversity through analysis of landscape ecology in Seoul cultivated area. *Korean Journal of Environmental Ecology*, **16**, 249–60.

Sorrie, B. (2005). Alien vascular plants in Massachusetts. *Rhodora*, **107**, 283–329.

Soule, M. E. (1991). Land use planning and wildlife maintenance: guidelines for conserving wildlife in an urban landscape. *American Planning Association Journal*, **57**, 313–23.

Spirn, A. W. (1984). *The Granite Garden: Urban Nature and Human Design*. New York: Basic Books.

Spirey, A. (2002). Built environment: rooftop gardens, a cool idea. *Environmental Health Perspectives*, **110**(11).

Starling, F. L. do R. M. (2000). Comparative study of the zooplankton composition of six lacustrine ecosystems in Central Brazil during the dry season. *Revista Brasileira Biololeogia* **1(60)**, 101–11.

State of the World's Cities 2006/2007. (2006). London: Earthscan.

Steinberg, D. A., Pouyat, R. V., Parmalee, R. W., and Groffman, P. M. (1997). Earthworm abundance and nitrogen mimeralization rates along an urban-rural land use gradient. *Soil Biology and Biochemistry*, **29**, 427–30.

Steiner, F. (1994). Regional planning in the United States: historic and contemporary examples. In T. J. Bartuska and G. L. Young (eds.), *The Built Environment: Creative Inquiry Into Design and Planning*. New Brunswick: CRISP Publications, 319–29.

Steiner, F. (2002). *Human Ecology: Following Nature's Lead*. Washington, DC: Island Press.

Steinitz, C. and McDowell, S. (2001). Alternative futures for Monroe County, Pennsylvania: a case study in applying ecological principles. In V. H. Dale and R. A. Haeuber (eds.), *Applying Ecological Principles to Land Management*. New York: Springer, 165–93.

Stubbs, J. and Clarke, G. (eds.) (1996). *Megacity Management in the Asia and Pacific Region: Policy Issues and Innovative Approaches*. Volume I. Manila, Philippines: The Asian Development Bank.

Sukopp, H. and Hejny, S. (eds.) (1990). *Urban Ecology: Plants and Plant Communities in Urban Environments*. The Hague, Netherlands: SPB Academic Publishing.

Sukopp, H., Numata, M., and Huber, A. (eds.) (1995). *Urban Ecology as the Basis of Urban Planning*. The Hague, Netherlands: SPB Academic Publishing.

Sukopp, H. and Werner, P. (1983). Urban environment and vegetation. In W. Holzner, M. J. A. Werger, and I. Ikusima (eds.), *Man's Impact on Vegetation*. The Hague, Netherlands: W. Junk Publishers, 247–60.

Sutcliffe, A. (ed.) (1980). *The Rise of Modern Urban Planning, 1800–1914*. New York: St. Martin's Press.

Swanson, F. J., Jones, J. A., Wallin, D. O., and Cissel, J. H. (1994). *Natural variability – implications for ecosystem management. In East Side Forest Ecosystem Health Assessment. Volume 2. Ecosystems Management: Principles and Applications*, Portland, OR: US Forest Service, Report GTR, 89–103.

Swenson, J. J. and Franklin, J. (2000). The effects of future urban development on habitat fragmentation in the Santa Monica Mountains. *Landscape Ecology*, **15**, 713–30.

Takaya, Y. (1987). *Agricultural Development of a Tropical Delta: A Study of the Chao Phraya Delta*. Honolulu, University of Hawaii Press.

Thayer, R. L., Jr. (2003). *LifePlace: Bioregional Thought and Practice.* Berkeley: University of California Press.

The Management Plan for the Greenbelt. (1995). *The Planning Framework – Le Plan.* Ottawa: National Capital Commission.

Theobald, D. M. and Hobbs, N. T. (1998). Forecasting rural land-use change: a comparison of regression- and spatial transition-based models. *Geographical and Environmental Modelling,* **2,** 65–82.

Theobold, D. M., Miller, J. R., and Hobbs, N. T. (1997). Estimating the cumulative effects of development on wildlife habitat. *Landscape and Urban Planning,* **39,** 25–36.

Tjallingii, S. B. (1995). *Ecopolis: Strategies for Ecologically Sound Urban Development.* Leiden: Backhuys Publishers.

Tonmanee, N. and Kuneepong, P. (2004). Impact of land-use change in Bangkok metropolitan and suburban areas. In G. Tress, B. Tress, B. Harms, *et al.* (eds.), *Planning Metropolitan Landscapes: Concepts, Demands, Approaches.* Wageningen, Netherlands: DELTA Series 4, 114–23.

Townsend, C. R., Harper, J. R., and Begon, M. (2000). *Essentials of Ecology.* Malden, MA: Blackwell Science.

Tress, G., Tress, B., Harms, B., *et al.* (eds.) (2004). *Planning Metropolitan Landscapes: Concepts, Demands, Approaches.* Wageningen, Netherlands: DELTA Series 4.

Trocme, M., Cahill, S., de Vries, H. (J. G.), *et al.* (eds.) (2003). *Habitat Fragmentation Due to Transportation Infrastructure: The European Review.* Brussels: European Commission, COST Action 341.

Troy, P. N. (1995). *Australian Cities.* Cambridge: Cambridge University Press.

Troy, P. N. (1996). *The Perils of Urban Consolidation.* Sydney: Federation Press.

Tuan, Y.-F. (1977). *Space and Place: The Perspective of Experience.* Minneapolis, MN: University of Minnesota Press.

Turner, B. (2005). *The Statesman's Yearbook 2005.* New York: Macmillan.

Turner, B. L., Clark, W. C., Kates, R. C., *et al.* (eds.) (1990). *The Earth as Transformed by Human Action: Global and Regional Changes in the Biosphere Over the Past 300 Years.* Cambridge: Cambridge University Press.

Turner, M. G. (2005). Landscape ecology: what is the state of the science? *Annual Review of Ecology and Systematics,* **36,** 319–44.

Turner, M. G., Gardner, R. H., and O'Neill, R. V. (2001). *Landscape Ecology in Theory and Practice.* New York: Springer-Verlag.

Turner, T. (1992). Open space planning in London: from standards per 1000 to green strategy. *Town Planning Review,* **63,** 365–86.

Ulrich, R. S. (1984). View through a window may influence recovery from surgery. *Science,* **224,** 420–1.

United Nations Population Division. (2001). *World Urbanization Prospects: The 2001 Revision.* New York: United Nations.

United Nations Population Division. (2005). *World Population Prospects: The 2005 Revision.* New York: United Nations.

Vallejo, R. and Alloza, J. A. (1998). The restoration of burned lands: the case of eastern Spain. In J. M. Moreno (ed.), *Large Forest Fires.* Leiden: Backhuys, 91–108.

van Bohemen, H. D. (2004). *Ecological Engineering and Civil Engineering Works: A Practical Set of Engineering Principles for Road Infrastructure and Coastal Management.* Delft, Netherlands: Directorate-General of Public Works and Water Management.

van den Bergh, C. J. M. and Verbruggen, H. (1999). Spatial sustainability, trade and indicators: an evaluation of the ecological footprint. *Ecological Economics,* **29**, 61–72.

van der Ree, R. and McCarthy, M. A. (2005). Inferring persistence of indigenous mammals in response to urbanization. *Animal Conservation,* **8**, 309–19.

Verboom, J. and Wamelink, W. (2005). Spatial modeling in landscape ecology. In J. A. Wiens and M. R. Moss (eds.), *Issues and Perspectives in Landscape Ecology.* Cambridge: Cambridge University Press, 79–89.

Vernez Moudon, A. (ed.) (1989). *Master-planned Communities: Shaping Exurbs in the 1990s.* Seattle, WA: University of Washington, College of Architecture and Urban Planning, Proceedings of Urban Design Program Conference.

Vigier, F. C. D. (1997). Housing as an evolutionary process: planning and design implications. In W. A. Allen, R. G. Courtney, E. Happold, and A. M. Wood (eds.), *A Global Strategy for Housing in the Third Millennium.* London: E & FN Spon, 89–106.

von Krosigk, K.-H. (2001). *Der Berliner Tiergarten.* Berlin: Markus Sebastian Braun.

von Stulpnagel, A., Horbert, M., and Sukopp, H. (1990). The importance of vegetation for the urban climate. In H. Sukopp (ed.), *Urban Ecology.* The Hague, Netherlands: SPB Academic Publishing, 175–93.

Vos, C. C., Baveco, H., and Grashof-Bokdam, C. J. (2002). Corridors and species dispersal. In K. J. Gutzwiller (ed.), *Applying Landscape Ecology in Biological Conservation.* New York: Springer, 84–104.

Wackernagel, M. and Rees, W. (1996). *Our Ecological Footprint: Reducing Human Impact on the Earth.* Gabriola Island, British Columbia, Canada: New Society Publishers.

Warner, S. B., Jr. (2001). *Greater Boston: Adapting Regional Traditions to the Present (Metropolitan Portraits).* Philadelphia, PA: University of Pennsylvania Press.

Warren, R. (1998). *The Urban Oasis: Guideways and Greenways in the Human Environment.* New York: McGraw-Hill.

Wein, R. (2006). *Coyotes Still Sing in My Valley: Conserving Biodiversity in a Northern City.* Edmonton, Alberta, Canada: Spotted Cow Press.

Wetzel, R. G. (2001). *Limnology.* Philadelphia: Saunders.

Wetzel, R. G. and Likens, G. E. (2000). *Limnological Analyses.* New York, Springer.

Wheater, C. P. (1999). *Urban Habitats.* London: Routledge.

White, R. R. (2002). *Building the Ecological City.* Cambridge, UK: Woodhead Publishing.

Whitehand, J. and Morton, N. (2004). Urban morphology and planning: the case of fringe belts. *Cities,* **21**, 275–89.

Wickham, J. D., O'Neill, R. V., and Jones, K. B. (2000). Forest fragmentation as an economic indicator. *Landscape Ecology,* **15**, 171–9.

Wiens, J. A. (2002). Riverine landscapes: taking landscape ecology into the water. *Freshwater Biology,* **47**, 501–15.

Wiens, J. A. and Moss, M. R. (eds.) (2005). *Issues and Perspectives in Landscape Ecology.* Cambridge: Cambridge University Press.

Wilcove, D. S., Rothstein, D., Dubow, J., *et al.* (1998). Assessing the relative importance of habitat destruction, alien species, pollution, overexploitation, and disease. *BioScience*, **48**, 607–15.

Williams, N. S. G., Morgan, J. W., McDonnell, M. J., and McCarthy, M. A. (2005). Plant traits and local extinctions in natural grasslands along an urban–rural gradient. *Journal of Ecology*, **93**, 1203–13.

Williams, R. (1983). Keywords: A Vocabulary of Culture and Society. New York: Oxford University Press.

Willis, K. G., Turner, R. K., and Bateman, I. J. (eds.) (2001). *Urban Planning and Management*. Cheltenham, UK: Edward Elgar Publishing.

Wilson, E. O. (1984). *Biophilia*. Cambridge, MA: Harvard University Press.

Wilson, E. O. (1992). *The Diversity of Life*. Cambridge, MA: Belknap Press of Harvard University Press.

Witham, J. H. and Jones, J. M. (1987). Deer–human interactions and research in the Chicago metropolitan area. In L. W. Adams and D. L. Leedy (eds.), *Integrating Man and Nature in the Metropolitan Environment*. Columbia, MD: National Institute for Urban Wildlife, 155–9.

Wong, S. (2004). Cities, Internal Organization of. *International Encyclopedia of Social and Behavioral Sciences*, 1825–29.

Woodlands New Community. (1973–74). Philadelphia: Wallace, McHarg, Roberts, and Todd.

World Bank. (2006). *Where is the Wealth of Nations? Measuring Capital for the 21st Century*. Washington, DC: World Bank.

Wu, F. (2006). Globalization and the Chinese City. New York: Routledge.

Wu, J. (2007). Making the case for landscape ecology: an effective approach to urban sustainability. *Landscape Journal*, in press.

Wu, J. and Hobbs, R. (2007). *Key Topics in Landscape Ecology*. Cambridge: Cambridge University Press.

Yamamoto, M. (1930). *Japan Geographic Compendium: Nine States Kyushu* (in Japanese). Tokyo: Gennosuke Inoue Publisher.

Yang, R. (2004). *Planning Outline for the Tourist Attraction System in Beijing City 2005–2025*. Beijing: Tsinghua University, Institute of Urban Planning and Design and Institute of Resource Protection and Tourism.

Yaro, R., Arendt, R., Dobson, H., and Brabec, E. (1990). *Dealing with Change in the Connecticut River Valley: A Design Manual for Conservation and Development*. Amherst, MA: Lincoln Institute of Land Policy.

Yu, K. and Padua, M. (eds.) (2006). *The Art of Survival: Recovering Landscape Architecture*. Mulgrave, Victoria, Australia: Images Publishing Group.

Zaitzevsky, C. (1982). *Frederick Law Olmsted and the Boston Park System*. Cambridge, MA: Harvard University Press.

Zipperer, W. C. and Foresman, T. W. (1997). Urban tree cover: an ecological perspective. *Urban Ecosystems*, **1**, 229–49.

Zonneveld, I. S. and Forman, R. T. T. (eds.) (1990). *Changing Landscapes: An Ecological Perspective*. New York: Springer-Verlag.

Index

Lightning Source UK Ltd.
Milton Keynes UK
UKOW06f2229090714

234835UK00004B/13/P

9 780521 670760